KV-019-007

POSTURE: SITTING, STANDING, CHAIR DESIGN AND EXERCISE

POSTURE: SITTING, STANDING, CHAIR DESIGN AND EXERCISE

By

DENNIS ZACHARKOW, P.T.

Posture and Seating Consultant
Formerly, Rehabilitation Specialist
Department of Physical Medicine and Rehabilitation
Mayo Clinic
Rochester, Minnesota

CHARLES C THOMAS • PUBLISHER
Springfield • Illinois • U.S.A.

PO 1509

Published and Distributed Throughout the World by

CHARLES C THOMAS • PUBLISHER
2600 South First Street
Springfield, Illinois 62717

This book is protected by copyright. No part of
it may be reproduced in any manner without
written permission from the publisher.

© *1988 by* CHARLES C THOMAS • PUBLISHER

ISBN 0-398-05418-5

Library of Congress Catalog Card Number: 87-18149

With THOMAS BOOKS *careful attention is given to all details of manufacturing
and design. It is the Publisher's desire to present books that are satisfactory as to their
physical qualities and artistic possibilities and appropriate for their particular use.*
THOMAS BOOKS *will be true to those laws of quality that assure a good name
and good will.*

A qualified health professional should be consulted before attempting
any of the exercises described in this book. These exercises are not
intended to be used in place of proper medical diagnosis and treatment.

Printed in the United States of America
SC-R-3

Library of Congress Cataloging-in-Publication Data

Zacharkow, Dennis.
    Posture : sitting, standing, chair design, and exercise by
Dennis Zacharkow.
        p.   cm.
    Bibliography: p.
    Includes index.
    1. Posture. 2. Sitting position. 3. Standing position. 4. Chair
design. 5. Exercise—Physiological aspects. I. Title.
    [DNLM: 1. Posture. WE 103 Z16p]
QP301.Z32 1988
613.7'8—dc19
DNLM/DLC
for Library of Congress   87-18149
                                                           CIP

Lanchester Library

*Dedicated to the memory of
my father,
George Zacharkow
and my sister,
Georgeanna Zacharkow*

PREFACE

Compared to primitive man living an outdoor life, civilized man has become a "standing-around, and a sitting-down animal, rather than a running-around one" (Drew, 1926). Modern man is subjected to "an altered environment to which the body adapts itself automatically" (Drew, 1926). Nearly all of modern man's activities encourage the forward position of the arms and head, with the tendency of gravity to pull the body forward and downward.

Cultures where other resting postures such as squatting or kneeling predominated also show a transition to the Western sitting posture with the simultaneous transition to industrialization (Helbig, 1978). With the transition to industrialization, there is actually "a demand for static and sedentary modes of living" (Barlow, 1946). As Bennett (1928) stated, "The most universal physical occupation of civilized human beings is sitting."

In regards to the chair, Aveling (1879) commented that "Of all the machines which civilization has invented for the torture of mankind, . . . there are few which perform their work more pertinaciously, widely, or cruelly, than the chair. It is difficult to account for the almost universal adoption, at least in this country, of such an unscientific article of furniture." Coghill (1941) referred to the chair as "the most atrocious institution hygienically of civilized life."

Leibowitz (1967) remarked that " . . . hundreds of variations upon the shape of the chair have been produced, many differing enormously in terms of how one must sit in them. Indeed, *we,* not the chair, have made the compromise. We have agreed to adjust our bodies to the dictates of chairs; only rarely do we find a chair that in its design has contracted to fulfill the requirements of the human body. In such ways have we permitted the forms and products of our culture to change our body alignments in order to satisfy *their* structural requirements. We have accustomed ourselves to habitual modes of use that are literally disfiguring."

Corlett (1981) warned that "We cannot afford to ignore posture, primarily because to do so creates such widespread misery, and secondly because the costs, both the social costs of unnecessary disease and the direct costs in lost productivity, are more than any modern industrial nation should be prepared to pay."

An important concept presented throughout this text is man's spontaneous search for postural stability when sitting and standing. In order to fully appreciate this concept, along with the close relationship of faulty standing and sitting postures, the first five chapters should be read in chronological order before proceeding to any of the later chapters on specific chair designs (office seating, motor vehicle seating, disabled seating, etc.).

In addition, Chapter Six on School Seating should also be read before proceeding to any of the other chapters on chair design. The concepts in this chapter are critical for *all seating applications*. In regards to school, Bennett (1928) stated that "civilization has imposed upon the child one of the most distinctly sedentary occupations yet devised." It is in the artificial environment of the school where the child's postural habits will be formed for life.

Seating and posture authorities from over one hundred years ago stressed the extreme importance of proper seating and postural habits for both schoolchildren and adults. The author has found their writings to be more insightful and pertinent today than most of the current literature. Reading and re-discovering their works has been a very inspiring and humbling experience. It is hoped that this text will help give the recognition to all these individuals of the late nineteenth century and early twentieth century, whose works are just as important and relevant today.

Dennis Zacharkow

Rochester, Minnesota

ACKNOWLEDGMENTS

I have always felt that spending the extra research time exploring and backtracking the literature from various disciplines is the most exciting and rewarding way to study a subject. Therefore, I am very grateful to the following librarians for all their help in obtaining many of the classic works used in this project: Sandie Renaux, Library Associate, and all the research librarians of the Rochester Public Library, Rochester, Minnesota; and Marjorie Ginn and Anita Thomas of the Plummer Library, Rochester, Minnesota.

My thanks to Bob Schrupp, P.T. and to Jim Youdas, P.T. for taking on the time-consuming task of reviewing my manuscript. Their input resulted in some very important changes in the text. The help and encouragement of Dr. Frank Rocco of Winona State University, Winona, Minnesota, Dr. James Tarlano of Dayton, Ohio, and Bill Westwood of Rochester, Minnesota were also much appreciated throughout this project.

I would also like to thank Marge Beauseigneur for typing the manuscript, Mark Curry for his work on the photographic prints, and Rob Barta of Ads and Art, Inc., Rochester, Minnesota for both his work on the adapted line illustrations and for designing the dust jacket.

Lastly, I am very grateful to many of the authors cited throughout this text for their generosity in sending me copies of their works, and to all the publishing houses, organizations, and individuals that granted me permission to reproduce material from their publications.

CONTENTS

POSTURE: SITTING, STANDING, CHAIR DESIGN AND EXERCISE

Chapter One

STANDING POSTURE

The evolution of man's erect biped stance from a quadruped posture has been "marked by a narrowing of the base of support and a progressive elevation of the center of gravity of the body as a whole. Both mitigate against stability" (Hellebrandt and Franseen, 1943). With a center of gravity placed high above a relatively small supporting base, gravity is the major deforming force affecting man's stance.

Man's posture has therefore been described as a constant struggle to remain erect against the force of gravity (Whitman, 1924; Wagenhäuser, 1978). According to Jones (1933), "When man became a biped and assumed the upright position, he immediately made an enemy of gravity, and he has been fighting this relentless foe ever since."

"One of the penalties that the human being is forced to accept in his being the highest type of mammal, is that in locomotion, with the body used as an erect biped, gravity is constantly operating to drag the organs downward out of their normal position, as well as to draw the upper part of the body downward and forward into positions which must mean strain and weakness" (Goldthwait, 1915).

IDEALIZED ERECT STANDING

In the idealized erect standing rest posture, the line of gravity is considered to be located in the midline between the following points (Basmajian, 1978; Woodhull et al., 1985) (Figure 1):

a. The mastoid processes.
b. A point just in front of the shoulder joints.
c. A point just behind the center of the hip joints.
d. A point just in front of the center of the knee joints.
e. A point approximately five to six centimeters in front of the ankle joints (Woodhull et al., 1985; Klausen, 1965; Hellebrandt et al., 1938).

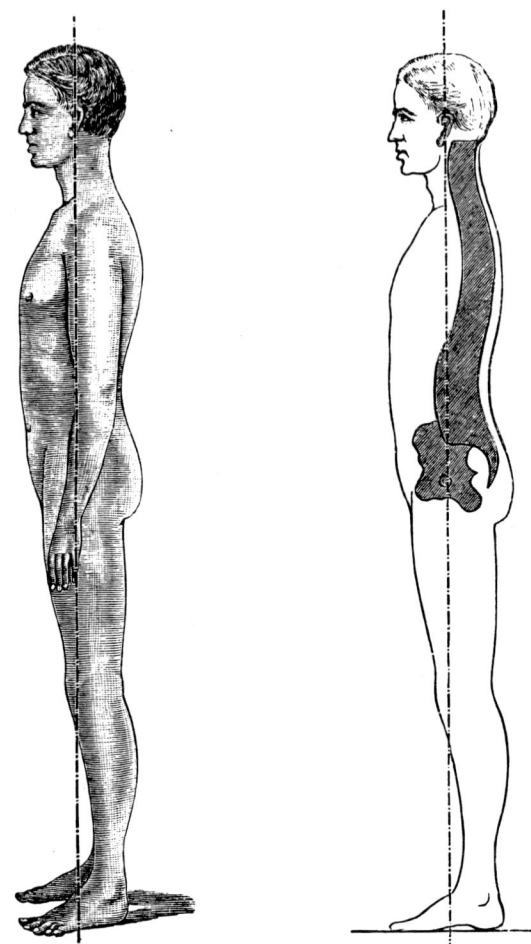

Figure 1. Line of gravity in idealized erect standing posture. From Staffel, F.: *Die Menschlichen Haltungstypen und ihre Beziehungen zu den Rückgratverkrümmungen.* Wiesbaden, J.F. Bergmann, 1889. Reproduced with permission of J.F. Bergmann Verlag.

The muscular activity required at the major body joints to achieve this idealized standing posture is as follows:

Ankle Joint

With the line of gravity passing in front of the ankle joints, the individual will tend to fall forwards from the ankles. This necessitates activity of the soleus muscles and possibly also the gastrocnemius muscles to counteract this anterior falling force (Åkerblom, 1948; Joseph and

Nightingale, 1952; O'Connell, 1958; Carlsöö, 1961, 1964; Okada, 1972; Soames and Atha, 1981; Okada and Fujiwara, 1983).

Knee Joint

With the line of gravity falling anterior to the center of the knee joint, gravitational forces will tend to extend the knee. Therefore, no muscle activity is apparently necessary for knee stabilization in idealized erect standing.

Hip Joint

With the line of gravity considered to fall just posterior to the hip axis, gravitational forces will tend to extend the hip and upwardly rotate the pelvis (Woodhull et al., 1985). The iliopsoas muscle has been found to be active in the erect posture, probably to resist both hyperextension of the hip joint and upward pelvic rotation (Snijders, 1969; Basmajian, 1978).

However, it is also important to note Carlsöö's comments (1972) regarding the hip's fixation in the erect standing rest position:

"... information on positional relationships between the gravitational line of direction and the hip's movement axis vary somewhat. This may be ascribed to varying methods of study, individual anatomical variations in test subjects or failure to accord sufficient consideration to postural sways. In any case, the trunk's vertical line and the hip's transverse movement axis pass very close to one another or even cross one another. It is not possible to say *a priori* if the joint's postural muscles should be sought among the hip's flexors or extensors."

"This muscular bracing of the hip in the standing rest position appears to occur intermittently in such a manner that certain muscles among both the joint's flexors and extensors may be in action as postural muscles for shorter or longer periods of time, depending on whether or not gravitational force acts to flex or extend the joint."

Carlsöö (1972) also noted that:

"If, from a standing position, the trunk is tilted forward slightly, about as much as when one stands at attention, the biceps femoris is activated. The reason for this muscle engagement is because gravitational force is displaced so far ventrally that it exerts a flexional effect on the hip, and the hip's extensors are activated to counteract this moment."

Spine

The line of gravity in idealized erect standing lies approximately one to two centimeters anterior to the center of the fourth lumbar vertebral body. By subjecting the spine to a constant forward bending moment, gravity therefore tends to flatten the lumbar lordosis, increase the thoracic kyphosis, and depress the trunk (Figure 2). This force is counteracted by slight but continuous activity from the erector spinae, the antigravity muscles of the trunk in idealized erect standing (Åkerblom, 1948; Kelly, 1949; Asmussen, 1960; Asmussen and Klausen, 1962; Carlsöö, 1964, 1972; Lindh, 1980; Forsberg et al., 1985).

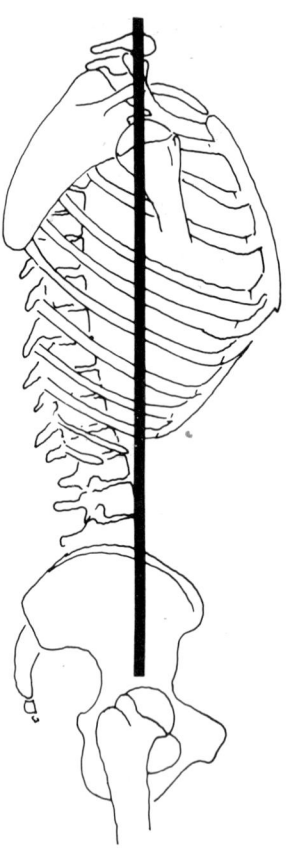

Figure 2. In idealized erect standing, the line of gravity will subject the spine to a constant forward bending moment. From Frankel, V.H., and Nordin, M.: *Basic Biomechanics of the Skeletal System.* Philadelphia, Lea and Febiger, 1980. Reproduced with permission of Lea and Febiger.

It is a common mistake to associate the idealized erect posture with a static concept of standing, instead of considering posture as "constant, continuous adaptation" (Campbell, 1935). Lee and Brown (1923) stressed the importance of continuous observation in order to understand an individual's habitual use of his body. Of particular importance is the way the load of the trunk is habitually carried.

Unfortunately, man's rapidly changing material environment has frequently resulted in a distorted carriage of the body. Man has become a "standing-around and sitting-down animal," with nearly all the activities of daily life encouraging a forward position of the arms and head (Drew, 1926). For the vast majority of individuals, daily work has a one-sided influence on the body, as the working object is always in front of the worker. "The result is often an inharmonious development of the body which makes itself known by bad carriage (round or crooked back, shoulders pulled forward, hanging head, etc.)" (Knudsen, 1920).

According to Keith (1923a), "It is not true, however, to say that our spines are not perfectly adapted to the upright posture; it would be more accurate to say that human spines were not evolved to withstand the monotonous and trying postures entailed by modern education and by many modern industries."

COMMON FAULTS IN THE CARRIAGE OF THE BODY: AN HISTORICAL PERSPECTIVE

Checkley (1890) considered the most common fault in carriage to involve a protruding lower abdomen, flattened chest, and round shoulders. Instead of the trunk being held erect, it has been allowed to "settle down" (Figure 3). In the correct standing position, Checkley (1890) commented on the forward carriage of the chest, and that the chin, chest, and toes should end in the same vertical line (Figure 4). According to Checkley (1890), in such a position "the body acquires its greatest ease, its greatest endurance and its greatest readiness." In Figure 5, Checkley (1890) illustrated beautifully the change in skeletal alignment from an habitual slumped posture to a posture with proper carriage of the body.

Checkley (1890) also commented that the correct standing position could be overexaggerated by the protrusion of the chest to an unnatural degree (Figure 6). He felt that the command "keep the shoulders back" could often lead to a "hollowed back" posture.

The most common postural fault described by Checkley (1890) is very

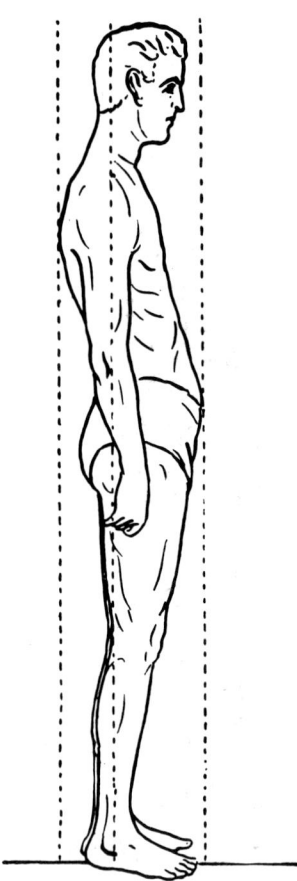

Figure 3. Most common standing postural fault, according to Checkley (1890). From Checkley, E.: *A Natural Method of Physical Training.* Brooklyn, William C. Bryant and Co., 1890.

similar to the round back posture described by Staffel (1889) a year earlier. Staffel (1889) characterized this posture as involving a forward shift of the pelvis, a prominent abdomen, a short lumbar lordosis just above the sacrum with the spine curving evenly backward above that point, drooping shoulders, a sunken thorax, and a forward inclination of the neck with the head thrust forward (Figure 7).

Lovett (1902) made an important statement regarding the classification of common faults in carriage which is still very relevant today:

"The difficulty with the classifications of faulty attitudes seems to be that the anteroposterior outline of the spine alone has been chiefly considered and but little or no attention has been paid to the relation of the feet, legs and pelvis to the spine and of the whole body to the

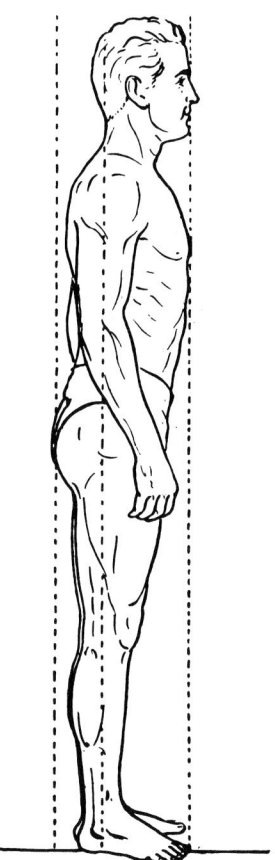

Figure 4. Correct standing posture, according to Checkley (1890). From Checkley, E.: *A Natural Method of Physical Training.* Brooklyn, William C. Bryant and Co., 1890.

perpendicular. And again no uniform system of measurement has been in use. Under these conditions have been only parts of the problem formulated. The general attitude is likely to be imperfectly represented by any drawing or observation made of the spine alone, because it is obvious that a backward curve of the upper part of the spine cannot occur without disturbing the normal relation of all the supporting structures below it. Equilibrium must be maintained, and the necessary adjustment involves feet, legs and pelvis in their relation to the spine and to the perpendicular."

Lovett (1902) considered round shoulders to be a general disturbance of the anteroposterior balance, involving the pelvis, legs, feet, and head, as well as the spine. He found the most frequent postural fault among seventy-two boys between the ages of fifteen and nineteen to be a round back posture with minimal lumbar lordosis. This posture also involved

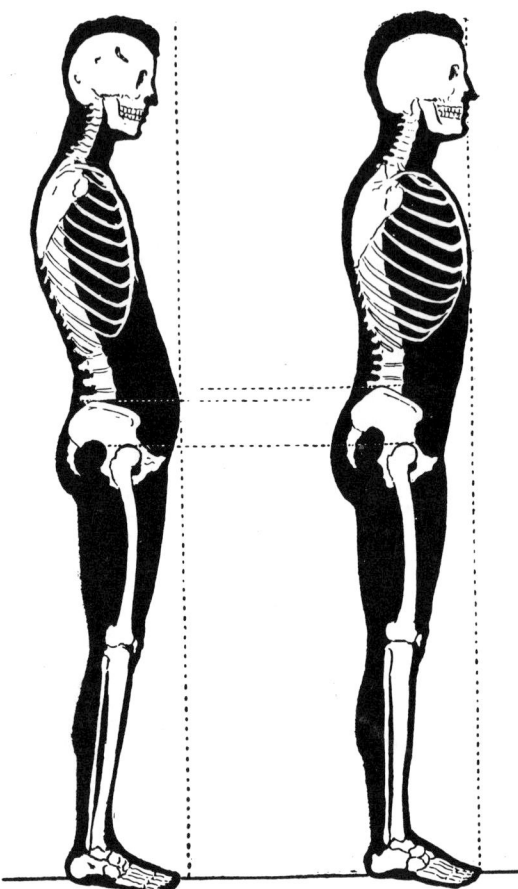

Figure 5. Change in skeletal alignment from an habitual slumped posture (left) to a posture with proper carriage of the body (right). From Checkley, E.: *A Natural Method of Physical Training.* Brooklyn, William C. Bryant and Co., 1890.

the pelvis deviating anteriorly and the hips hyperextended, along with round shoulders.

Skarstrom (1909) described the most common postural fault associated with poor general muscle tone and development as follows:

> "... the hips are carried excessively forward, the lumbar curve is low, not very marked, sometimes almost obliterated; while the thoracic is pronounced and long, often encroaching on the upper lumbar region. The entire forward position of the hips leads to decreased pelvic obliquity, the plane of the pelvis approaching too nearly the horizontal and the sacrum too nearly the vertical, wedged in between and spreading the ilia."

Goldthwait (1915) differentiated between the proper carriage of the

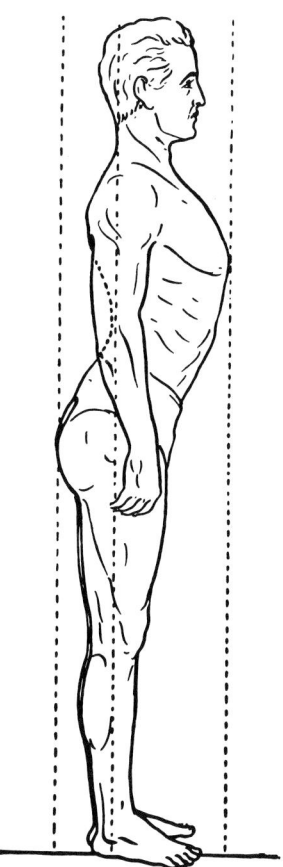

Figure 6. Exaggerated standing posture, distorting the spine and chest. From Checkley, E.: *A Natural Method of Physical Training.* Brooklyn, William C. Bryant and Co., 1890.

body and the commonly observed slouched attitudes, especially in regard to the effect on the abdominal and thoracic cavities:

"When used rightly, or fully erect, the feet, knees, hips, spine, shoulders, head and all the portions which represent the frame of the body, are used in balance, with the greatest range of movement possible without strain. In this position the chest is held high and well expanded, the diaphragm is raised, and the breathing and heart action are performed most easily. The abdominal wall is firm and flat, and the shape of the abdominal cavity resembles an inverted pear, large and rounded above and small below (Figure 8). The ribs have only a moderate downward inclination. The subdiaphragmatic space is ample to accomodate the viscera. In this position, also, there is no undue pressure upon, or interference with, the pelvic viscera or with the large ganglia at the back of the abdomen and in the pelvis."

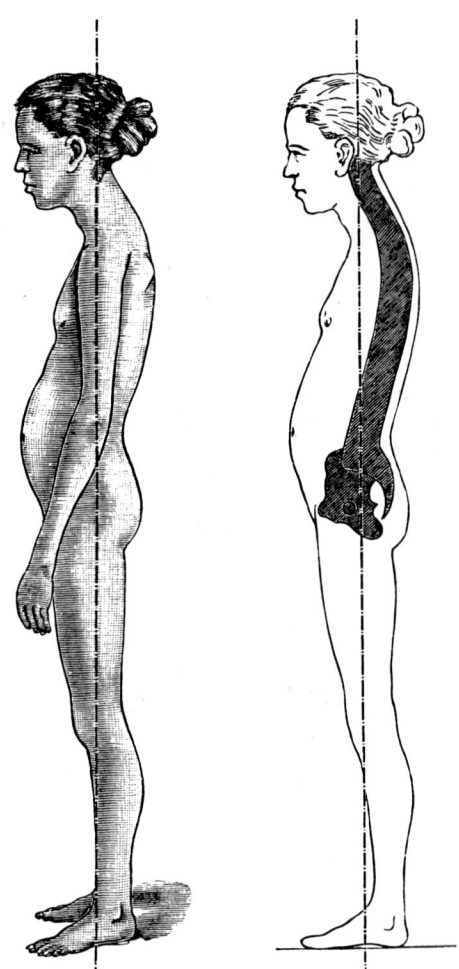

Figure 7. Round back posture, according to Staffel (1889). From Staffel, F.: *Die Menschlichen Haltungstypen und ihre Beziehungen zu den Rückgratverkrümmungen.* Wiesbaden, J.F. Bergmann, 1889. Reproduced with permission of J. F. Bergmann Verlag.

Goldthwait (1915) went on to describe the slouched postures commonly observed:

a. Standing

"In this position the chest is necessarily lowered, the lungs are much less fully expanded than normal, the diaphragm is depressed, the abdominal wall is relaxed, so that with the lessened support of the abdominal wall, together with the lowering of the diaphragm, the abdominal organs are necessarily forced downward and forward" (Figures 9 and 10).

b. Sitting

"In sitting it is perfectly possible to sit so that the trunk is in practically the same shape and with the different parts in practically the same relation, as they are when the body is used fully erect. At the same time in sitting it is very common to have the body markedly drooped, so that the body is rounded forward with the lumbar spine entirely reversed in its curve, with the ribs low, so that the thorax practically telescopes into the abdominal cavity."

c. Lying

"It is perfectly possible to produce practically the same effect upon the shape of the body and upon the thoracic and abdominal cavities that is present when the patient is sitting in the slouched position, if when lying down several pillows are placed under the head and shoulders, as is so frequently seen."

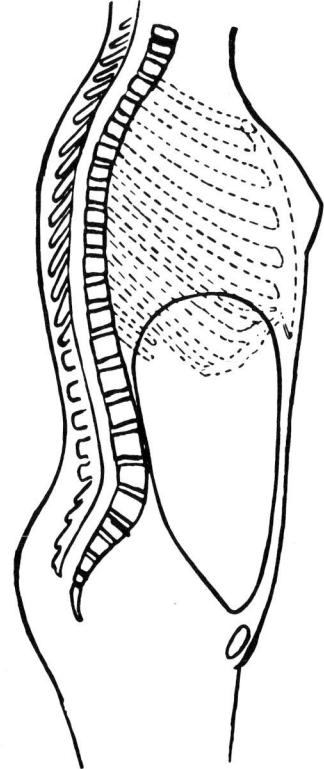

Figure 8. Proper position and shape of the thoracic and abdominal cavities. From Williams, J.T.: Visceral ptosis. A review. *Boston Medical and Surgical Journal*, *172:*13–18, January 7, 1915. Reproduced with permission of The New England Journal of Medicine.

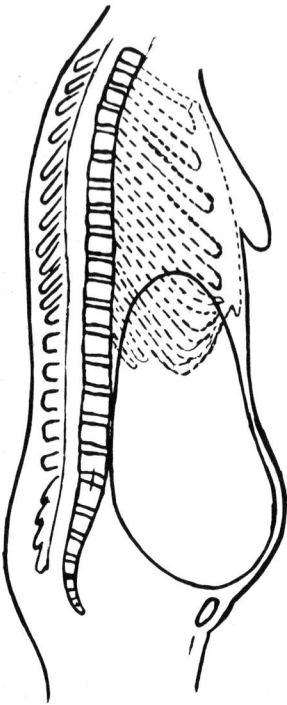

Figure 9. Position and shape of the thoracic and abdominal cavities in slouched postures. The chest is flat and the ribs are lowered, with a more acute costo-vertebral angle. The abdominal wall is relaxed, with a protruberant lower abdomen. From Williams, J.T.: Visceral ptosis. A review. *Boston Medical and Surgical Journal, 172:*13–18, January 7, 1915. Reproduced with permission of The New England Journal of Medicine.

Goldthwait stressed that:

"The body as a whole should be considered and not just the chest or the abdomen or the feet or the back or any one part, and it should be considered with reference to use in the different positions in the routine of life, especially those which are maintained for long periods, the occupational postures."

Mosher (1913, 1914, 1919) stressed that proper posture is largely dependent upon the position of the feet in standing, and upon the position of the pelvis in sitting.

Mosher (1919) referred to two postural habits that would result in a slouched posture with the characteristic round shoulders, forward thrust head, flat chest, and relaxed abdomen:

1. "Standing and walking with body weight resting more heavily upon the heels than upon the balls of the feet" (Figure 11).

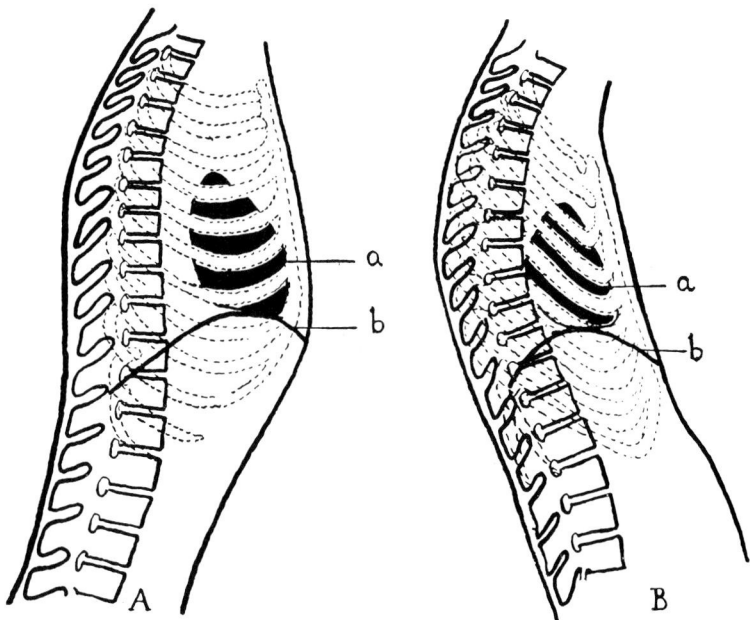

Figure 10:
  A. With good body mechanics, the diaphragm (b) is higher anteriorly under the heart (a) than at its attachment to the spine. Note in particular the greater depth and fullness in the lower rib region.
  B. With poor body mechanics, the anterior part of the diaphragm has sagged. With the lowering of the diaphragm, the depth of the chest has decreased and the abdominal wall is relaxed. The angulation of the ribs is more nearly vertical.
  From Goldthwait, J.E., Brown, L.T., Swaim, L.T., and Kuhns, J.G.: *Body Mechanics in the Study and Treatment of Disease.* Philadelphia, Lippincott, 1934. Reproduced with permission of J.B. Lippincott Company.

"In standing with the body weight upon the heels, the knees relax, the pelvis drops at the back, and the lumbar curve of the spine straightens, carrying the trunk backward beyond its center of gravity. To regain lost equilibrium the shoulders and head are moved forward for ballast and the ribs drop."

2. "Sitting with pelvis rocked back and chest doubled upon the abdomen and head hanging heavily upon its spinal and scapular muscles" (Figure 12).

A classic book by Bancroft (1913) entitled *The Posture of School Children* merits a detailed review. Bancroft considered the most common postural fault in standing to be a collapsed posture referred to as the fatigue position. In the fatigue position, "the neck and head droop forward, and the upper part of the trunk sinks backward."

Bancroft advocated the vertical line test to judge correct and incorrect standing posture. With this test, a line is dropped from the front of the

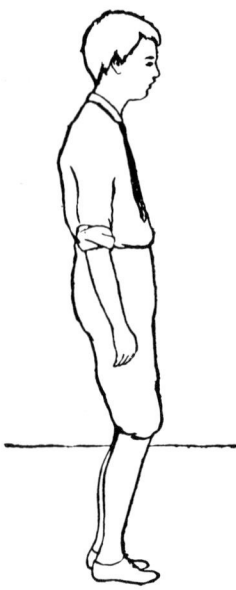

Figure 11. Standing posture with the body weight over the heels. From Cornell, W.S.: *Health and Medical Inspection of School Children.* Philadelphia, F.A. Davis, 1912. Reproduced with permission of F.A. Davis Company.

Figure 12. One of the sitting postures that produces round shoulders, flat chest, a forward thrust head, and a relaxed abdomen. From Mosher, E.M.: *Health and Happiness.* New York, Funk and Wagnalls, 1913. Reproduced with permission of Harper and Row, Publishers, Inc.

ear to the forward part of the foot. In proper posture, this line should parallel the long axes of the main segments of the body (neck and head, and trunk). With the fatigue position, the axes of these main segments of the body do not form one continuous vertical line, but are broken into a zigzag. As a result, "a straight line indicating the axis of the head, neck, and upper back slopes downward and backward to the region of the shoulders instead of being vertical; the axis of the lower trunk joins this upper line at an angle and slopes in the opposite direction—downward and forward to the hips" (Figure 13).

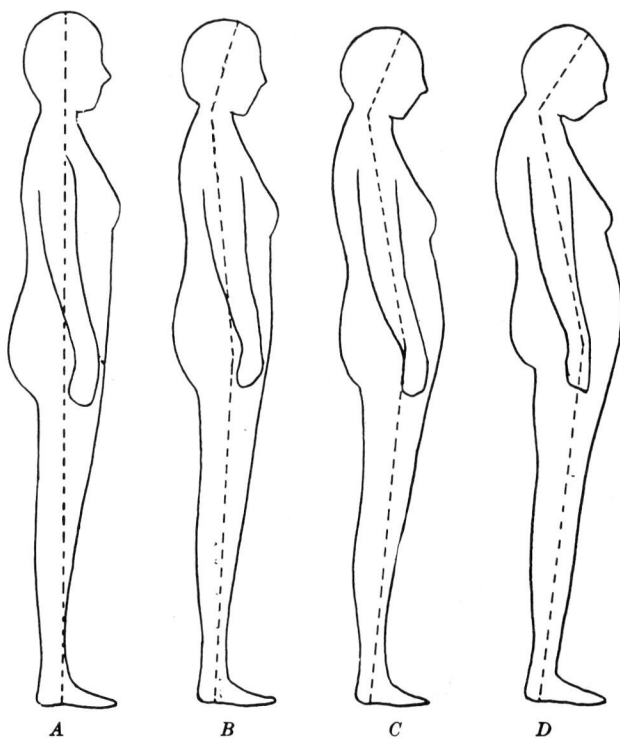

Figure 13. As the standing posture deteriorates into a greater fatigue position, the axes of the body segments are broken into a zigzag. From Drew, L.C.: *Individual Gymnastics,* 3rd. ed. Philadelphia, Lea and Febiger, 1926. Reproduced with permission of Lea and Febiger.

With the vertical line test, Bancroft considers it very important to place the vertical line in front of the ear (rather than through the mastoid process) and in the forward part of the foot (between the middle of the arch to the ball of the foot). This is considered necessary to emphasize the forward carriage of the weight, since in the fatigue position with the upper part of the trunk having sunk backward, most people habitually carry the body weight too far backward.

In regards to postural correction, Bancroft stresses the importance of training the muscular sense of correct position. As most people carry their body weight too far backward, the upper part of the trunk must be brought forward so that the chest is over the toes. The individual is taught to stretch the arms strongly sideways at shoulder level, with the palms turned downward. Holding the arms in this position, the individual sways forward from the ankles so that the body weight is "nearly or quite over the balls of the feet, not, however, rising on the toes, but keeping the heels on the ground."

Bancroft feels that the advantage of this method is that it treats the body as a whole. It takes advantage of the natural mechanical reactions of the body, so that if the upper part of the trunk is placed in position by swaying the body weight forward while stretching the arms out to the sides, the rest of the body naturally falls in line.

Bancroft considers phrases such as "lift the chest," "abdomen in," and "shoulders back" to be useless for postural correction. She feels that the use of these phrases will result in the formation of other postural faults.

The only command Bancroft found useful for reinforcing the moving forward of the upper part of the trunk is "chest over toes." However, she stresses that the swaying forwards to bring the chest over the toes should come from the ankle joints, and should not be misinterpreted as a bending at the waist or hips.

Several comments are made by Bancroft pertaining to trunk muscle tone and posture. Regarding the abdominal muscles, she feels that "there is no part of the body that suffers more from the relaxation of prolonged or poor sitting positions."

In regards to the back muscles, Bancroft states that "When one considers that nearly all occupations with the hands or arms are in front of the body, drawing the shoulders forward, it is not surprising that the posterior or back muscles, which antagonize this forward action and hold the shoulder blades in place, are kept unduly stretched, so that they lose their tone and allow the shoulders to droop forward. If, in addition to this, the spine be bowed over with an exaggerated curve in the dorsal region, the shoulders naturally slip forward still more, yielding to the pull of gravitation."

Bancroft also comments on trunk muscle imbalance and incoordination:

"Equally important with the cultivation of strength in weak muscles is the establishment of correct habits of coordination in the various muscle groups. To illustrate briefly, in the habitual fatigue position the muscles on the front of the body are too much contracted, and those in the back too much relaxed. So habitual has this relation become, that it feels to be right, and in anything the individual does, the faulty relation between these groups will be maintained."

McKenzie (1915) stressed the proper relation between the thorax and the pelvis in correct carriage of the body (Figure 14). Sutton (1925) considered the sagging abdomen to be the start of the postural deviations that would result in a relaxed, slouched standing posture.

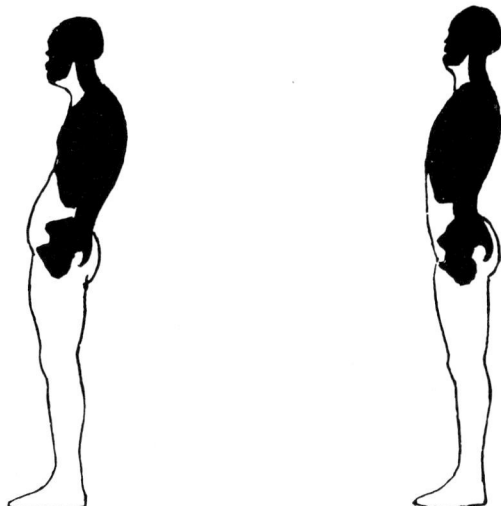

Figure 14:
    Left: Note the improper axial relationship of the head, thorax, and pelvis in the slumped, fatigue posture.
    Right: The proper axial relationship of the head, thorax, and pelvis, according to McKenzie (1915).
    From McKenzie, R.T.: *Exercise in Education and Medicine,* 2nd ed. Philadelphia, Saunders, 1915. Reproduced with permission of W.B. Saunders Company.

According to Kellogg (1927):

"The slumped posture is the natural result of the fact that our house chairs, the seats occupied by children in schools and by adults in churches, theaters, trains, street cars and other public places, are rarely constructed in such a way as to support the body in a normal posture when it is relaxed. Consequently, the moment relaxation occurs, gravitation takes control and the body is pulled into whatever position the supporting structure may cause it to assume.

The universality of this condition among civilized people, in the opinion of the writer, gives to this question of posture, and especially the sitting posture, great importance as a health factor. The standing posture is in general simply the natural result of the sitting posture. If the habitual sitting posture is such as to hold the body in proper form, the same poise will be maintained in the erect position."

In normal posture, Kellogg (1927) considered the most anterior portion of the skeleton to be the lower end of the sternum. However, "if the upper part of the trunk is moved backward in relation to the Vertical Axis, the dorsal curve will be increased, for the reason that the head and shoulders must be carried forward in order to maintain the vertical balance. The natural effect of this change in posture will be to lower the chest. As a result of this, the diaphragm will be depressed, the viscera crowded down, the lower abdomen will be made more prominent."

Kellogg (1927) considered the poise of the body to be determined by four essential factors (Figure 15):

1. the Vertical Balance
2. the Pelvic Obliquity
3. the Chest Ratio
4. the Head Angle

The Vertical Balance

"When the standing body is in normal Vertical Balance, a line dropped from a point an inch or two in front of the ear will pass through the center of the plane of the obliquity of the pelvis and will strike the foot at the instep. This line may be called the Vertical Axis of the body, although it lies a little anterior to the gravitational center of the erect body. It serves as a line of reference in analyzing the posture of a person in the standing position."

The Pelvic Obliquity

By Pelvic Obliquity, Kellogg means the inclination of the plane of the brim of the pelvis from the horizontal. The Pelvic Obliquity is marked by a line passing from the pubis to the promontory of the sacrum.

The Chest Ratio

"The Chest Ratio expresses the relation between that portion of the chest that is in front of the Vertical Axis and that portion which is behind this line. This relation is determined thus: A line is drawn from the point of intersection of the Vertical Axis with the plane of the pelvis and the lower end of the sternum, which is normally the most prominent

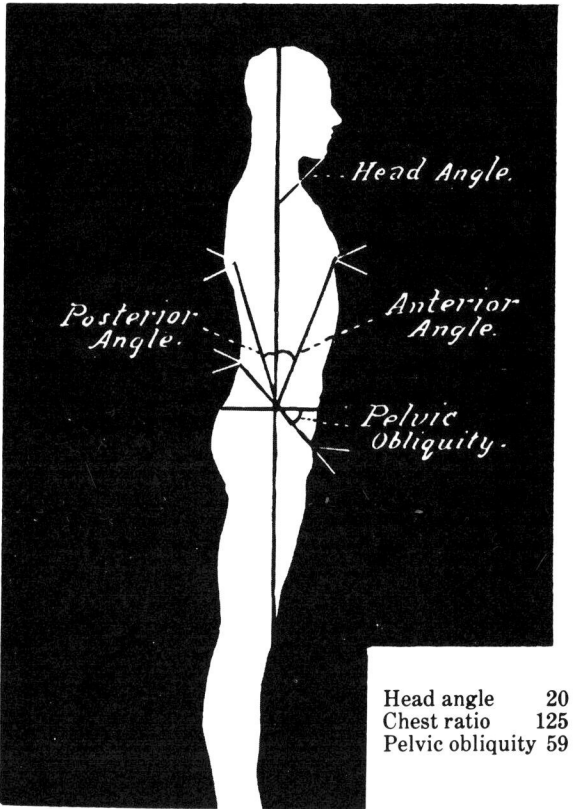

Head angle   20
Chest ratio   125
Pelvic obliquity 59

Figure 15: Four essential factors in the poise of the body, as illustrated in the shadow-graph of a Swedish athlete. From Kellogg, J.H.: Observations on the relations of posture to health and a new method of studying posture and development. *The Bulletin of the Battle Creek Sanitarium and Hospital Clinic, 22:*193–216, 1927.

portion of the chest. Another line is drawn from the same starting point to a point at the back of the chest just opposite the lower end of the sternum. The two angles thus formed are measured and the anterior is divided by the posterior. Normally, the anterior angle is always larger than the posterior, never smaller."

The Head Angle

"The angle formed by the chin with the Vertical Axis is measured by a line passing from the chin through the upper end of the sternum to the Vertical Axis. This measures the degree of forward carriage of the head. A large head angle is associated with an exaggerated posterior cervico-dorsal curve."

Figure 16 illustrates the faulty carriage Kellogg found most frequently in individuals of sedentary habits: a large posterior chest angle, a large head angle, and a decreased pelvic obliquity.

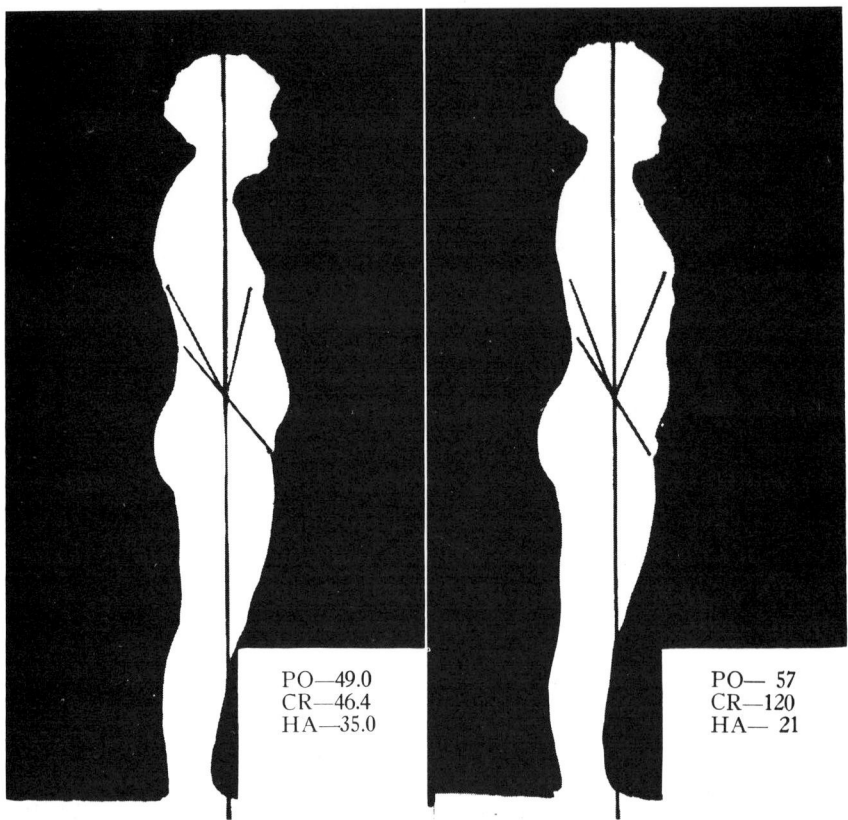

PO—49.0
CR—46.4
HA—35.0

PO— 57
CR—120
HA— 21

Figure 16.
   Left: The faulty carriage found most frequently in individuals of sedentary habits.
   Right: The same individual after postural correction. From Kellogg, J.H.: Observations on the relations of posture to health and a new method of studying posture and development. *The Bulletin of the Battle Creek Sanitarium and Hospital Clinic, 22:*193–216, 1927. Note: PO = Pelvic obliquity CR = Chest ratio HA = Head angle.

Rathbone (1934) considered the long single curve with sway back in standing to owe its origin in many cases to a poor sitting posture. "The back is curved throughout its length during the sitting; and, when the person habituated to that stretch of his spinal muscles rises, all he has to do to stand is to thrust his hips forward and not change the relationship of the spinal segments one bit" (Rathbone and Hunt, 1965) (Figure 17).
   In regards to this round sway back standing posture, Rathbone (1934)

commented that "Unfortunately we are seeing a great deal of it in high school years, among boys as well as girls; and no other form of posture anomaly demands such studied attention at present."

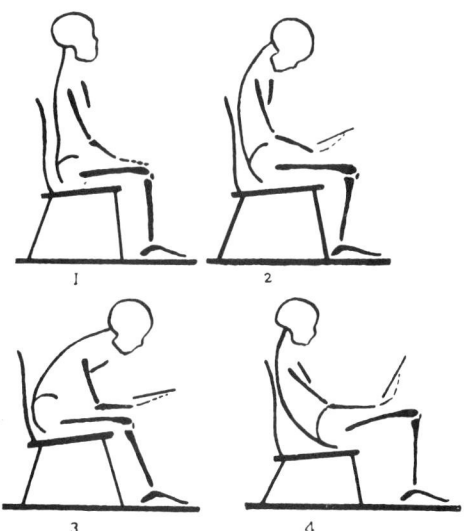

Figure 17. Sitting postures (2,3,4) contributing to a sway back standing posture. From Drummond, W.B.: *An Introduction to School Hygiene.* London, Edward Arnold, 1915. Reproduced with permission of Edward Arnold Publishers Ltd.

With this posture, "The lumbar curve is not increased; instead it is obliterated or reversed. There is a sharp hyperextension at the lumbosacral juncture due to the swaying back of the upper part of the body, but this bend cannot be called an increase in the lumbar curve." Rathbone (1934) found the erector spinae muscles to be weak in individuals with this posture, along with usually a decreased pelvic inclination.

Rathbone (1934) also observed this posture to be very common among individuals having serious emotional difficulties. Apparently, many of these individuals "lose the ability to stand in complete extension and assume a position of more or less flexion." Earlier, James (1922) commented that in depression "the flexors tend to prevail."

Rathbone (1934) also felt that proper chair design could improve standing posture. "While the chair is holding the trunk in extended position, the neuromuscular system is being patterned in a desirable posture which can carry over into standing and into movements."

Hawley (1937) also commented upon the effect of prolonged poor sitting postures on the individual's standing attitude:

"Indeed, in some sitting attitudes, slouching into a relaxed posture with the weight of the trunk supported in part by the coccyx and sacrum instead of by the ischial tuberosities causes the pelvis to assume a horizontal position. This in turn involves a flattening or reversal of the lumbar lordosis and, because most chairs push the shoulders forward, the relaxed sitting posture generally involves an abnormal degree of thoracic kyphosis with abducted scapulae and a forward position of the head.

Furthermore the increased kyphosis of the thoracic spine with depression of the ribs and a forward position of the head tend to become habitual, thereby increasing the difficulties of proper standing posture and affecting adversely the functioning of the organs of the thoracic, abdominal and pelvic cavities.

Moreover, the standing posture is likely to suffer due to an abnormal shortening of the anterior musculature and ligaments of the thoracic and cervical regions and a loss of tone of the abdominal muscles. Thus the standing posture commonly seen in sedentary individuals shows a forward head, depressed thorax, round back, forward shoulders and a prominent abdomen."

Howland (1936) characterized the round sway back posture as involving a decreased lumbar lordosis, along with a distinct hyperextension at the lumbosacral joint, the whole upper back swaying backward from this point. Howland (1936) felt that this posture was often mislabeled as involving an increased lumbar lordosis due to the presence of large hips or a protruding abdomen.

Knudsen (1947) considered the thoracic spine to be the region that suffered first and also most frequently from prolonged improper sitting, resulting in decreased mobility. He described the long round back posture as involving a long curve from the middle of the cervical spine to the lower lumbar region. "The erector spinae are not sufficiently developed and strong to perform the big work required for keeping the trunk upright throughout the day. They let it sink forward so that it is supported by ligaments more than by muscles" (Knudsen, 1947).

Knudsen (1947) described the long round back posture as involving a decreased pelvic inclination, with the normal lumbar curve almost straightened except for a sharp bend at the lumbosacral joint:

"In that way the weight of the trunk is partly carried backward and the line of gravity would fall too far behind the axis through the hip joints, if the head and the upper part of the dorsal spine were not carried forward as a counterbalance, thus rounding the upper part of the back.

The muscles will alter according to this position of the skeleton. As the normal lumbar curve is almost straightened the erector spinae of this region are lengthened; the same is the case with those of the dorsal spine, whereas those of the neck are shortened. In front the abdominal muscles are shortened as the sternum and the pubic symphysis are nearer one another than normally. In the long round back the shoulders are nearly always brought too far forward, and, therefore, the pectoral muscles are shortened."

Anderson (1951) referred to the approximation of the sternum to the pelvis in the slouched standing posture as a "postural depression." He implicated sitting in a kyphotic position as a major cause of this slouched standing posture, as it results in tightness of the rectus abdominis and overstretching of the back extensors (Figure 18).

Figure 18. A kyphotic sitting posture resulting in a "postural depression," with approximation of the thorax to the pelvis. From Barlow, W.: *The Alexander Technique.* New York, Warner Books Edition, 1980. Reproduced with permission of Alfred A. Knopf, Inc.

Interestingly, sit-ups are often advocated for correcting a slouched standing posture with its characteristic protruding lower abdomen. Sit-up exercises, however, will actually strengthen the tight rectus muscle that is already causing a "postural depression," approximating the thorax to the pelvis. Overstretching of the back extensors will also be reinforced with these exercises.

Wiles (1937) also felt that the action of the rectus abdominis was misunderstood in standing posture. Since in upright standing the thorax is the least fixed point compared to the pelvis, the major effect of contracting the rectus muscle will be to bring the thorax closer to the pelvis anteriorly. Wiles (1937) also commented that loss of tone in the upper lumbar and lower thoracic erector spinae would allow the entire weight of the upper part of the body to come on to the spine, resulting in a thoracolumbar kyphosis.

Schoberth (1962) mentioned that prominent features of the relaxed or fatigued standing posture included a sunken thorax and protruding abdomen. The stabilization of this posture was considered to involve more passive mechanisms. He described a backward rotation of the pelvis, with the gravity line of the trunk shifted further behind the hip joints. With this posture, there is apparently a relaxation of the iliopsoas, with the hip joints stabilized by the stretched iliofemoral ligaments (Snijders et al., 1976).

More recently, Salminen (1984) studied the standing postural faults of 370 Finnish schoolchildren, from ages eleven to seventeen. A functional postural fault in the anteroposterior plane was found in 109 individuals (29.5 percent). The most common anteroposterior fault was a long roundback posture (49.5 percent).

Long roundback was defined by Salminen (1984) as:

> "A back where the spine had a long accentuated kyphosis continuing to the lumbar region, resulting in straightening of lumbar lordosis in the upper and middle parts of the lumbar spine. The stomach was pushed forward and lordosis was present only in the lower region of the lumbar spine. The hip joint seemed to be somewhat in extension and the whole upper trunk was inclined to some extent backwards. The head, cervical spine, and shoulders were pushed forward to retain balance. Posture could be retained with little muscular effort by leaning onto ligaments."

In regards to this common long roundback posture where the gravity line of the upper trunk is shifted backwards, Salminen makes reference to an earlier Finnish study by Tawast-Rancken (1960). According to Salminen, in this study of approximately three hundred primary and secondary school pupils, Tawast-Rancken (1960) described "mostly kyphotic postures, in which the trunk inclines backward and in which little muscle effort is required." Hyperlordotic (hollow back) postures were rarely found, and Tawast-Rancken (1960) felt that in many studies long roundback postures and hyperlordotic postures are confused.

Hamstring tightness was found in over forty percent of the students in Salminen's study (1984), and was most frequently associated with the long roundback posture as opposed to other anteroposterior deviations. Knudsen (1920, 1947) and Vidal (1984) also found an association between hamstring tightness and a kyphotic standing posture with a decreased pelvic inclination.

THE SEARCH FOR POSTURAL STABILITY

According to Kelly (1949):

"Posture is a distinct problem to humans because the skeleton is fundamentally unstable in the upright position. A four or even a three-legged chair or stool can be quite stable. But who ever heard of a two-legged piece of furniture? The two-legged human body presents a continuous problem in maintaining balance, a problem augmented because the feet are a very small base of support for a towering superstructure. And as though this were not problem enough, the trunk, head, and arms are supported from the hips upward by a one-legged arrangement of the spine."

Hellebrandt and Franseen (1943) commented that the "architectural design of the human is not conducive to stability. A segmented structure, its center of gravity is placed high above a relatively small supporting base."

Standing is in reality movement upon a stationary base, with postural sway being inseparable from man's upright stance (Hellebrandt, 1938). Postural sway is always the greatest in the anteroposterior plane. Thomas and Whitney (1959) reported that during postural sway, "rotation of the trunk relative to the lower limbs occurs in addition to, and generally in the same direction as, rotation of the whole body about the ankle pivots."

Man's postural behavior can be interpreted as spontaneous attempts at attaining relative stability of the body's segmented structure (Branton, 1969). The Minimal Principle of Nubar and Contini (1961) postulates that "the individual will, consciously or otherwise, determine his motion (or his posture, if at rest) in such a manner as to reduce his total muscular effort to a minimum consistent with imposed conditions, or 'constraints.'" Through the introduction of passive mechanisms, the antigravity posture-maintaining musculature can to some extent be relaxed (Schoberth, 1962).

As Kelly (1949) commented:

"Individuals too commonly fall into postural habits where the shoulders and upper trunk are carried backward and the hips forward. In such positions they hang on the ligaments which cross the front of the hip joint and on those which hold the various spinal bones in a column, because in this posture the antigravity muscles of the trunk are required to expend little energy. The strain falls largely on the ligaments. These ligaments are able to withstand strain for short periods without damage or complaint. But over longer periods, as is the case with poor habitual standing and walking positions, or through long hours of sitting, the ligaments may fail to protect joints adequately.

If a person's habitual posture involves much leaning on ligaments, he does not sway as readily as he otherwise would, because he is already balanced at one extreme of his total range of possible movement. Hence, he develops local strains more rapidly."

Symmetrical Stabilization

In idealized erect standing posture, the hip joints are not in maximum extension. Radiographic investigations by Åkerblom (1948) on twenty-five subjects showed that the amount of further hip extension possible from erect standing may be as much as fifteen degrees, with the average being six degrees. In normal standing posture, Åkerblom considered the trunk to be balanced over the hip joints in a position of unstable equilibrium, with the line of gravity usually falling just posterior to the center of the hip joints.

According to Ahlback and Lindahl (1964), "It is sometimes stated that the hip joint in the erect position is fully extended and that it is the iliofemoral ligament that provides the stabilization of the joint. In the cases in which radiographic measurements were made in the erect position it was found that there was a flexion of the hip joint."

With the line of gravity usually falling slightly behind the hip axis in erect standing, gravitational forces will tend to slightly extend the hip and decrease the pelvic inclination (Woodhull et al., 1985). By leaning the upper trunk backwards as in the slumped, fatigue posture, one will tilt the pelvis backward and fully extend the hip joint (Brunnstrom, 1972; Braus, 1921). The hip joints will then be passively stabilized in extension by tension of the iliofemoral ligaments.

Stabilization of the hip in full extension through tension of the iliofemoral ligament can also increase knee stabilization in extension. According to Meyer's theory (1853), as explained by Brunnstrom (1972):

"... in complete hip extension when the iliofemoral ligament becomes tensed, this ligament causes the femur to internally rotate with respect to the pelvis (and with respect to the tibia). This internal rotation of the femur at the knee constitutes a locking mechanism, since the shape of the femoral condyle is such that flexion can take place only if the femur first externally rotates somewhat with respect to the tibia. Meyer furthermore points out that stability of the knee is also affected by the iliotibial tract which is attached to the anterior surface of the lateral condyle of the tibia. As the hip extends fully, the iliotibial tract is put on a stretch and thus stabilizes the knee against flexion."

According to Benninghoff (1939), "When the iliofemoral ligament is put under tension by tilting the pelvis backwards the upper part of the band tends to rotate the femur inwards and thus holds the bones in the position of the terminal rotation. The pelvis, when tilted backwards thus fixes the knee joint in a position in which it is protected against bending."

Luciani (1915) felt that passive stabilization of the hip and knee due to tension of the iliofemoral ligament could also contribute to stabilization at the ankle joint:

> "The line of gravity of the whole body falls on the ground in a plane somewhat anterior to the line between the two tibio-astragalic articulations, and the body tends to fall forwards. This is avoided by the fact that the plane of flexion in this joint is very oblique with that of the other side; the two planes of flexion form an angle of 60° open to the front. In order that flexion at these two joints should be possible, it is therefore necessary for the two knees to be moved apart from each other, and flexed. When flexion of the knees is prevented, falling forward owing to flexion of the tibio-astragalic articulations is also prevented. As the fixation of the hip-joint determines the fixation of the knee, the fixation of this joint leads to the fixation of the ankle. Here again the gastrocnemius, soleus, posterior tibial, and posterior peroneal muscles also take part in maintaining fixation."

As the body is predominantly an open-chain system of links, additional stability can be imparted through temporary closed chains of the body segments (Dempster, 1955b). Standing with the hands in the pockets or the arms crossed over the chest are two common examples (Dempster, 1955b; Schoberth, 1962).

The common habit of standing with the hands in the pockets was implicated as facilitating a slouched, fatigue posture by Bancroft (1913), Goldthwait (1909), and Browne (1916).

Asymmetrical Stabilization

Smith (1953) noted that immobile periods of standing are "habitually prolonged in certain individuals, and it is well known that some occupations demand, on occasion, a form of standing which is continuously immobile." He observed asymmetrical standing attitudes to occur four times as often as symmetrical attitudes.

Schoberth (1962) commented that the striving for the most economic use of muscle power causes an asymmetry of the human posture. In the commonly observed asymmetrical position, the body weight is carried on one leg, with the other leg being placed slightly forward and out to the side, acting as a prop and bearing very little weight (Figure 19). As with the symmetrical slumped, fatigue posture, stabilization in extension at the hip and knee of the weight bearing leg will result from tension in the iliofemoral ligament and iliotibial tract (Phelps and Kiphuth, 1932). By increasing the distance between the feet, additional stability is achieved as the size of the standing base is increased and the center of gravity is lowered (Glassow, 1932).

Figure 19. Asymmetrical standing posture, with the body weight over the right leg. From Mosher, E.M.: *Health and Happiness.* New York, Funk and Wagnalls, 1913. Reproduced with permission of Harper and Row, Publishers, Inc.

With the asymmetrical standing posture, McKenzie (1915) considered the strain to be borne by the ligaments of the hip and spine for prolonged periods. "Such cases are nearly always accompanied by rounding of the shoulders, flattening of the chest, protrusion of the abdomen, and rotation of the vertebrae—all signs of muscular fatigue and ligamentous strain."

According to Phelps and Kiphuth (1932), in the asymmetrical attitude, "there is less muscular effort necessary to maintain stability than in any other standing position. However, habitual maintenance of this position in individuals who stand a great deal of the time is likely to produce ligamentous strains eventually because of the asymmetrical distribution of weight."

Vierordt (1862) recorded the oscillations of the head in different standing positions, in order to determine the most stable standing posture. With Vierordt's technique, a pen was attached to the head by a cap, which traced on paper fixed horizontally from above the anteroposterior and lateral oscillations. The fewest oscillations were found with the asymmetrical standing posture.

Mosher (1913) illustrated the asymmetrical skeletal alignment that occurred when standing on the right leg with the left leg directed diagonally forward, a posture she referred to as "right foot twist" (Figure 20). With this posture, it will be noted that the pelvis tilts down to the left, the spine assumes a long C-curve convexity to the left, the right shoulder is lowered, and the head tilts to the right.

A study on standing weight distribution by Marsk (1958) found that individuals working with the right hand had a marked tendency to use the left foot as the supporting foot.

According to Marsk (1958):

"It seems, thus, that the use of the right or left hand by the individual in question has a bearing influence upon the selecting of the standing-foot; those working with the right hand being more inclined to use the left foot for that purpose and those working with the left hand more inclined to use the right foot.

Judging from the results, the assumption of a crossed adjustment arm-leg therefore finds support. The cause of this fact seems to be that basic principles of statics and dynamics act in favour of using the contra-lateral leg as the standing-leg. The most stable equilibrium — doubtless of fundamental importance when working in a standing position — is then achieved by establishing a diagonal adjustment between arm and leg."

Mosher (1914) commented that sitting with one leg "tucked under," or habitually leaning to one side when writing, would result in a spinal attitude similar to the asymmetrical standing posture (Figures 21 and 22). Mosher (1913) also stressed the importance of learning a different way to shift the body weight from side to side with prolonged standing, thereby preventing the loss of body symmetry. With one foot placed a short step forward in advance of the other, Mosher (1913, 1914) advocated swaying the entire trunk gently forward until the body weight rests on the ball of the forward foot (Figures 23 and 24). When alternating the

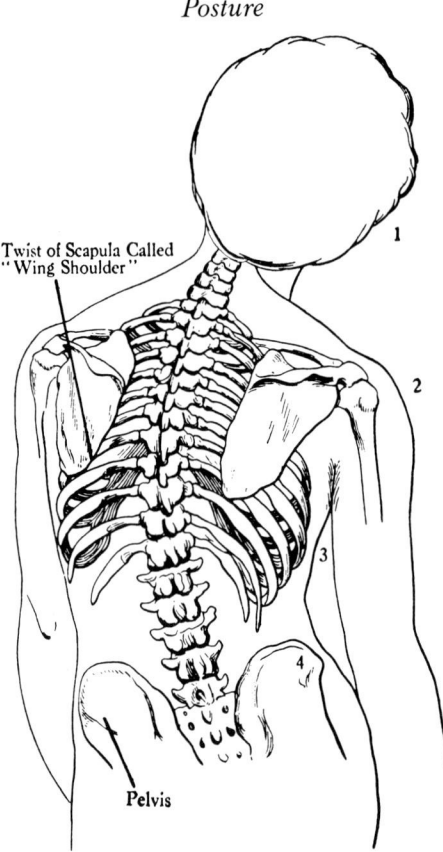

Twist of Scapula Called
"Wing Shoulder"

Pelvis

Figure 20. Asymmetrical skeletal alignment that occurs in the asymmetrical standing posture of Figure 19, with the body weight over the right leg. Note in particular: the head is tilted towards the right shoulder (1); the right shoulder is lowered (2); the shortened trunk on the right (3); the raised pelvis on the right (4). The spine assumes a long C curve convexity to the left. From Mosher, E.M.: *Health and Happiness.* New York, Funk and Wagnalls, 1913. Reproduced with permission of Harper and Row, Publishers, Inc.

weight bearing leg, the other foot would be moved forward to take the body weight.

Lee and Wagner (1949) also stressed, "Never stand with the weight on one foot unless it is a one-foot-forward position and with the weight on the forward foot."

Energy Expenditure

The Minimal Principle of Nubar and Contini (1961) postulates that "in all likelihood the individual will, consciously or otherwise, determine his motion (or his posture, if at rest) in such a manner as to reduce

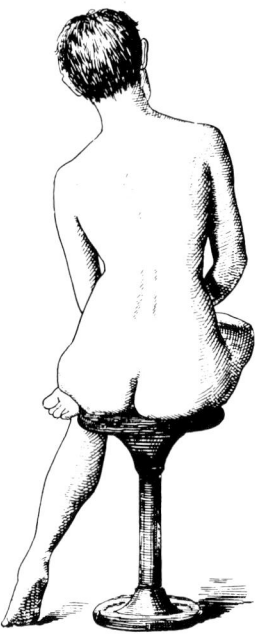

Figure 21. Sitting with one leg "tucked under." From Bradford, E.H., and Stone, J.S.: The seating of school children. *Transactions of the American Orthopaedic Association, 12:* 170–183, 1899.

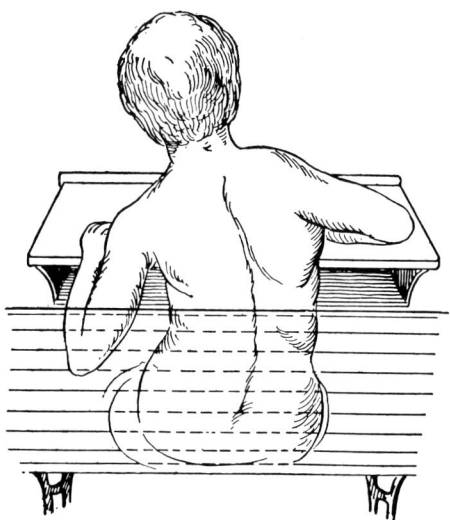

Figure 22. Leaning to one side when writing. From Pyle, W.L. (Ed.): *A Manual of Personal Hygiene,* 5th ed. Philadelphia, Saunders, 1913. Reproduced with permission of W.B. Saunders Co.

Figure 23. Proper way of shifting the body weight with prolonged standing, according to Mosher (1913, 1914). From Mosher, E.M.: *Health and Happiness.* New York, Funk and Wagnalls, 1913. Reproduced with permission of Harper and Row, Publishers, Inc.

Figure 24. A lateral view of shifting the body weight to the forward foot. From Cornell, W.S.: *Health and Medical Inspection of School Children.* Philadelphia, F.A. Davis, 1912. Reproduced with permission of F.A. Davis Company.

his total muscular effort to a minimum consistent with imposed conditions, or 'constraints.'"

In the symmetrical slumped, fatigue posture and the asymmetrical standing posture previously described, the antigravity postural musculature is required to expend a minimum of energy, with the strain falling largely on the ligaments (Kelly, 1949). A reduced energy expenditure can be expected as tension acting on the ligaments does not increase the energy demands of the body (Tichauer, 1967).

According to McCormick (1942), the first reported investigation of the metabolic cost of various standing postures was by Zuntz and Katzenstein in 1891. Taking as a base line the sitting metabolic rate of the one male subject studied, they found a three percent increase in metabolic rate when standing in the asymmetrical attitude, a six percent increase in easy relaxed standing, and a twenty-five percent increase when standing at attention.

McCormick's (1942) investigation of the metabolic cost of maintaining a symmetrical standing posture indicated that the body alignment associated with the minimum metabolic increase involved:

1. The knees pushed back to the limit of hyperextension.
2. The hips thrown forward to the limit of extension.
3. The thoracic curve increased to the maximum.
4. The head forward.
5. The upper part of the trunk tilted slightly backward over the hip joints.

McCormick (1942) made some important remarks in regards to the question of whether a standing alignment with minimum energy expenditure should be recommended as an ideal posture to assume:

"In addition to the metabolic cost of a given alignment there are a number of factors which must be considered in determining ideal posture. The appearance of the alignment, the amount of strain caused by it on joints and ligaments, its effect on the functioning of the internal organs, its efficiency as a starting position and basic underlying pattern for movement are a few of these factors.

The alignment associated with minimum metabolic increase—where the knees are pushed back, the hips thrust forward, the trunk tilted back, the upper back rounded, and the head dropped forward—does not seem to measure up to these other requirements. This alignment seems to be maintained primarily by means of ligamentous action, and it is quite possible that strain would be put upon these ligaments which, as the tissues lose their elasticity, might produce such disabilities as sacroiliac strains. It is also possible that in this alignment the internal organs are either compressed or allowed to slip downward and their functioning may be somewhat impaired. Finally, very few individuals would agree that this alignment is acceptable aesthetically.

Undoubtedly the ideal alignment should be an efficient one, but that does not mean that there should be a minimum expenditure of energy. *Efficiency implies a maximum return for a minimum output,* not simply a minimum output. The alignment which was found to be associated with minimum metabolic cost brings practically no returns in terms of other factors involved, and if authorities could agree on anything, they would doubtless agree that this alignment cannot be considered to be good posture."

STABILIZATION IN PREGNANCY AND OBESITY

The common view regarding the postural changes associated with pregnancy is that with enlargement of the uterus and the increased weight of the fetus, the center of gravity is shifted forward and there is an increased anterior tilt of the pelvis. An increased lumbar lordosis is necessary to maintain the upright posture and prevent the individual from falling forwards. A large increase in erector spinae muscle activity will therefore be required to maintain an upright standing posture.

However, there is actually more evidence to support other adjustments of the body segments to achieve postural stabilization with pregnancy. As Appleton (1946) stated, "The characteristic standing attitude of the pregnant woman must again be accepted as one that is in general terms appropriate, even though individual differences may be seen that reflect previous habit or a relative weakness of the trunk musculature. The pelvis is usually tilted less forwards; by this means, as well as by a plantar flexion of the ankles, the mid-thoracic region is carried backwards relatively to the feet and the centre of gravity of the body is thus prevented from being displaced too far forwards by the enlarged abdomen."

Fries and Hellebrandt (1943) found that with the increase in the anteriorly placed load as pregnancy progresses, the individual compensates by leaning backward beyond her normal gravitational center (Figure 25). The major postural adjustments are made by "elevating the head, extending the cervical spine, stabilizing the knee joint, and leaning backward from the ankle."

Therefore, the increased anterior load of pregnancy causes a "temporary overcompensation which displaces, backward, the average location of the

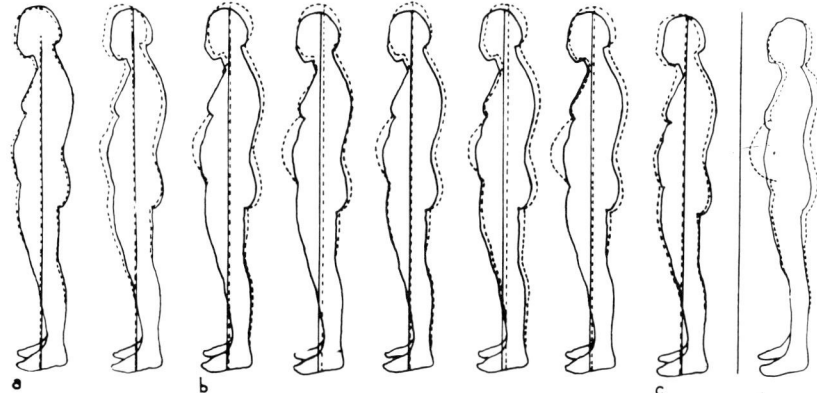

Figure 25. Tracings of superimposed photographs demonstrating the postural changes accompanying pregnancy. In a,b, and c, successive observations of one subject are compared in turn (dotted line) to a photograph taken early in pregnancy (solid line).
   a. First trimester
   b. Biweekly observations of the third trimester
   c. Six weeks post partum
   In part d, a photograph of a different subject taken 24 hours ante partum (dotted line) is superimposed upon a photograph taken 20 days after parturition (solid line).
   From Fries, E.C., and Hellebrandt, F.A.: The influence of pregnancy on the location of the center of gravity, postural stability, and body alignment. *The American Journal of Obstetrics and Gynecology,* 46:374–380, 1943. Reproduced with permission of The C. V. Mosby Company.

vertical projection of the center of gravity in the sagittal plane." This backward displacement of the center of gravity would protect against "an acute unbalancing force which might lead to a forward fall, endangering the fetus" (Fries and Hellebrandt, 1943).

Fries and Hellebrandt (1943) also reported that the exaggerated lumbar curve commonly thought to occur in pregnancy was not seen. More recently, Snijders et al. (1976) measured the dorsal contour of the spines of sixteen pregnant women. They found that in every case the spine was straighter immediately before childbirth compared to after childbirth.

These postural changes associated with pregnancy would reinforce the Minimal Principle postulated by Nubar and Contini (1961). There will be a decreased expenditure of energy with these postural adaptations compared to the strong static contraction of the erector spinae that would be required if an excessive lumbar lordosis was the posture adopted to prevent the individual from falling forwards.

With obesity, it is also easy to presume an increased lumbar lordosis, as the soft tissue contours of a protruding abdomen and large buttocks can be misleading. Howorth (1946) described the posture of an obese person as involving a compensation for the protruding abdomen. The compensation involves leaning the upper trunk backwards, resulting in a long thoracolumbar kyphosis.

In individuals with an excess of abdominal fat, Kellogg (1927) referred to the forward carriage of the hips, along with the posterior displacement of the upper trunk in relation to the vertical axis (Figure 26).

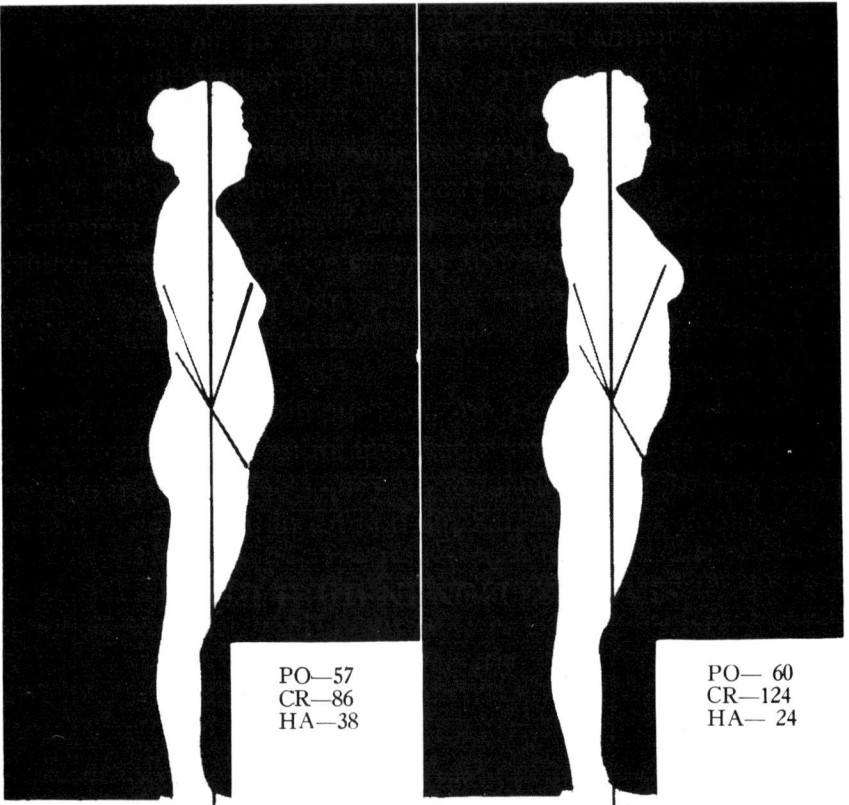

PO—57
CR—86
HA—38

PO— 60
CR—124
HA— 24

Figure 26.
    Left: The forward carriage of the hips and the posterior displacement of the upper trunk, observed by Kellogg (1927) in most individuals with an excess of abdominal fat.
    Right: The protruding lower abdomen is greatly lessened with proper postural correction. From Kellogg, J.H.: Observations on the relations of posture to health and a new method of studying posture and development. *The Bulletin of the Battle Creek Sanitarium and Hospital Clinic, 22:*193–216, 1927. Note: PO = Pelvic obliquity CR = Chest ratio HA = Head angle.

STABILIZATION IN HIGH HEELS

The common view regarding the posture in high heels is that there will be a forward displacement of the center of gravity. An increase in activity of the erector spinae, along with an increased lumbar lordosis, will be necessary to keep the trunk erect and prevent the individual from falling forwards.

However, with the use of a balance apparatus to determine the location of the center of gravity, Reynolds and Lovett (1910) found that "high-heeled shoes tip the body back as a whole without making any appreciable change in the lumbar curve." They mention that a forward displacement of the center of gravity when standing would lead to an increased demand on the posterior musculature to maintain the erect position. With high heels, instead of this increased demand on the back musculature, Reynolds and Lovett (1910) found that "the strain on the posterior musculature is relieved by motion of the center of gravity backward, through movement of the body backward as a whole, chiefly from the ankle joint " (Figure 27).

The tendency for the individual to lean backward when wearing high heels was also found in the Antioch College studies (1931) on the effects of women's shoes on body mechanics (Figure 28). Compared to a group of women accustomed to wearing low-heeled shoes (heel height not exceeding one inch), the following standing postural faults were found to be two to three times more frequent among a group of women who habitually wore high-heeled shoes (heel height exceeding 1½ inches): a sway back posture, an increased thoracic kyphosis, and an increased relaxation of the abdominal muscles (Antioch College, 1931).

Bendix et al. (1984) recently examined the effect of high heels on eighteen healthy women. They found that with high heels the lumbar lordosis and the forward pelvic inclination were both decreased. In order to counteract the tendency of the body to fall forward with high heels, the line of gravity was found to shift backward by extension of the ankle joints, and a shifting of the legs and trunk backwards.

Effect of Heels on Normal Foot Physiology

Stewart (1945) considered the heels of all shoes to be "utterly destructive of normal foot physiology." He felt that heels should be omitted altogether in shoe design.

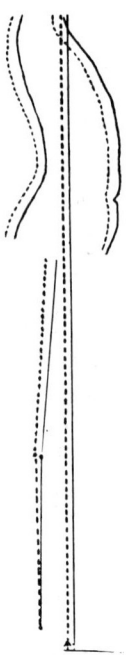

Figure 27. The standing posture induced by high heeled shoes (dotted line) compared to the normal standing posture with the unshod foot (solid line). From Reynolds, E., and Lovett, R.W.: An experimental study of certain phases of chronic backache. *The Journal of the American Medical Association, 54:*1033–1043, March 26, 1910. Copyright 1910, American Medical Association. Reproduced with permission.

Stewart (1945) described the adverse effect of heels as follows:

"The muscles of the non-weight bearing foot appear to be toneless. The instant weight is borne on the foot, one may feel reflex or postural tone develop in the abductors hallucis and minimi digiti, and in the adductor hallucis and to a lesser degree in the other plantar muscles. The insertion of a heel even as low as that of the average man's shoe causes a decrease in the tone of these postural muscles by about a half, and there develops a slight increase in the tension on the plantar fascia. The use of a two or three inch heel completely relieves the intrinsic postural muscles of the foot of their tone, and markedly increases tension in the plantar fascia; i.e., the postural reflex mechanism of the foot is negated by heels, and passive tension on non-contractile fascia is substituted for muscle tone. When one considers the fact that many individuals never go unshod, and that even their bedroom slippers are heeled, one can then appreciate the fact that the postural use of the muscles of the foot are rarely if ever used, and they are constantly straining on tissue with an essentially different physiology."

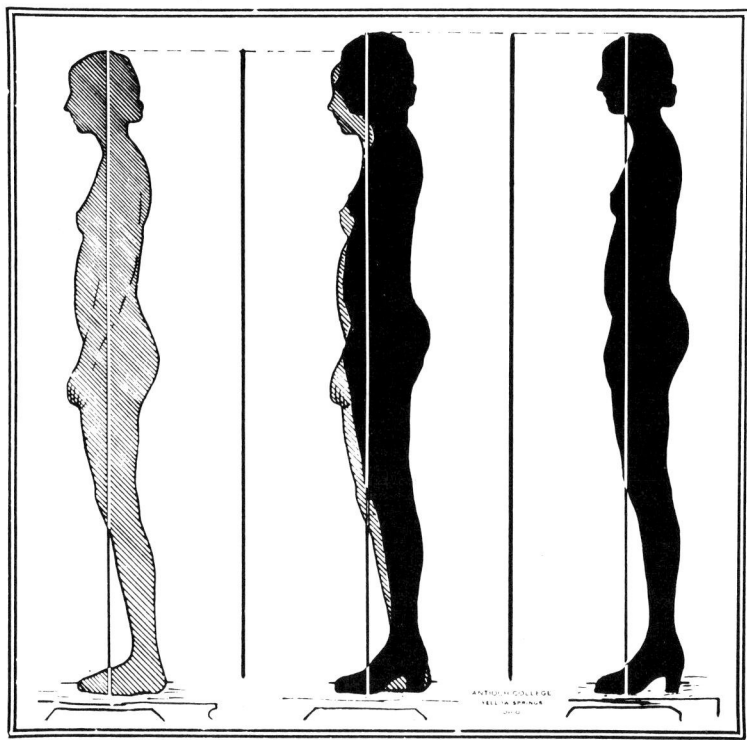

Figure 28. The figures on the left and right sides of this illustration are the same subject, barefoot (left) and in high-heeled shoes (right). The center figure is a composite of the two, showing the change in posture with high heeled shoes. From *The Antioch College Studies of the Effects of Modern Shoes upon Proper Body Mechanics.* Yellow Springs, Antioch College, 1931. Reproduced with permission of Antioch University.

TRUNK STABILIZATION

In all animals below man, the shoulders and arms serve to support the trunk. However, man's assumption of an upright, bipedal posture demanded an elaboration of the mechanism of trunk and visceral support. As Keith (1923b) stated:

"It was with the evolution of the plantigrade posture of the human body that the postural compression of the abdominal viscera reached its highest point. The arms no longer assisted in supporting the body—the reverse became the case—the erect trunk had to afford a firm basis of action for the arms. For this purpose the postural complex of muscles which surround and enclose the abdominal cavity is called continuously into action.

The moment we assume a sitting or standing posture there is set going a reflex mechanism which throws the muscles of the belly wall into a state of postural tone."

Keith (1923b) also noted that:

"Under modern conditions of life, this postural mechanism of man's body is particularly liable to break down."

Stockton (1913) stressed that:

"A certain amount of intra-abdominal pressure is necessary if the viscera of that region are to be held in their proper places and proper relations. This is possible only when the body is erect in sitting and standing, when the chest is kept habitually raised to its normal position, and when the abdominal muscles are strong, and are not allowed to relax, pouch out, and thus favor the descent of the organs" (Figures 29 & 30).

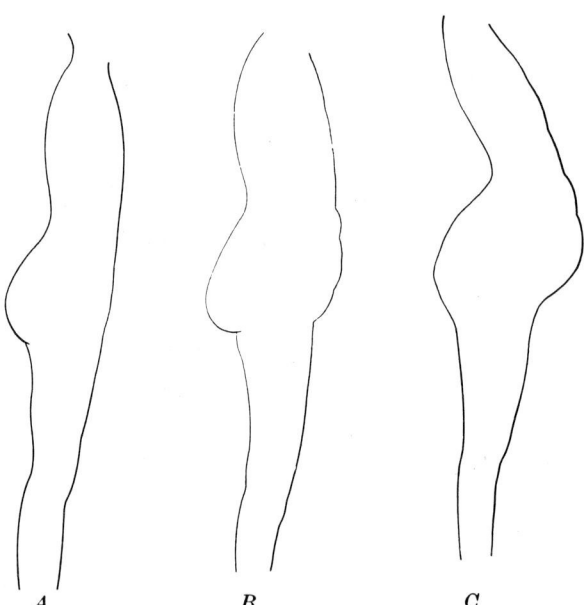

A   B   C

Figure 29. With relaxation of the postural abdominal muscles (B and C), there will be a protrusion of the lower abdomen and an improper axial relationship of the thorax and pelvis.

From Bowen, W.P.: *Applied Anatomy and Kinesiology,* 2nd ed. Philadelphia, Lea and Febiger, 1919. Reproduced with permission of Lea and Febiger.

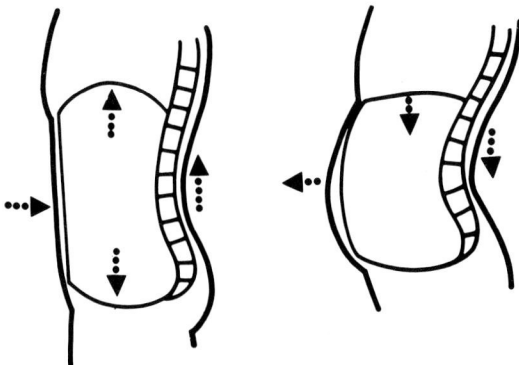

Figure 30. The proper intra-abdominal pressure (left illustration) will help restore the correct axial relationship of the thorax and pelvis, along with "elongating" the lumbar spine. From Macnab, I.: *Backache.* Baltimore, Williams and Wilkins, 1977. Reproduced with permission of Ian Macnab, M.D.

Trunk Muscle Activity

Upright Erect Posture

Besides the slight but continuous activity of the erector spinae with the upright erect posture, certain specific abdominal muscle activity is also present (Keith, 1923b; Floyd and Silver, 1950; Ono, 1958; Teshima, 1958; Strohl et al., 1981; DeTroyer, 1983). These critical postural abdominal muscles can be grouped together as the "lower abdominal" muscles. Included here are the lower fibers of the transversus abdominis, the lower anterior fibers of the internal obliques, and the lower (lateral) fibers of the external obliques (Figure 31). These lower abdominal muscles should be considered as part of man's normal postural antigravity muscles.

The role that these lower abdominal muscles play in upright erect standing involves:

1. Compression and support of the abdominal viscera (Ono, 1958; Hatami, 1961; Kendall et al., 1971).
2. Maintaining the proper axial relationship of the thorax and pelvis. The lower (lateral) fibers of the external obliques are considered the most important regarding this function (Kendall et al., 1971).

Bancroft (1913) referred to the protrusion of the abdomen in the slumped posture, produced by the backward sagging of the thorax: "In correcting the position, and bringing forward the upper spine, there is a

Posture

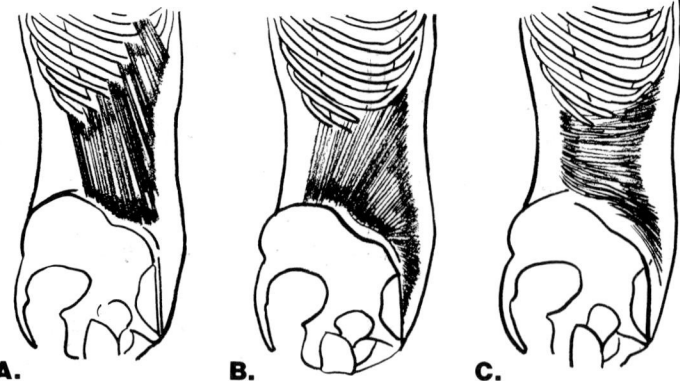

Figure 31. The postural lower abdominal muscles.
  A. External oblique abdominal muscle.
  B. Internal oblique abdominal muscle.
  C. Transversus abdominal muscle.
Adapted from Knudsen, K.A.: *A Textbook of Gymnastics,* 2nd ed. *Volume One. Form-Giving Exercises.* London, J. & A. Churchill Ltd., 1947. With permission of Longman Group Ltd.

tightening of the abdominal muscles, but these act on the front of the pelvis as a fixed point, and bring the thorax forward rather than draw the abdomen backward."

   3. An important role in inspiration, as explained by DeTroyer (1983):

   "The persistent activity of the abdominal muscles during inspiration may also confer some mechanical advantage onto the respiratory muscles. By opposing the descent of diaphragmatic dome, this inspiratory muscle activity will prevent excessive shortening of the diaphragm, thus maintaining the diaphragmatic muscle fibers on a more advantageous portion of the length tension curve and preserving their increased ability to generate pressure. In addition, by decreasing the compliance of the abdominal compartment, this inspiratory activity may also improve the diaphragm's ability to lift the lower ribs, and hence, to expand the lower rib cage, in much the same way that an externally applied abdominal compression helps the diaphragm in expanding the rib cage in seated quadriplegic subjects. As a result, the abdominal muscle tone which naturally exists in the standing position may allow the rib cage to expand more for a given activation of the diaphragm and inspiratory rib cage muscles. In so doing, the abdominal muscles may take on a substantial fraction of the load imposed on the inspiratory muscles."

   Goldthwait et al. (1934) referred to the fullness of the lower rib region and the high diaphragmatic position at the start of inspiration with good lower abdominal muscle tone.

Co-Contraction of Erector Spinae and Lower Abdominals

Rathbone (1934) stressed that the lower abdominal muscles should be grouped with the extensors of the body, as they "take part in all trunk movements, due to a reflex connection with the other muscles of the torso. They are intimately connected in the reflex of extension and are contracted whenever the body assumes the completely extended position."

Rathbone (1936) also commented upon the relationship between weak abdominal muscles and weak erector spinae: "It has been our observation for several years that the lower abdominal wall and the erector spinae muscles are associated reflexly in all vigorous extensions of the human body. Therefore, with sagging abdominal walls one would expect to find weak backs."

This relationship between the erector spinae and lower abdominals has also been noted in more recent studies. Woodhull-McNeal (1986) found a positive correlation between activity in the erector spinae and external obliques when standing. Deusinger and Rose (1985) reported a high degree of correlation between back extensor force and external oblique EMG activity with prone back extensions. Carlsöö (1980) also reported strong activity of the lower (lateral) external obliques during strength testing of the back extensors.

Trunk Muscle Activity with Slumped, Fatigue Posture

Asmussen (1960) has shown that with erect upright standing posture there will be continuous activity from the erector spinae, as the line of gravity of the head, arms, and trunk is anterior to the center of L4, the vertebra that lies most ventrally in the lumbar lordosis. As gravity will tend to pull the spine forwards, the erector spinae are the active antigravity muscles of the trunk, counteracting this flexion force.

However, Klausen and Rasmussen (1968) found that the activity in the erector spinae ceases and is replaced by activity in the upper rectus abdominis with a slight leaning backwards of the upper trunk, and the corresponding shifting of the line of gravity slightly posterior. This leaning backwards of the upper trunk is characteristic of the most common anteroposterior postural fault, the slumped, fatigue posture. It is also characteristic of the standing posture adopted with obesity, high heels, and during the later stages of pregnancy.

This change in trunk muscle activity with a backward leaning trunk posture was observed earlier by Bowen (1917). Floyd and Silver (1955) also reported relaxation of the erector spinae with a backward displacement of the center of gravity of the body. This was observed with a slight backward swaying movement, occurring mainly at the ankle joints.

Besides the relaxation of the erector spinae with a backward leaning trunk posture, Okada (1972) also reported relaxation of the erector spinae with the asymmetrical slouched standing posture.

Back Pain Implications

Bancroft (1913) stated that in individuals in whom the slumped, fatigue standing posture had become habitual, "the muscles on the front of the body are too much contracted, and those on the back too much relaxed."

Martin's description (1912) of the relaxed, slouched posture of an individual prone to ptosis included the following: a flat lumbar spine, descent of the ribs with a flat chest and a tight upper abdominal wall, a bulging lower abdominal wall from muscular weakness and visceral pressure, and a forward bend of the neck. In these individuals, Martin (1912) found the erector spinae to be "woefully neglected and atrophied."

Martin (1912) brought up several questions regarding the relationship between the strength and endurance of the trunk musculature and the occurrence of this posture:

> "Is it because the anterior group of muscles or flexors are proportionately more powerful than the extensors for the task they have to perform?"
> "Is it because when the back muscles or body extensors are favored or allowed to remain dormant for a long time by the individual assuming frequently the position of relaxation, these muscles will not only become weakened but will be obliged when they do act, to overcome and take up the slack of the stretched fascia, aponeurosis and ligaments, and this in time will lead to their exhaustion and to the development of still further deformity while they recuperate, and thus another vicious circle will be established?"

Collins et al. (1982) found that in tasks requiring increased activity from the erector spinae muscles, individuals with chronic low back pain showed less activation of these muscles than individuals without back pain. Pope et al. (1985) found low back pain patients to be flexor overpowered.

Concerning low back pain and pregnancy, Mantle et al. (1977) reported that 48 percent of a group of 180 women experienced backache during pregnancy. The most frequent aggravating factor for backache was found to be standing. For over 60 percent of these women, the onset of low back pain was around the six month of pregnancy. It is during the third trimester of pregnancy when the gravity line will be shifted the furthest posteriorly (Fries and Hellebrandt, 1943). With pregnancy, Fries and Hellebrandt (1943) mentioned that the flexor muscle groups are involved in postural control against gravity.

An important one year prospective study on low back pain in a general population by Biering-Sørensen (1984a,b) revealed that the two most important risk factors for first time occurrence of low back trouble in men were:

1. Poor isometric endurance of the back muscles.
2. Hypermobility of the lumbar spine in flexion.

According to Biering-Sørensen (1984b), the finding that good back muscle endurance should protect to some extent against low back trouble "seems to be in agreement with the fact that the back muscles maintain the erect posture of the spine throughout the day which must require a certain isometric back muscle endurance. Such endurance is probably also essential considering many manual handling procedures, including lifting and load carrying."

One characteristic apparently in common with standing postures involving a backward leaning trunk posture is the relaxation of the erector spinae. Habitual standing postures where the erector spinae are no longer the antigravity trunk muscles could result in disuse, overstretching, rapid fatigue, and a decreased isometric endurance of the erector spinae, along with increased stress on the ligaments and discs (Jokl, 1982). This would be accentuated when combined with prolonged, slouched sitting postures.

These same habitual slouched standing and sitting postures will also have a tendency to result in weakness and overstretching of the lower abdominal postural antigravity muscles, along with a decrease in the normal resting intra-abdominal pressure (Stockton, 1913). With ambulation, the forward position of the hips and backward position of the upper trunk characteristic of slouched postures will result in the body weight landing heavily on the heels (Kellogg, 1927; Cureton and Wickens, 1935). The decrease in normal resting intra-abdominal pressure along with the posterior displacement of the line of gravity may therefore possibly increase the spinal trauma with walking and other activities.

Proper Postural Correction

Gurfinkel et al. (1981) considered the position of the trunk to be the "principal parameter of regulation" for stabilization of the vertical posture. The axial relationship of the thorax and pelvis, and the way the load of the trunk is habitually carried are therefore of major importance both in postural analysis and postural correction. In order to restore the normal thorax-pelvis relationship and bring the upper trunk forward, and over the hips, the key area for spinal correction is the lower thoracic and upper lumbar spine (T9 thru L1). As will be discussed in the chapter on school seating, this is the region of the spine where prolonged improper sitting will have the most harmful effect. This is also considered to be the spinal region where the erector spinae are the weakest, and where extension needs to be emphasized (Rathbone, 1934; Wiles, 1937).

Activation of the erector spinae of this region and movement of the thorax forward should result in a spontaneous contraction of the lower abdominals, probably from segmental reflexes at the lower thoracic and upper lumbar levels. With the upper trunk moved forward into its proper position, there will be a lengthening of the rectus abdominis. The lower end of the sternum, as opposed to the lower abdomen, will be the most anterior part of the torso (Checkley, 1890; Kellogg, 1927; Frost, 1938). The chest, lower ribs, and diaphragm will all be raised to their most optimal position (Goldthwait et al., 1934). As the upper trunk is brought over or slightly anterior to the hips, as opposed to being carried posterior to the hips, the hamstrings will be activated as a postural anti-gravity muscle contributing to hip stabilization (Carlsöö, 1961, 1972).

Chapter Two

UNSUPPORTED SITTING POSTURES

A COMPARISON OF SITTING TO STANDING

Compared to standing, there will be a decreased physiological load on the individual when sitting (Grandjean, 1980a; Konz, 1983). Therefore, as a prolonged posture, sitting will be less fatiguing than standing.

Weight Bearing of Lower Limbs

As opposed to supporting the entire body weight in standing, most of the weight bearing function has been taken off the lower extremities in sitting. In standing, the fatigue from this weight bearing function is not due to muscle fatigue, as the muscle activity is only slight to moderate. Instead, it is due to the stress placed upon the ligaments and skeletal system. For example, with most individuals the integrity of the long arch of the foot in static standing is from ligamentous support and not muscle activity (Basmajian, 1978).

Cardiovascular Adaptations

The cardiovascular demands of standing are greater than for sitting. Ward et al. (1966) investigated several cardiovascular parameters of twenty healthy individuals while supine, five minutes after changing from a supine to standing posture, and five minutes after changing from a supine to sitting posture.

Due to the influence of gravity, they found that standing resulted in a marked peripheral pooling of blood. Compared to the supine position, the stroke volume decreased 45 percent and the heart rate increased 36 percent. There was a 27 percent decrease in the cardiac output.

The postural change from supine to sitting resulted in approximately half as much peripheral pooling of blood as the change from supine to standing. From supine to sitting, the stroke volume decreased 20 percent, and there was an 18 percent increase in heart rate. As a result, cardiac output fell only 10 percent.

More recently, Shvartz et al. (1982) studied the hemodynamic responses during prolonged sitting. With data from six healthy young men, they found that after five hours of quiet sitting there was a 19.4 percent increase in venous pooling in the calf, and a 15.6 percent decrease in calf blood flow.

Even though there is a greater increase in venous hydrostatic pressure in the lower leg with standing compared to sitting, one rarely stands completely still for prolonged periods (Pollack and Wood, 1949). The "venous muscle pump" is easily activated when standing, as the calf muscles will contract when walking, and even with involuntary postural sway (Hellebrandt et al., 1940).

With sitting, however, the calf muscles are usually relaxed and the "venous muscle pump" will not be activated. Therefore, even though the cardiovascular effects of standing are greater than for sitting, swelling and discomfort of the lower leg can be a major problem from prolonged sitting (Winkel, 1981; Pottier et al., 1969).

Energy Consumption

If the energy consumption when lying down is taken as 100 percent, standing will result in an 8 to 10 percent increase in energy consumption, and stooping will result in a 50 to 60 percent increase in energy consumption. Sitting, however, will only result in a 3 to 5 percent increase in energy consumption (Grandjean, 1973).

Stability

In standing, the center of gravity of the trunk is higher than in sitting, and the base of support involves only the feet. In sitting, not only is the center of gravity of the trunk lowered, but the base of support is enlarged, extending from the feet to the buttocks (Asatekin, 1975; Carlsöö, 1972). When applicable, the base of support will be expanded to include the projection of the backrest surface on the ground (Figure 32). The increased stability of the body when sitting with proper support for the buttocks, feet, and back will increase one's capacity for precision tasks or fine movements (Ayoub, 1972, 1973; Asatekin, 1975).

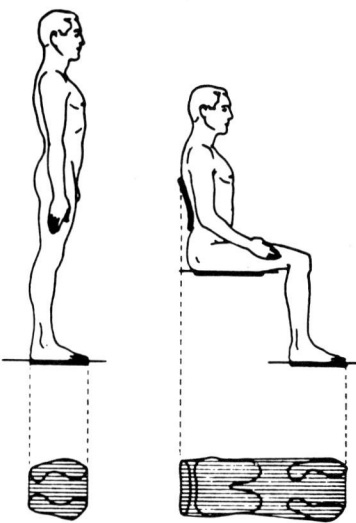

Figure 32. The base of support standing, compared to sitting in a chair with a backrest. From Asatekin, M.: Postural and physiological criteria for seating — a review. *M.E.T.U. Journal of The Faculty of Architecture, 1:*55–83, Spring 1975. Reproduced with permission.

On the other hand, there is a greater potential for pelvic instability in sitting compared to standing. In a relaxed standing posture, a passive locking mechanism is available at the hip joints from ligamentous support at full hip extension. With sitting this passive locking mechanism is not available, as the hip joints are in a mid-position. As a result, muscular stabilization or other external support is needed to stabilize the trunk over the hips (Meyer, 1873; Åkerblom, 1948; Coe, 1983).

Major Disadvantages of Sitting

"Today, the sitting position is the most frequent body posture in industrialized countries: we sit in the car or the train on the way to or from work, we sit most of the time at the work place, and in the evening we again sit in front of the television set. It can be stated without exaggeration that the sitting position is characteristic for modern times" (Grandjean and Hünting, 1977).

Unfortunately, sitting is probably also the unhealthiest of all the prolonged postures of the human body (Helbig, 1978). Whether due to poorly designed chairs or workstations, musculoskeletal factors, or improper movement patterns, a slouched kyphotic sitting posture predominates among observed sitting postures. However, compared to poor standing postures, poor sitting postures will usually always be accompanied by a greater degree of spinal flexion.

As a result, a prolonged, slouched sitting posture with a kyphotic lumbar spine has been frequently implicated as a major cause of low back pain (Keegan, 1953; Kottke, 1961; Cyriax, 1975; McKenzie, 1981). In contrast to a lordotic sitting posture, the slouched sitting posture will stress the posterior fibrous wall of the discs and posterior ligaments of the back, as well as cause a greater pressure increase within the discs. Overall, depending on how kyphotic the sitting posture, there will be an increased potential for pain and stress to the lower back, upper back, and neck.

This prolonged, slouched sitting posture has also been implicated as impairing both respiratory and digestive functioning (Goldthwait, 1909, 1915; Schurmeier, 1927; Bunch and Keagy, 1976; Goldthwait et al., 1952). This posture can constrict the abdominal and thoracic cavities, and increase the pressure on the abdominal viscera.

> "In this position the chest is necessarily lowered, the lungs are much less fully expanded than normal, the diaphragm is depressed, the abdominal wall is relaxed, so that with the lessened support of the abdominal wall, together with the lowering of the diaphragm, the abdominal organs are necessarily forced downward and forward" (Goldthwait, 1915).

EFFECT OF SITTING ON PELVIC INCLINATION AND LUMBAR LORDOSIS

When an individual goes from a standing to a relaxed, unsupported sitting position, the pelvis rotates backwards and there is a subsequent change of the lumbar lordosis into a kyphosis (Figure 33). This pelvic rotation is due in part to the tension of the hip extensors as the hips are flexed (Keegan 1953, 1964; Carlsöö, 1972). (Figure 34).

However, the major pelvic rotation upon sitting does not begin until after the buttocks are resting on the seat. This backward rotation of the pelvis is mainly due to the posterior rocking over the ischial tuberosities that occurs as the gravity line of the trunk comes to lie posterior to the ischial tuberosities (Åkerblom, 1948) (Figure 35B+C).

The amount of backward pelvic rotation that occurs when going from a standing to a relaxed, unsupported sitting posture has been investigated by Andersson et al. (1979) and Åkerblom (1948). Based on data from eight individuals, ages 21 to 44, Andersson et al. (1979) reported an average pelvic rotation of 28 degrees. Åkerblom (1948) reported an aver-

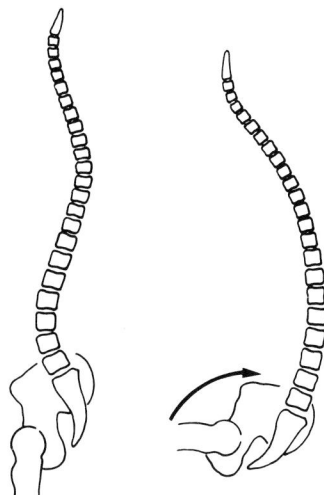

Figure 33. When changing from a standing position (left) to a relaxed, unsupported sitting position (right), the pelvis rotates backward and there is a subsequent change of the lumbar lordosis into a kyphosis. From Grandjean, E.: *Fitting the Task to the Man,* 3rd ed. London, Taylor and Francis, 1980. Reproduced with permission of Taylor and Francis Ltd.

age pelvic rotation of 35 degrees, from a study involving 32 individuals. Schoberth (1969) stressed that the shape of the spine in sitting "depends directly on the position of the pelvis."

Even though the thighs have changed from a vertical to a horizontal position when going from a standing to a relaxed, unsupported sitting posture, the actual hip flexion that occurs is not 90 degrees. It is not unusual to find only 50 to 60 degrees of actual hip flexion in a relaxed, unsupported sitting posture (Åkerblom, 1948; Schoberth, 1962; Carlsöö, 1972).

The lumbar flexion or kyphosis that occurs in relaxed, unsupported sitting is necessary in order for the individual to assume an upright posture after the pelvis has rotated backwards (Strasser, 1913; Åkerblom, 1948). This flexion involves mainly the lower three lumbar segments (Andersson et al., 1979; Schoberth, 1962; Åkerblom, 1948).

Based on data from twenty-five individuals, ages 5 to 41, Schoberth (1962) found an average total flexion of 30.4 degrees from lumbar segments L3-L4, L4-L5, and L5-S1 when going from a standing to a relaxed, unsupported sitting position. Overall, Andersson et al. (1979) found an average decrease in the lumbar lordosis of 38 degrees, of which 28 degrees was due to pelvic rotation.

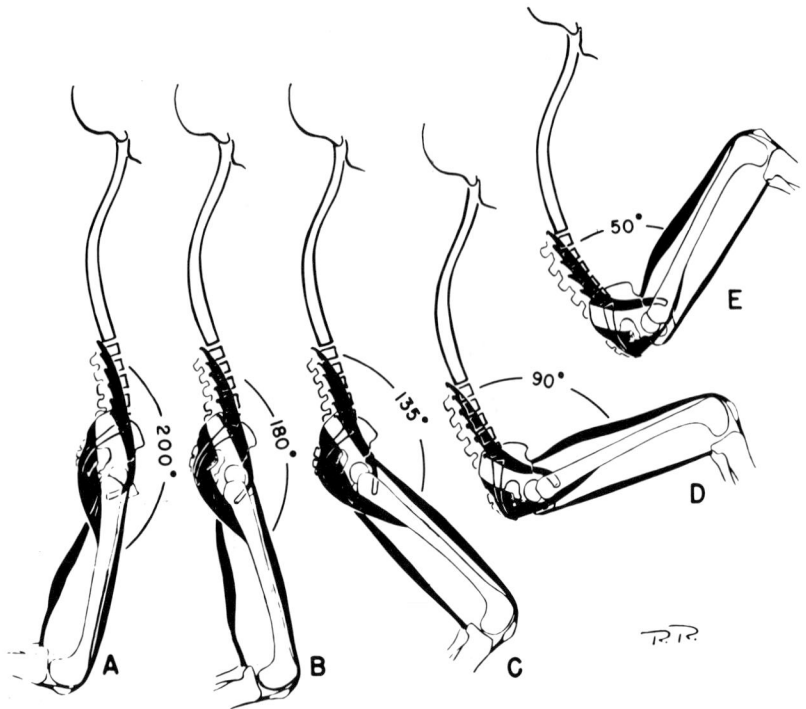

Figure 34. The pelvic rotation when sitting is partly due to the tension of the hip extensors as the hips are flexed. Note in particular D and E. From Keegan, J.J.: Alterations of the lumbar curve related to posture and seating. *The Journal of Bone and Joint Surgery, 35-A:*589–603, 1953. Reproduced with permission of The Journal of Bone and Joint Surgery.

The relaxed, kyphotic sitting posture just described is not the only unsupported sitting posture capable of being assumed by the individual. For example, one can counteract the backward pelvic rotation and lumbar kyphosis with an active tightening of the erector spinae musculature, resulting in either a straight or lordotic sitting posture (Figure 36).

The actual unsupported sitting posture assumed depends on various factors such as the mobility of the hips, the mobility of the spine, the individual's habit patterns, and the individual's fatigue level.

THREE BASIC UNSUPPORTED SITTING POSTURES

Schoberth (1962) described three basic unsupported sitting postures. He differentiated between these three postures based on the center of gravity of the trunk and the percentage of body weight transmitted by the feet to the floor. These three sitting postures are most easily observed

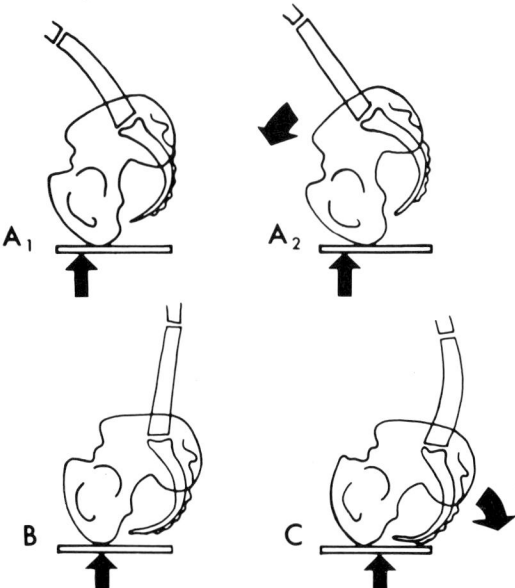

Figure 35. The location of the gravity line of the trunk (vertical arrows), and the posture of the pelvis and lumbar spine in different unsupported sitting postures.

A1. Anterior sitting posture with little or no pelvic rotation, but with kyphosis of the lumbar spine.

A2. Anterior sitting posture with a forward rotation of the pelvis and no kyphosis.

B. Relaxed middle sitting position.

C. Posterior sitting position with a backward rotation of the pelvis and kyphosis of the lumbar spine.

From Andersson, B.J.G., Örtengren, R., Nachemson, A.L., Elfström, G., and Broman, H.: The sitting posture: an electromyographic and discometric study. *Orthopedic Clinics of North America*, 6:105–120, 1975. Reproduced with permission of W.B. Saunders Company.

when sitting on a flat surface without a backrest, the feet flat on the floor with the thighs horizontal and the lower legs vertical (Figure 37).

In the middle position, the center of gravity of the trunk is above the ischial tuberosities, and the feet transmit approximately 25 percent of the body weight to the floor (See Figure 35 B). When sitting relaxed in this posture, the lumbar spine is either in a slight kyphosis or straight. However, with an active contraction of the erector spinae, a more upright middle position will result, with the lumbar spine changing to either a straight or lordotic posture. The more lordotic the upright posture, the more the pelvis will rotate forward, with a corresponding anterior shift of the trunk's gravity line.

In the anterior position, the center of gravity of the trunk is anterior to the ischial tuberosities and the feet transmit more than 25 percent of the

Figure 36.
    A. Erect unsupported sitting posture with proper axial relationship of head, thorax, and pelvis.
    B. Kyphotic unsupported sitting posture with marked backward pelvic rotation.
    From Goldthwait, J.E., Brown, L.T., Swaim, L.T., and Kuhns, J.G.: *Body Mechanics in the Study and Treatment of Disease.* Philadelphia, Lippincott, 1934. Reproduced with permission of J. B. Lippincott Company.

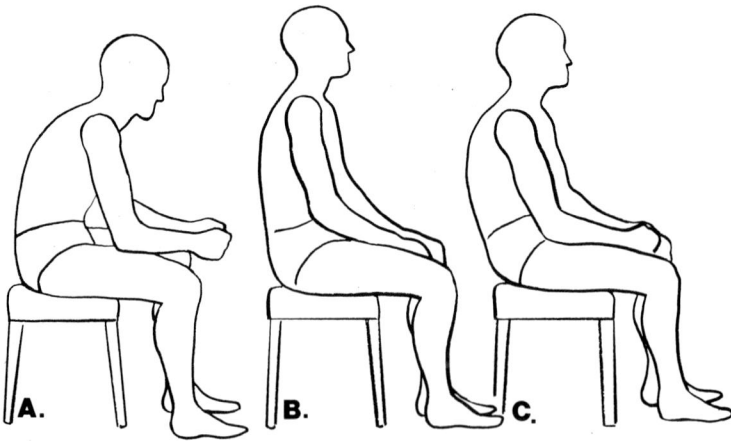

Figure 37. Three basic unsupported sitting postures. A. anterior B. middle C. posterior.

body weight to the floor. This forward leaning posture (as when leaning forward over a horizontal desk) can be assumed from the middle position in either of two ways:

1. With little or no pelvic rotation, but maximal flexion of the spine (See Figure 35-A1).

2. By a forward rotation of the pelvis, keeping the lumbar spine either in slight kyphosis, straight, or in lordosis (See Figure 35-A2).

The degree of straightening or lordosis of the lumbar spine in this posture would depend on several factors including the extent of conscious activation of the erector spinae musculature and the degree of hip mobility.

In the posterior sitting position, the center of gravity of the trunk is above or behind the ischial tuberosities, and the feet transmit less than 25 percent of the body weight to the floor. The posterior position is obtained from the middle position by a backward rotation of the pelvis, resulting in a kyphosis of the lumbar spine (See Figure 35-C). In this posterior sitting posture, the greater the backward rotation of the pelvis, the greater the posterior shift of the trunk's gravity line behind the ischial tuberosities.

The shape of the lumbar spine is usually the same in the most frequently observed anterior and posterior sitting positions. The lumbar spine is in a marked kyphosis and the erector spinae muscles are relaxed, with the spine being supported by the posterior ligaments (Åkerblom, 1948; Floyd and Silver, 1955; Carlsöö, 1972).

DISC PRESSURE STUDIES

The lumbar disc pressure is considerably lower in standing compared to unsupported sitting postures (Andersson et al., 1974d, 1975; Fiorini and McCammond, 1976; Okushima, 1970). Of all the unsupported sitting postures, the disc pressure is the lowest in the lordotic upright posture and the highest in the kyphotic anterior sitting posture (Figure 38).

The following factors are considered responsible for the change in disc pressure from standing to unsupported sitting:

1. Compared to erect upright standing, in relaxed unsupported sitting the pelvis is rotated backwards with a flattening or reversal of the lumbar lordosis. The gravity line of the upper body, already anterior to the lumbar spine in erect standing, will shift further forward. This will result in a long lever arm for the force exerted by the weight of the trunk, producing an increased torque in the lumbar spine. If the trunk is bent forward, this torque will increase even further (Lindh, 1980) (Figure 39).

Figure 38. Mean values of normalized lumbar disc pressure in standing and various unsupported sitting postures. From Andersson, B.J.G., Örtengren, R., Nachemson, A., and Elfström, G.: Lumbar disc pressure and myoelectric back muscle activity during sitting. 1. Studies on an experimental chair. *Scandinavian Journal of Rehabilitation Medicine,* 6:104–114, 1974. Reproduced with permission.

With active contraction of the erector spinae and a more upright sitting posture, the disc pressure will be reduced as compared to a relaxed middle or posterior sitting position. This is because as the backward pelvic rotation and lumbar flexion are reduced, the lever arm for the force exerted by the weight of the trunk will be shortened (Lindh, 1980).

2. In the normal lordotic standing posture, the intervertebral compressive force is shared between the discs and facet joints. Approximately 16 percent of this compressive force is carried by the facet joints when standing (Adams and Hutton, 1980). The facet joints will not take any of this load in kyphotic sitting postures, resulting in higher compressive loads on the intervertebral discs.

3. A further reason for the increased disc pressure with unsupported kyphotic sitting postures would be the greater deformation of the disc in these postures, compared to the normal physiological shape of the disc in lordosis (Andersson et al., 1974d).

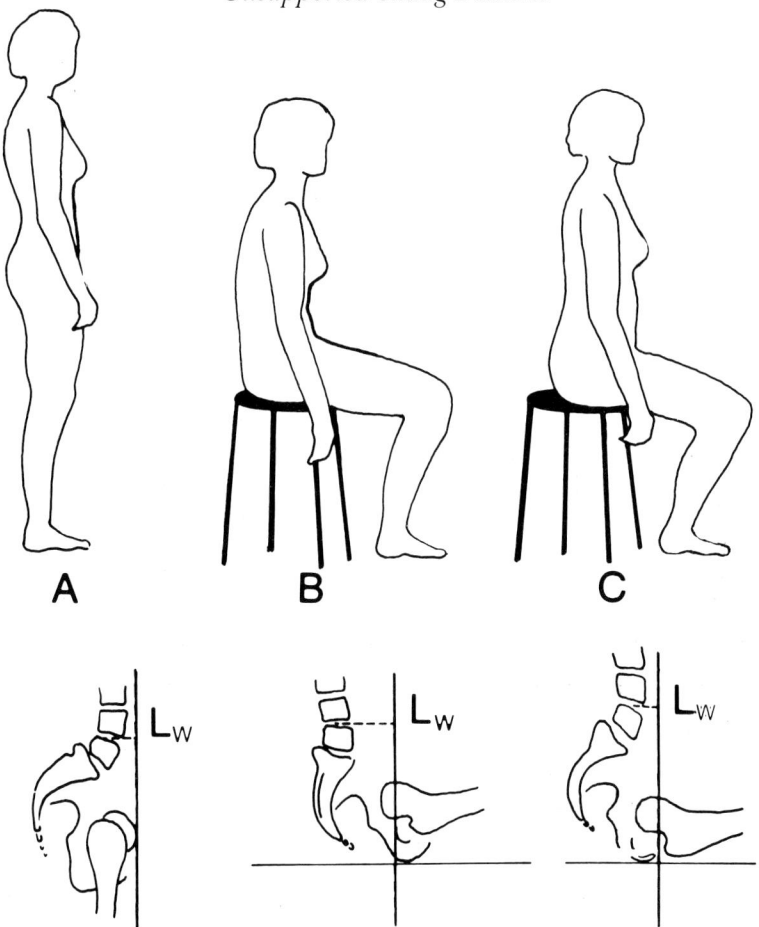

Figure 39. Compared to erect upright standing (A), the line of gravity for the upper body, already ventral to the lumbar spine, shifts further ventrally during relaxed unsupported sitting as the pelvis is tilted backward and the lumbar lordosis flattens (B). This creates a longer lever arm (Lw) for the force exerted by the weight of the upper body. During erect sitting the backward pelvic tilt is reduced and the lever arm shortens (C), but it is still slightly longer than during erect upright standing. From Frankel, V.H., and Nordin, M.: *Basic Biomechanics of the Skeletal System.* Philadelphia, Lea and Febiger, 1980. Reproduced with permission of Lea and Febiger.

4. In addition, a drop in the normal resting intra-abdominal pressure when sitting with lax lower abdominal muscles would also increase the spinal loading and disc pressure (Frymoyer and Pope, 1978; Armstrong, 1965). This is an important factor that is often overlooked.

USCLE ACTIVITY IN UNSUPPORTED SITTING

All unsupported sitting postures are basically unstable without further external support (Meyer, 1873). This is due to the pelvic instability inherent in unsupported sitting (Coe, 1983). The hip joints are in an intermediate position, and "the upper part of the body cannot be locked relative to the thighs by any form of passive checking mechanism" (Åkerblom, 1948). The balance is therefore maintained by the muscles of the hip joint and trunk.

Middle Sitting Position

Sitting with the center of gravity of the trunk directly over the ischial tuberosities is a position of unstable equilibrium since the ischial tuberosities, with their narrowed, curved surface, provide only a linear support (Helbig, 1978; Meyer, 1873). The equilibrium will resemble that of a rocking chair on a pair of very short rockers (Hope et al., 1913). In addition, sitting with a lordosis in this middle position cannot be held for a prolonged period of time by many individuals due to the continuous static work of the erector spinae muscles.

Posterior Sitting Position

An individual can slump into a posterior sitting posture, which will relax the back musculature (Schoberth, 1962; Carlsöö, 1972; Andersson et al., 1974a). Stability will be improved due to the additional supporting surface provided by the coccyx, sacrum, and posterior buttocks. With the gravity line now shifted posterior to the ischial tuberosities, the psoas major will become the main antigravity muscle (Keagy et al., 1966).

Leaning back more than a few degrees without external support (such as a backrest or backward placement of the hands) becomes a very unstable posture since there is minimal weight bearing on the legs. To maintain such a posture also requires increased activity from the rectus abdominis muscle and the neck musculature (Asatekin, 1975; Cotton, 1904). Stretching the arms and legs forward can also help the individual to barely maintain this posture (Meyer, 1873; Åkerblom, 1948).

Neck Musculature

In a lordotic upright sitting posture, the gravity line of the head passes anterior to the cervical spine, thereby requiring slight to moderate activity of the posterior neck musculature to counteract the tendency for the head to incline forwards (Steen, 1966).

With a slumped, kyphotic sitting posture, the gravity line of the head will pass further anterior to the cervical spine, and there will be an increased demand placed on the posterior neck musculature (Jones et al., 1961; Gray et al., 1966; Bunch and Keagy, 1976). An increase in neck muscle activity will also be required to keep the head erect and the gaze horizontal (Figure 40).

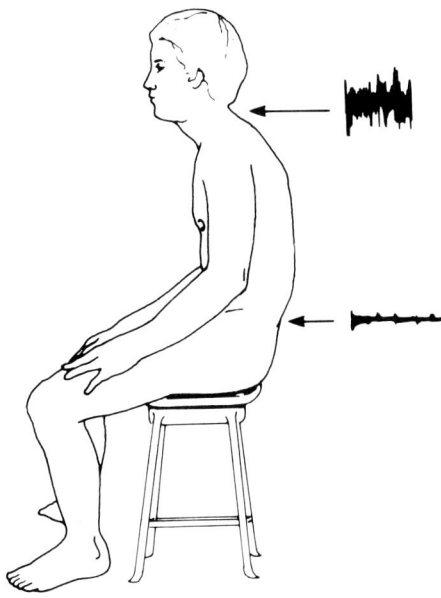

Figure 40. Increased demand placed on the posterior neck musculature in a slumped, kyphotic sitting posture. From Bunch, W.H., and Keagy, R.D.: *Principles of Orthotic Treatment.* St. Louis, Mosby, 1976. Reproduced with permission of W. H. Bunch, M.D.

The greater the slump and thoracolumbar kyphosis, the greater will be the forward thrust of the head, resulting in a marked increase in activity from the upper trapezius and other posterior neck musculature (Gray et al., 1966). A greater than 50 percent increase in muscle tension at the back of the neck has been reported when going from an erect to a slumped sitting posture (Gray et al., 1966).

The alteration in the shape of the cervical spine in a slumped, kyphotic sitting posture would probably resemble the contour described by Ingelmark (1942). From radiological examinations of sixteen patients with pain in the middle and lower neck and trapezius muscle tenderness, Ingelmark (1942) found an absence of the normal cervical lordosis at the C4 to C7 level, and a hyperlordosis above the C4 level.

Anterior Sitting Position

Compared to a middle sitting position, the stability is improved as the individual leans forward into an anterior sitting position. This is due to the increased supporting surface provided by the upper posterior thighs and the increased body weight on the feet with this posture. However, without external support, as the gravity line is shifted anterior to the ischial tuberosities, the erector spinae and hip extensor muscles must contract to prevent the trunk from falling forwards (Cotton, 1904; Åkerblom, 1948; Schoberth, 1962; Andersson et al., 1974a). With extreme spinal flexion, the erector spinae will relax and only hip extensor activity will be required to maintain this posture (Åkerblom, 1948; Floyd and Silver, 1955; Floyd and Roberts, 1958; Carlsöö, 1972).

If there is external support, the erector spinae and hip extensor muscle activity can both be relieved, and the anterior sitting position can become the most stable (although not physiologically the most beneficial) unsupported sitting posture. Examples of such external support are as follows (Meyer, 1873):

1. The hands and forearms are supported on the thighs.
2. The anterior trunk is supported by the edge of a table.
3. The arms are supported on a table.

As the gravity line of the head is also moved further anterior to the cervical spine with the anterior sitting postures, there will also be an increased stress placed on the posterior neck musculature.

FREQUENTLY OBSERVED
UNSUPPORTED SITTING POSTURES

Although the slightly lordotic upright sitting posture is the most beneficial from a physiological point of view, it is not the unsupported sitting posture most frequently observed. First of all, as will soon be discussed, a large percentage of the population is unable to achieve the lordotic upright sitting posture due to age-related factors. Secondly, for many individuals the lordotic upright sitting posture will be the most fatiguing due to the sustained static contraction required of the erector spinae muscles (Lundervold, 1951b; Schoberth, 1962; Floyd and Silver, 1955; Floyd and Roberts, 1958).

According to the Minimum Principle of Nubar and Contini (1961), "the individual will, consciously or otherwise, determine his motion (or his posture, if at rest) in such a manner as to reduce his total muscular effort to a minimum consistent with imposed conditions, or 'constraints.'" As will be discussed in more detail in the next chapter, the individual will most frequently assume "closed-chain" positions that result in increased stability and decreased muscle activity and energy expenditure (Dempster, 1955a,b).

Two of the most commonly observed unsupported sitting postures are therefore as follows:

1. Anterior sitting posture with the individual resting his forearms on his thighs (Meyer, 1873) (See Figure 37A).

In this position, since the weight of the trunk is taken by the arms, the strain is reduced in the back and hip extensor musculature. The decreased energy expenditure associated with this posture has been verified by Hanson and Jones (1970). Among eleven male college students, they reported an average decrease in heart rate from 87 to 66 beats per minute when the individual changed from an erect upright posture to an anterior sitting position with the trunk weight resting on the arms. In one individual, there was actually a decrease in heart rate of 30 beats per minute between these two postures.

2. Posterior sitting posture with the legs crossed (Schoberth, 1962). In addition, the fingers are frequently interlocked around the uppermost knee (Figure 41). The locking effect from crossing the legs will help regain the stability that is lost in the posterior sitting posture from the decreased weight bearing on the feet. The lower leg will be securely anchored by the upper leg (McConnel, 1933). The muscle activity required from the psoas major to maintain this posture should also decrease, as it will have a better distal stabilization with the legs crossed (Smidt et al., 1983). The close approximation of the uppermost knee to the arms after the legs are crossed results in an easy to assume resting place for the hands, which will further increase trunk stability.

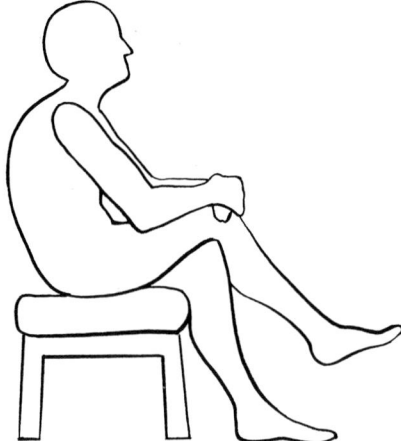

Figure 41. Posterior sitting posture with the legs crossed and the fingers interlocked around the uppermost knee.

FACTORS PREVENTING A LORDOTIC SITTING POSTURE

Decreased Hip Mobility

One of the most important factors influencing the sitting posture one assumes is the state of mobility of the hips (Le Floch and Guillaumat, 1982). Decreased hip mobility will make it impossible to achieve a lordotic upright sitting posture.

Tight Hamstrings.

The hamstring muscles, when tight, will pull the pelvis into a posteriorly rotated position. In order to keep the trunk upright, a marked flexion of the lumbar spine will be necessary (Floyd and Roberts, 1958; Stokes and Abery, 1980; Brunswic, 1984a,b). Hamstring tightness is not only found in the adult population, but it is also very prevalent among school-age children and adolescents (Kendall and Kendall, 1948; Milne and Mierau, 1979). (See Chapter Six on School Seating.)

The effect of the hamstrings on lumbar flexion will also be related to the angles of hip flexion and knee extension assumed by the individual on a specific seat. Brunswic (1984a,b) found a roughly linear relationship between the angles of the hips and knees when sitting and the percentage of lumbar flexion. An increase in hip flexion or in knee extension both increased the percentage of lumbar flexion in a ratio of 1:2. An increase in knee extension of 20 degrees would therefore correspond to an approximate 10 degree increase in hip flexion.

According to Brunswic (1984a):

> "These results can be explained anatomically by the role of the hamstring muscles as a posterior rotator of the pelvis. The tension of the hamstrings imparted by the flexion of the hips or the extension of the knees rotates the pelvis backwards. In order to maintain an upright trunk, the posterior rotation of the pelvis is compensated for by an increased flexion of the lumbar spine."

Degenerative Changes

Hip mobility may also be limited due to degenerative changes in the joint, restricting hip flexion. The individual will then be forced to flex the lumbar spine in order to sit upright (Rosemeyer, 1973).

Limited hip mobility in abduction will also impair one's ability to achieve a lordotic upright sitting posture (Mandal, 1982). This is because less hip flexion is possible with the hip in an adducted position.

Decreased Back Extension Mobility

As a result of keeping the thoracolumbar spine flexed day after day, some individuals may have lost the ability to extend their spines sufficiently. These individuals often sleep, sit, stand, and exercise in postures involving thoracolumbar spinal flexion.

The ability to achieve a lordotic sitting posture is also related to structural changes of the spine associated with aging. Milne and Lauder (1974) found that a lumbar lordosis was absent in an increasingly large proportion of men and women as age increased beyond sixty years.

Improper Movement Patterns

During the school years, improper movement patterns may develop from exercises that reinforce thoracolumbar flexion (instead of thoracolumbar extension) with hip flexion. Examples of such exercises are toe touches and full sit-ups (Zacharkow, 1984b).

Sit-ups can overly strengthen the upper rectus muscle and overstretch the back extensors (Anderson, 1951). Toe-touching exercises may promote hypermobility in spinal flexion, especially when done with very tight hamstrings. These types of exercises all stress thoracolumbar flexion mobility, the antithesis of the spinal mobility needed for a lordotic upright sitting posture.

This improper movement pattern can often be observed in the anterior sitting posture a student adopts at a horizontal desk (Cotton, 1904) (Figure 42). Goldthwait (1909), Mosher (1914, 1919), and Schurmeier (1927) have all stressed that in an anterior sitting posture the trunk should be kept straight and not flexed. The forward flexion should occur at the hips and not the spine.

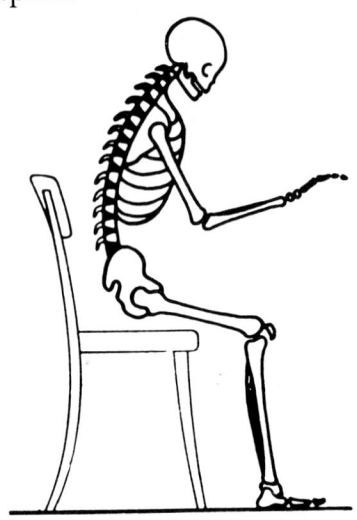

Figure 42. Improper movement pattern when assuming an anterior sitting posture, with flexion occurring at the spine instead of at the hips. From Rizzi, M.: Entwicklung eines verschiebbaren rückenprofils für auto-und ruhesitze. In Grandjean, E. (Ed.): *Proceedings of the Symposium on Sitting Posture.* London, Taylor and Francis, 1969, pp. 112–119. Reproduced with permission of Taylor and Francis Ltd.

PERCENTAGE OF POPULATION ABLE TO ACHIEVE A LORDOTIC UPRIGHT SITTING POSTURE

Due to the many factors just described, there will be many variations in the posture of the lumbar spine in the upright sitting position. According to Åkerblom (1948), "Some of the curves are very little different from those obtained in the relaxed position, while on the other hand there are a few cases in which the curves are hardly to be distinguished from those obtained in the standing position. However, they usually show an intermediate position between standing and maximal ventriflexion."

Hooton (1945) in his survey of body measurements for seat design concludes that the lumbar lordosis "tends to be flattened practically to the vanishing point in most subjects when they sit erect."

Other investigators have been able through their research to specify the percentage of a specific population able to achieve a lordotic upright sitting posture, as seen in Table I. In regards to the lack of a sitting lumbar lordosis in many schoolchildren, Schoberth (1969) explained that in children age six, the shape of the vertebral column has not been completely developed. The lumbar lordosis develops in conjunction with the caudal and ventral displacement of the first sacral vertebra during ages eight to twelve, which leads to curvature of the sacrum. Before this time period, the sacrum exhibits a relatively extended shape.

Table I

ABILITY TO
ACHIEVE LORDOTIC UPRIGHT SITTING POSTURE

Investigator	Population Studied	% Able to Achieve Lordotic Upright Sitting Posture
Schoberth (1962)	1035 Schoolchildren ages 6-14	30.5%
Branton (1984)	114 Railway Employees mean age: men - 46 years women - 38 years	71.1%
Institute for Consumer Ergonomics (1983 b)	758 Elderly ages 65 and over	62%
Institute for Consumer Ergonomics (1983 b)	502 Disabled Individuals ages 16-64	48%

Chapter Three

INSTABILITY OF SITTING

As discussed in the preceding chapter, the sitting position is basically unstable without additional external support. This is because the ischial tuberosities, with their rounded shape resembling the rockers of a rocking chair, provide only a linear base of support (Meyer, 1873). Also, in the sitting position the hip joints are in an intermediate position and the trunk cannot be locked relative to the thighs by ligamentous restraint (Åkerblom, 1948; Meyer, 1873; Coe, 1983). As a result, muscle activity is necessary for fixation of the trunk when sitting without additional stabilizers.

SITTING IN A CHAIR

A common misconception, however, is to consider sitting in a chair as a static activity, as opposed to a dynamic activity. According to Branton (1966), the sitting body is "not merely an inert bag of bones, dumped for a time in the seat, but a live organism in a dynamic state of continuous activity."

Branton's (1969) mechanical model of the sitting body from the waist down depicts four degrees of freedom to move, even with the feet planted firmly on the floor (Figure 43):

1. Rocking of the pelvis over the ischial tuberosities.
2. Flexion and extension at the pelvic-femoral joint.
3. Flexion and extension at the knee joint.
4. Flexion and extension at the ankle joint.

In the sitting posture, the hip, knee, and ankle joints are near the midpoint of their range of motion, and therefore they are in a state of maximum mobility. Branton (1966, 1969) mentioned that even if an individual appears to sit still, his body is continuously moving. The freedom of the pelvis to move, which will be present in all sitting postures when the upper sacrum is not supported by a backrest, will result in "continuous hunting" or relatively fast oscillatory movements of the pelvis rocking over the ischial tuberosities.

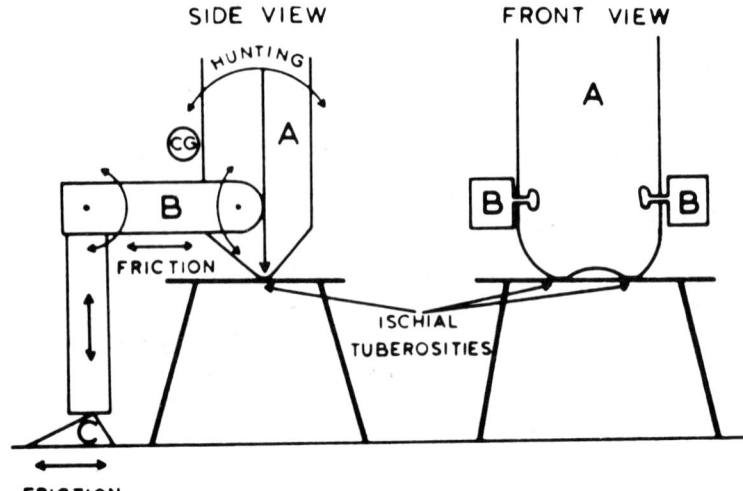

Figure 43. Branton's mechanical model of the sitting body from the waist down. A. trunk B. femur C. foot

    CG. Center of gravity of sitter.

    Hunting = oscillatory movements of the pelvis rocking over the ischial tuberosities.

    From Branton, P.: Behaviour, body mechanics and discomfort. In Grandjean, E. (Ed.): *Proceedings of the Symposium on Sitting Posture.* London, Taylor and Francis, 1969, pp. 202–213. Reproduced with permission of Taylor and Francis Ltd.

Therefore, Branton (1966) hypothesized that there is a continual need for postural stability when sitting, so that the seated person "spontaneously takes up such postures as will allow him to sit stably, while relieving his brain and muscles from greater exertion than would be necessary otherwise."

"If the seat does not allow postures which are both stable and relaxed, the need for stability seems to dominate that for relaxation and postures are adopted which rigidify the body internally in compensation. In other words, if seat features fail to stabilize him, the person must stabilize himself, e.g. by crossing his legs, or by supporting his head on his hand. This may be at some extra cost in muscle work." (Branton, 1966).

WAYS OF ACHIEVING STABILITY

Closed Chains of Body Segments

Dempster (1955b) compared the dynamic human body to an open-chain system of links that rotate about the joint centers. He described how certain joint motions may be stabilized:

"The fingers of the two hands may be interlocked to interconnect the right and left upper limb links; the legs may be crossed for seated stability; the arms may be crossed or placed on the hips. In such actions as these, temporary approximations to closed chains are effected.

Link chains may be cross-connected as in crossing the knees (vis., pelvis and right and left thighs) or in placing the hand on the same or opposite shoulder. To the extent that these temporary closed chains approximate a triangular linkage, there is a degree of stability imparted even without muscular actions, but this is still approximate because of the interposed soft tissues. The closer the links approximate a closed triangular, or pyramidal, pattern, the less muscles are called upon for stabilizing action at joints. One may recognize many rest positions involving this principle: crossed arms, hands in pockets, or such sitting positions as crossed knees, ankle on opposite knee, elbow on knee, or head in hand" (Dempster, 1955b).

Temporary closed chains may also include certain environmental objects. The most commonly observed example is when the arms are supported on a desk (Meyer, 1873).

Branton and Grayson (1967) observed many closed-chain positions in the five thousand observations they made on sitting posture in train seats. Through time-lapse film, the following sequence of postures was observed to occur with one man a total of twelve times throughout the entire five hour train ride:

"Starting well back, legs apart, he gradually slides into slumped posture, sometimes using the arms to prop himself up. As the arms seem to fail to stop the forward movement of the pelvis, he crosses his knees. Next he stretches his legs forward, and ends up with his body and legs in almost a straight line from the neck down, as near to horizontal as can be. After a very short spell in this position he raises himself back up, only to begin another journey down the slope. Each of these sequences takes between 10 and 20 minutes" (Branton and Grayson, 1967).

The train seat slowly and repeatedly ejected the sitter for two main reasons:

1. A poor anthropometric fit of the chair and its stabilizing features (backrest, armrests, upholstery) with the sitter.

2. Movement of the train (bumps, accelerations, vibrations, stops).

In order to determine the normal movements made by a person sitting in a chair, Dillon (1981) filmed individuals in pairs as they engaged in conversation over a one hour period. The pattern of sitting behavior observed was for the individual to start sitting in an upright posture, and then gradually to adopt a more slumped posture. During the slumping process, a repeated crossing and uncrossing of the legs was observed, which caused the buttocks to "walk" forward on the seat. Eventually, the individual would lift himself back to an upright posture in a single movement, but he would then start the slumping process over again.

Crossing the Knees

The frequent observation of crossing the knees as a closed-chain position merits further discussion. Take as an example a chair with an inclined backrest, horizontal seat, and a slippery seat cover. When the individual leans against the backrest, there will be a force tending to push the buttocks forward on the seat, and slowly eject the sitter (Åkerblom, 1948).

Using Branton's mechanical model of sitting (1969), there will be four degrees of motion observed as the individual's buttocks slide forward on the seat, with the feet firmly on the floor:

1. Upward rotation of the pelvis (rocking over the ischials).
2. Extension at the hip joints.
3. Flexion at the knee joints.
4. Dorsiflexion at the ankle joints.

The first spontaneous attempt to stop the forward slide will usually be to cross the legs. If the right knee is crossed over the left knee, the right hip will be "locked" in a position near the end range of flexion-adduction, the right knee will be "locked" over the left thigh, and the right ankle will be removed from a weight bearing position. The increased weight bearing over the left thigh with the crossed knee posture will further resist the forward slide, as the left leg and foot will be securely anchored to the floor (McConnel, 1933).

In addition, due to the pull of the right hip abductors as the right knee is crossed over, the individual's pelvis and trunk will rotate slightly to the left as the right buttock "walks" forward on the seat. This asymmetrical posture will further help resist the forward slide, as the individual's hip joints are no longer in the same anteroposterior plane as the chair seat and backrest.

Dillon's (1981) observations of repeated crossing, uncrossing, and opposite crossing of the legs can be explained by the accelerated fatigue from this asymmetrical sitting posture. Not only will the muscles fatigue faster when asymmetrically loaded, but there will also be an increased pressure imposed on one buttock. This will result in an acceleration of ischemia and discomfort, necessitating another position change (Dempsey, 1962; Schoberth, 1962).

Another commonly observed closed-chain position would involve stretching the legs forward into extension. This posture will "lock" the knees and ankles, and the lower extremities will become rigid posts pushing against the floor to resist the forward slide (Branton, 1969).

Chair Features as Stabilizers

Certain chair features are critical to help support the seated individual so that he can maintain a stable posture whether working or relaxing. The aim of proper support for all types of chairs and workstations should be *optimal stability with minimal restraint* (Branton, 1966, 1970; Shipley, 1980; Hedberg et al., 1981; Schaedel, 1977; Dempsey, 1962; Zacharkow, 1984a; Do et al., 1985; Grandjean, 1984c; Nakaseko et al., 1985).

Important stabilizers include the backrest, armrests, and the type of seat cushion and upholstery cover. The proper chair dimensions are also critical in providing essential stabilization for the sitter. These include the proper seat inclination, seat height, seat depth, and seat to backrest angle.

The most frequent train seat posture observed by Branton and Grayson (1967) in their five thousand observations involved having the back and arms well supported by the backrest and armrests, and the feet flat on the floor. This posture would allow the individual to sit stably, along with the greatest reduction in muscular exertion. Overall, Branton and Grayson (1967) found that the sitting postures where the chair provided the greatest stabilization were held five to ten times as long as the least supported postures.

Examples of Improper Stabilizers

Soft Cushioning. If cushioning is too soft, the individual will not receive proper support from the seat, and an increase in muscle activity will be needed to stabilize the sitter (Kohara, 1965; Kohara and Hoshi, 1966; Branton, 1966).

On the relationship between seat cushioning and stability, Branton (1966) commented that:

> "... a state can easily be reached when cushioning, while relieving pressure, deprives the body structure of support altogether and greatly increases instability. The body then 'flounders about' in the soft mass of the easy chair and only the feet rest on firm ground. Too soft and springy a seat would therefore not allow proper rest, but may indeed be very tiring because increased internal work is needed to maintain any posture."

Vertical Backrest. A properly designed backrest should support some of the individual's body weight, with its main functions being to provide trunk stabilization and relaxation of the back muscles (Kroemer, 1982).

As early as 1873, Meyer critiqued the vertical backrest as providing no support at all for the vertebral column. Such a backrest prevents an individual from leaning back and obtaining any back support unless he first slides his buttocks forward on the seat. Similar comments were later made by Staffel (1884) and Strasser (1913).

By sliding the buttocks forward on the seat, the individual's thigh-trunk angle will widen and there will be less tension in the hip extensors. The opening of the thigh-trunk angle will help reduce the stress on the lumbar spine, diaphragm, and abdominal viscera (Bradford and Stone, 1899; Keegan, 1953, 1962, 1964; Asatekin, 1975; Corlett and Eklund, 1984).

Unfortunately, when sliding the buttocks forward to obtain support from a vertical backrest, support is obtained by the upper thoracic spine around the scapular region. The pelvis and lumbar spine remain unsupported (Figure 44). Unless the individual strongly contracts the erector spinae to hold the lumbar spine in a straight or lordotic posture, a kyphotic posture will result (Strasser, 1913). The resulting posture will tend to upwardly rotate the pelvis even more, and the individual will usually cross his legs in an attempt to achieve some stabilization (Figure 45).

In addition, as a result of sliding the buttocks forward, the body weight on the seat will now be localized to the gluteal region posterior to the ischial tuberosities. This will quickly lead to ischemia and discomfort from the distortion of muscle and other tissues (Babbs, 1979; Asatekin, 1975).

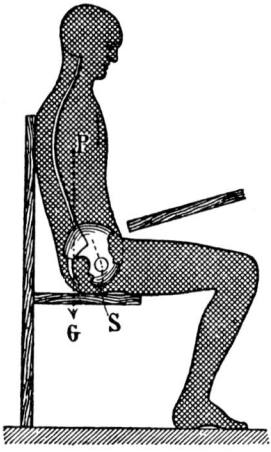

Figure 44. Posture assumed on a chair with a high vertical backrest, in order to obtain back support. From Burgerstein, L., and Netolitzky, A.: *Handbuch der Schulhygiene.* Jena, Verlag Von Gustav Fischer, 1895. Reproduced with permission of Gustav Fischer Verlag.

Figure 45. Crossing the legs in a chair with a high vertical backrest, in an attempt to obtain further stabilization. From Drew, L.C.: *Individual Gymnastics,* 5th ed. Philadelphia, Lea and Febiger, 1945. Reproduced with permission of Lea and Febiger.

The first 10 degrees of backrest inclination are considered critical in order to properly stabilize the trunk (Andersson et al., 1975). Electromyographic studies have shown how a marked decrease in back muscle activity will occur as the trunk is supported by a slightly inclined backrest (Figure 46). This finding would help explain the often observed tendency for individuals to tilt a chair with a vertical backrest on its rear legs (Bennett, 1928; Travell, 1955).

Unfortunately, many chairs today continue to be designed with high vertical backrests. Many health professionals and teachers are probably still requesting that one should learn to sit up straight in a chair with a high vertical backrest, and not slump.

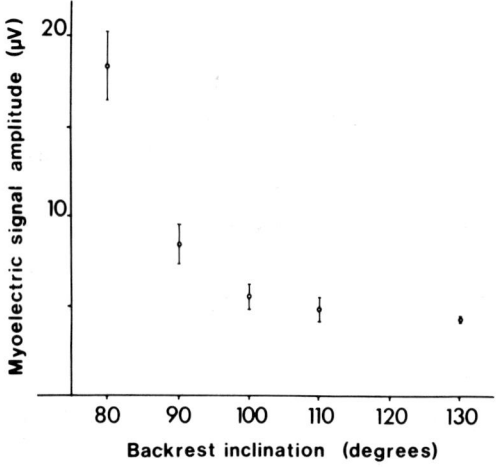

Figure 46. Effect of backrest inclination on myoelectric back muscle activity at the L3 level. Note in particular the change from 90 degrees to 100 degrees of backrest inclination. From Andersson, B.J.G., Örtengren, R., Nachemson, A.L., Elfström, G., and Broman, H.: The sitting posture: an electromyographic and discometric study. *Orthopedic Clinics of North America,* 6:105–120, 1975. Reproduced with permission of W. B. Saunders Company.

Chapter Four

COMFORT, FATIGUE, AND PRODUCTIVITY

The principal source of postural stress when sitting is the prolonged static muscular workload needed to support various body segments against the force of gravity (Pheasant, 1984; Östberg, 1984; Coe, 1979, 1983). Therefore, the goal of a good chair should be to support the sitter so that he can maintain a stable posture and thereby minimize the static contraction of postural muscles (Bhatnager et al., 1985). A stable sitting posture is a necessary condition for seat comfort (Kohara, 1965; Kohara and Sugi, 1972).

When proper body support is neglected, discomfort and fatigue may soon follow, along with the potential for a considerable amount of inefficiency at work (Branton, 1970; Schaedel, 1977). Stecher (1911) commented upon how improper seating can result in restlessness and inattention among schoolchildren. In regards to school seating, Lincoln (1896) considered discomfort to be "an indirect cause of deformity, as it invariably leads a child to take improper positions."

Traffic accidents and death may result from the distractions due to the body pressure discomfort caused by improper automobile seating (Hockenberry, 1982; Diebschlag and Müller-Limmroth, 1980).

FATIGUE

Bitterman (1944) proposed the following definition of fatigue: "a reduction in efficiency resulting from continued work and reversible by rest."

Earlier, Goldthwait (1909) emphasized the waste of energy and reduction in efficiency that would result from the strain of maintaining a poor posture. He stressed that:

> "... there is a certain definite amount of energy available for expenditure with each individual; that this energy can be expended in many different ways, but that if expended for one thing it no longer is available for other efforts, except after periods of recuperation. The form of the expenditure, whether in mental or physical expression, is

immaterial, and the waste of this energy is not only undesirable, but is usually harmful."

According to Goldthwait, when maintaining proper posture the muscular forces would be used with the least waste and the viscera would be most favorably situated for function. As a result, the greatest amount of energy will be available for "whatever function the individual may choose or be forced to perform."

Recently, Kuorinka (1982) also discussed how the energy available for productive work would be greatly diminished if also required for maintaining a poor posture.

Garner (1936) credits Halfort from Berlin as making the first direct association of fatigue with faulty posture in 1848. According to Garner (1936), as a direct result of fatigue, an individual's "ability to perform is reduced; his output is diminished; the quality of his production is lowered; his mental aptitude is reduced; his susceptibility to disease is increased in direct ratio to the lowering of his vitality; he is much more liable to commit errors, and his likelihood to sustain personal injuries is greatly enhanced."

Clark (1954) stated that a good chair can reduce fatigue by providing the proper stabilizing features to reduce unnecessary static muscular work and give some degree of relaxation to the sitter. He stressed that body stabilization was particularly important in moving vehicles.

Roberts (1963) correlated the fatigue of passengers with the physical work they do during a journey:

"Every time a vehicle sways, rounds a curve, goes over a bump, accelerates or slows, the passenger is thrown slightly off balance. He then expends energy, usually without thinking about it, in using his muscles to regain stability."

Roberts added that as another source of fatigue, "the design of the seat or some other feature of the accomodation may cause the passenger to assume and hold an unsatisfactory posture, in the maintenance of which the muscles employed may not work at their most efficient rate."

Grandjean's (1968, 1970, 1980a) general fatigue diagram illustrates how the degree of fatigue for an individual is an accumulation of all the different stresses of the day. Recuperation must be balanced out over the twenty-four hour cycle in order to avoid adverse effects on the individual's well-being and efficiency (Figure 47).

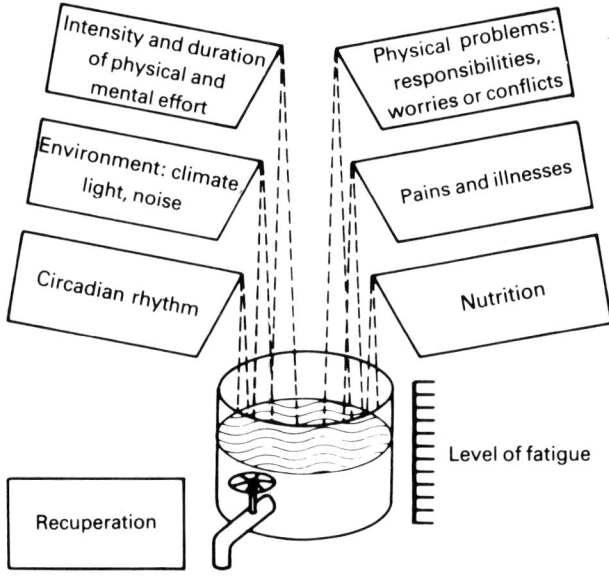

Figure 47. Grandjean's general fatigue diagram. From Grandjean, E.: *Fitting the Task to the Man,* 3rd ed. London, Taylor and Francis, 1980. Reproduced with permission of Taylor and Francis Ltd.

Dainoff (1984a) commented upon Grandjean's dynamic fatigue model as applied to working at a video display terminal (VDT):

"In this approach, the individual is seen as generating a kind of generalized fatigue as a result of coping with a variety of demands, both physical and psychological. The rate of fatigue buildup is proportional to the cumulative intensity of the demand. Thus, the discomfort experienced by working at a keyboard which is too high is added to the discomfort experienced from performing a repetitive data entry task under high productivity pressure. Normally, fatigue recovery occurs by means of work-rest breaks during the day, and through rest and relaxation after work. If, however, the cumulative buildup of fatigue across the typical workday is sufficiently and consistently large enough, normal recovery mechanisms are not sufficient, fatigue begins to accumulate from previous workdays, and the individual enters a state of chronic fatigue or stress. Presumably, this generalized state of debilitation is manifested in a variety of physical symptoms which are superimposed on any specific localized symptoms of strain or chronic trauma (e.g., wrist disorders, back disorders)."

COMFORT

The concept of comfort in regard to seating is very difficult to define, as can be seen by the following descriptions.

Hertzberg (1958):

"Although we tend to speak of comfort and discomfort as if they were two states of consciousness, for testing purposes it is more realistic to consider that there is only one, discomfort, and that 'comfort' is only the absence of discomfort, just as cold is the absence of heat. It is the neutral sensory condition. Thus one cannot 'provide comfort' in a seat design, one can only 'eliminate sources of discomfort.' In this sense of comfort, the implication of euphoria or active pleasure is ignored.

Until more objective methods are developed, the length of time that a sample of subjects can endure a given position is so far the simplest, though not the only way of assessing the discomfort resulting from it.

All words used to describe the quality of a seat or the degree of bodily discomfort ('good,' 'bad,' 'better,' 'mild,' 'severe,' etc.) are relative, the only theoretical absolute being the state of 'zero discomfort,' i.e., the absence of pain.

Before one can say that a given seat or body support is 'better' or 'worse' than another, one must specify the expected duration of occupancy. What is 'good' for two hours may be 'bad' for four hours, and so on."

Branton (1969):

"Most investigators assume that the term comfort denotes a feeling or an affective state, which varies subjectively along a continuum from a state of extreme comfort through indifference to a state of extreme discomfort. We find it most difficult to envisage deriving extreme feelings of well-being merely from sitting however good the chair may be. In our view, therefore, the possible continuum extends only from indifference to extreme discomfort. The absence of discomfort denotes a state of no awareness at all of a feeling and does not necessarily entail a positive affect. Similarly, the absence of pain does not necessarily entail the presence of pleasure."

Corlett (1973):

Comfort "can be seen as arising from the summation of all the bodily sensations, each having a random distribution of levels. In any given situation, if the balance is such as not to draw attention to any one sensation, a person can be said to be comfortable. If a sensation distracts attention from the task in hand, then a state of discomfort can be said to exist."

Pile (1979):

"What will be considered 'comfortable' by a user depends very much on the way a seat is used and on how long it is used.

A reclining chair found comfortable for a nap will not be suitable for dining. A classroom chair will not (and should not) encourage sleep. Comfort can in the end, only be judged in relation to intended use."

Measurement Methods for Comfort

Shackel et al. (1969) described four basic methods for assessing chair comfort, to be used in seat design:

1. Anatomical and physiological factors.
2. Observations of body position and movement.
3. Observation of task performance.
4. Subjective methods.

Anatomical and Physiological Factors

A basic misconception regarding sitting and comfort is to directly apply linear anthropometric body dimensions to the chair design. These measurements are taken with the trunk vertical and with 90 degree angles at the hips, knees, ankles, and elbows (Figure 48).

Figure 48. The basic anthropometric position, with the trunk vertical and 90 degree angles at the hips, knees, ankles, and elbows. From NASA: *Anthropometric Source Book. Volume One: Anthropometry for Designers.* NASA Reference Publication 1024, 1978.

However, this is a posture assumed by very few people when sitting (Lueder, 1984; Mandal, 1981, 1982, 1984; Barkla, 1961). At best, it can only be tolerated for a few minutes.

Many standards and brochures for video display terminal (VDT) operators have chair and workstation recommendations based on anthropometric models, advocating 90 degree sitting angles with an upright (vertical) trunk posture. Yet an important field study by Grandjean et al. (1982a, 1983a, 1984) revealed that only approximately 10 percent of VDT operators actually sit upright. The majority of operators were found to sit with trunk inclinations of 10 to 20 degrees from upright.

Direct application of anthropometric model dimensions ignores the fact that sitting is a dynamic activity. The comfort of a chair will be influenced by the sitter's activity in that chair, or the work task requirement (Lueder, 1983; Drury and Coury, 1982; Kaplan, 1981; Shipley, 1980; Pile, 1979). In addition, Branton (1970) and Schaedel (1977) have emphasized the need to consider the effect of upholstery on anthropometric dimensions, as the loaded in-use contours of the seat and backrest are the critical factors in chair comfort, not the free, unloaded contours.

There are, however, various physiological and anatomical measurements that are applicable to chair comfort, and should be applied to chair and workstation design in order to reduce the potential for musculoskeletal disease.

Intervertebral Disc Studies. Some studies have demonstrated the beneficial effect that a lumbar support, inclined backrest, and armrests will have on decreasing lumbar disc pressures when sitting (Andersson et al., 1974d, 1975; Guillon et al., 1985). Eklund and Corlett (1984) used changes in body height as a measure of disc compression. They were able to compare the influence of various chair designs on the rate of shrinkage or disc compression.

EMG Studies. These studies have demonstrated how different chair features can either reduce or increase the myoelectric activity of various postural muscles. Andersson et al. (1974a,c,d; 1975) showed how an increase in backrest inclination can reduce erector spinae muscle activity. Burton (1984) and Yamaguchi et al. (1972) also studied the effect of different chair designs and seat to backrest angles on back muscle activity.

Lundervold (1951a,b; 1958) showed how certain chair features that contribute to sitting instability, such as a flexible backrest or freely moveable casters, could result in an increase in back muscle activity. Kvarnström (1983) and Granström et al. (1985) demonstrated the beneficial effect of proper arm support on reducing trapezius muscle activity and the static load on the neck, shoulders, and arms. In regards to a VDT workstation, Weber et al. (1984) reported a reduction in trapezius muscle activity when using a keyboard with a forearm-wrist support, compared to working at the same keyboard height without a forearm-wrist support.

Blood Flow Studies. The effect of chair design on lower extremity venous blood flow and foot swelling is considered an important physiological assessment for discomfort (Trendafilov et al., 1967; Winkel and Jørgensen, 1986). Pottier et al. (1967,1969) and Morimoto (1973) used a plethysmograph to measure the effect of various chair seat heights and thigh compression on venous blood flow and foot swelling.

Pressure Measurements. The pressure distribution over the seat and backrest with various chair designs is considered to be a very critical factor in sitting comfort (Hertzberg, 1958, 1972; Dempsey, 1962; Lay and Fisher, 1940; Kohara and Sugi, 1972; Babbs, 1979; Shipley, 1980; Kamijo et al., 1982). Hertzberg (1958, 1972) studied the comfort—discomfort factor in regards to the shape of the seat. His unpublished studies demonstrated that a properly contoured and cushioned seat would reduce the extreme point pressures under the ischial tuberosities that occur with a hard, flat seat. Compared to a hard, flat seat, Kosiak et al. (1958) found a wider pressure distribution under the buttocks with a contoured seat, and also with the use of foam seat cushioning of one and two inches in thickness. Marchant (1972) compared the deformation of polyether foam to latex foam when compressed with a 200 mm in diameter indentor.

Kamijo et al. (1982), Diebschlag and Müller-Limmroth (1980), and Shields (1986) all measured the beneficial change in pressure distribution over the seat and backrest that would occur when using a lumbar support. Jürgens and Helbig (1973) and Helbig (1978) analyzed the pressure distribution on the seat, backrest, and feet with an experimental chair that was adjusted to various seat and backrest angles. They also studied both anterior and posterior sitting postures.

Jürgens (1969) and Bendix et al. (1985a) measured the pressure on the backrest when using a chair with an inclined seat surface compared to a chair with a declined seat surface. Swearingen et al. (1962) recorded the beneficial effect that certain chair features such as armrests and a backrest would have on reducing body weight from the seat.

Observations of Body Position and Movement

Bhatnager et al. (1984, 1985) observed the postural changes of four seated individuals over a three hour period as they inspected slides of printed circuit boards from three different screen heights. They found that the frequency of postural change or fidgeting increased by more than 50 percent over the three hours of observation. The total frequency of posture changing was found to be a sensitive indicator of postural stress and discomfort.

Grandjean et al. (1960) recorded the body movements of individuals while sitting in different chairs. They found among subjects tested that positive answers on questions related to seat comfort corresponded to fewer body movements on the chair, whereas negative answers corresponded to frequent body movements.

Branton (1966) correlated postural change with the need for stability. If the seat fails to stabilize the sitter, he will spontaneously take up such postures that will satisfy the need for body stability. As fatigue from muscle exertion sets in, another posture will be adopted as the individual seeks to minimize fatigue and stabilize his body segments.

According to Branton's (1966) instability hypothesis, "The greater the instability, the less likely is muscular relaxation and hence the greater is discomfort." The postures held longest were found by Branton (1966) to be the most stable.

Aveling (1879) thought it appropriate to quote the following passage from Cowper's *Sofa* in his discussion of sitting discomfort:

> But restless was the chair; the back erect
> Distressed the weary loins that felt no ease,
> The slippery seat betrayed the sliding part
> That pressed it, and the feet hung dangling down,
> Anxious in vain to find the distant floor.

Cantoni et al. (1984) performed a posture analysis and evaluation at the switchboard control room of a telephone company before and after the introduction of an ergonomically designed chair and VDT workstation. Compared to the old work place, at the new ergonomically designed workstation they found a considerable increase in the duration of sitting postures with the spine resting on the backrest of the chair. The duration of kyphotic sitting postures, which was very extensive at the old work place, was very minimal at the new workstation. The results showed a highly significant decrease in the number of postural changes at the new workstation compared to the old work place. Among the four individuals studied, the mean percent decrease in the number of postural changes per hour ranged from 44 percent to 71 percent.

In a study involving fourteen subjects, Pustinger et al. (1985) reported fewer episodes of motion and fewer health complaints at an adjustable VDT workstation compared to a fixed VDT workstation. Increased movement was also found to have an adverse effect on productivity. Mark et al. (1985) also found considerably less movement at properly adjusted VDT workstations.

Shephard (1974) has drawn attention to the fact that the frequency of postural changes can also be influenced by cultural factors. For example, a woman might cross her legs when sitting because she was taught to do so, as opposed to spontaneously adopting this posture for greater stability in the chair.

The frequency of postural changes or fidgeting may also be affected by other factors such as temperature, humidity, boredom, general stress associated with the specific work situation, and the daily rhythm of the individual, where one may be more mentally and physically active at certain times of the day (Jürgens, 1980; Pheasant, 1986).

Rathbone and Hunt (1965) stressed the importance of the chair holding the trunk in an extended position, with proper support just below the shoulder blades, at the thoracolumbar junction. They felt that such a chair, with the proper seat height and depth, would result in greater comfort and a minimum of fidgeting, as the spinal muscles would not fatigue readily, and the respiration and circulation would not be hindered.

Observations of Task Performance

Several recent studies have shown how chairs and workstations that minimize worker discomfort can have a positive effect on productivity.

A study by Dainoff et al. (1982) compared the data-entry performance of thirteen trained typists at a VDT workstation with good ergonomic design to one with poor ergonomic design. At the workstation with good ergonomic design, the chair had a gas cylinder height adjustment for the seat, an adjustable backrest inclination, and an adjustable lumbar support. A wrist rest and copy holder were also part of the workstation. At the poorly designed workstation, the chair had a non-adjustable seat height, a flexible backrest, and no lumbar support. There was no copy holder or wrist rest, and the workstation was deliberately maladjusted in regards to proper keyboard height and screen viewing angle. A high level of glare was also present on the VDT screen. Results from this study showed fewer musculoskeletal complaints at the workstation with good ergonomic design, along with a 24.6 percent improvement in performance. The performance improvement was mainly due to an increase in keystroke rate (Dainoff, 1983, 1984b; Secrest and Dainoff, 1984).

Among data entry operators at an airline computer center in Singapore,

Ong (1984) found that an improved working environment, which included an improved chair design and working posture, significantly reduced visual and muscular complaints. The use of document holders and footrests were among the improvements made at the new workstation.

Ong (1984) reported a significant improvement in work efficiency and performance at the new workstation. The average keying speed increased from 9480 keystrokes per hour to 13,002 keystrokes per hour. In addition, the average monthly error declined from 1.54 percent to 0.11 percent.

Springer's study (1982a, 1982b, 1983) at a major corporation with VDT workstations found that employees preferred a comfortable, user adjustable workstation with the ability to adjust the seat and backrest positions of the chair, the work surface height, and the terminal screen and keyboard positions. Compared to the control workstation, the two most preferred workstations resulted in a 10 percent performance improvement for dialogue tasks, and a 15 percent performance improvement for data entry tasks.

Rohmert and Luczak (1978) investigated the performance and physiological strain at a "postal video letter coding workstation" after certain design improvements. Among the eight individuals studied they reported an increased performance of 80 to 120 codings per hour, along with a decrease in heart rate and myoelectric activity of the extensor digitorum with the ergonomic improvements. The workstation improvements included better foot supports, the addition of inclined arm supports, and the replacement of flat keyboards with inclined, angular keyboards.

An interesting study by Riskind and Gotay (1982) indicated that an individual's physical posture can have carry-over effects on motivated behavior. In a laboratory setting at two different universities, twenty undergraduate students were placed in either a slumped, kyphotic sitting posture, or an upright, lordotic chair sitting posture. The individuals who were previously placed in a slumped, kyphotic posture later showed significantly lower persistence in a standard learned helplessness task, an insoluble geometric puzzle. These results suggested that "the self perception of being in a more slumped-over physical posture predisposes a person to more speedily develop self-perceptions of helplessness later, following exposure to problems that the person finds to be insoluble."

Subjective Measurement Techniques

These measurements are taken during and after controlled sitting trials.

General Comfort Rating. Shackel et al. (1969) developed a scale of eleven statements to elicit from subjects at various intervals during a sitting trial their sensation of comfort-discomfort. The subjects are instructed to draw a horizontal mark anywhere on the eleven item scale to indicate their comfort rating. The eleven statements comprising the scale are as follows:

> I feel completely relaxed.
> I feel perfectly comfortable.
> I feel quite comfortable.
> I feel barely comfortable.
> I feel uncomfortable.
> I feel restless and fidgety.
> I feel cramped.
> I feel stiff.
> I feel numb (or pins and needles).
> I feel sore and tender.
> I feel unbearable pain.

Body Area Comfort Rating. One technique for assessing postural discomfort involves a diagram of the body divided into various regions (Corlett and Bishop, 1976; Corlett, 1981). At several intervals throughout a sitting trial, the subjects are first asked to indicate their overall level of discomfort on a five or seven point scale with "no discomfort" and "extreme discomfort" at the scale extremes. After the general rating, the body map is used (Figure 49). The individual first indicates the body part or parts that are the most painful or uncomfortable. Then, the next most uncomfortable body parts are indicated, and so on.

Results from certain studies regarding discomfort in specific body parts have important implications in chair design. Slechta et al. (1959a) evaluated the design characteristics of the C-118 pilot seat in regards to comfort. This pilot seat has armrests, is adjustable in seat height and backrest inclination, and has a fixed seat inclination of 9 degrees. The seat and back cushions are made of foam rubber, with a leather upholstery cover.

Hourly questionnaires were given to seventeen subjects during a voluntary sitting session of seven hours maximum duration. For all body regions the average time for onset of discomfort was 220 minutes, with the most discomfort experienced in the buttocks and back. Six of the seventeen subjects felt that the seat could be improved by providing better cushioning for the buttocks. Four of the seventeen subjects felt that

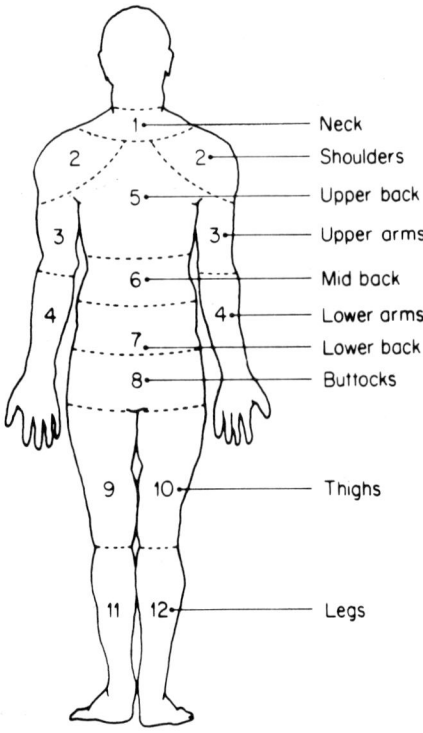

Figure 49. Body map. From Corlett, E.N.: Pain, posture and performance. In Corlett, E.N., and Richardson, J. (Eds.): *Stress, Work Design, and Productivity.* Chichester, Wiley, 1981, pp. 27–42. Reprinted by permission of John Wiley and Sons Ltd.

the back cushion should give better support to the lower back.

Another study by Slechta et al. (1959b) involved the evaluation of the C-124 crew seat. Design characteristics of this seat included an adjustable seat height, a fixed seat inclination of 2 degrees, a fixed backrest inclination of 10 degrees from vertical, and no armrests. The seat cushion was of foam rubber and the back cushion consisted of layers of glass fiber. Both cushions had a canvas upholstery cover. For all body regions, the average time of onset of discomfort was 134.5 minutes. Marked discomfort was experienced in the buttocks and back, with buttocks discomfort appearing after about one hour.

Fourteen of the seventeen subjects expressed the need for a firmer seat cushion, nine subjects felt the back cushion should give better lower back support, eleven subjects felt the backrest angle was too vertical and that it should be adjustable, and sixteen of the seventeen subjects expressed a need for armrests.

Wachsler and Learner (1960) re-analyzed data obtained in an earlier study regarding the relative comfort of six Air Force pilot and crew seats (Slechta et al., 1957). They found that individuals tended to rate the overall comfort of a seat mainly on the comfort of their backs and buttocks.

Oshima (1970) also found in his experiments on chair design for a bus passenger seat that the back and buttocks were the body parts that received the greatest subjective complaints with increasing sitting time. Wotzka et al. (1969), in their investigation of auditorium seats, found that students' main complaints of pain from the seats in four auditoriums involved the back and buttocks. In a study of chairs used by FAA air traffic controllers, the two most frequent body areas listed for discomfort from sitting were the lower back and buttocks (Kleeman, 1980). Discomfort in the lower back was felt by 29.7 percent of the air traffic controllers, and buttocks discomfort was felt by 16.3 percent of the controllers.

Chair Feature Checklist. This is a method where the sitter can specifically rate those features of a chair which might produce local discomfort (Shackel et al., 1969; Drury and Coury, 1982) (Figure 50).

Direct Ranking. This technique simply involves the individual sitting in several chairs and ranking them in order of comfort (Shackel et al., 1969).

Shackel et al. (1969) stressed in regards to seat comfort "the importance of the user's subjective assessment, and its essential primacy as the ultimate criterion of comfort against which other more convenient and perhaps more objective methods may be validated."

Some researchers take issue with relying too heavily on subjective criteria of comfort for chair design. Oxford (1973) finds it inadvisable to attach final importance in chair design to the comfort ratings of sitters. According to Oxford, although lumbar support is an essential physiological requirement for chair design, many individuals are unaware when it is not present. Many individuals are also not aware that a firm seat is physiologically better than a soft seat.

Barkla (1964) found that while comfort ratings after thirty minutes could reliably discriminate between different settings of an experimental chair, ratings after a five minute sitting session could not reliably discriminate between different settings.

Fiedler and Fiedler (1977) feel that people's knowledge regarding the optimum healthy sitting posture is completely inadequate. Bennett (1928) commented upon how a slumped sitting posture may be more comfortable for an individual whose habitual sitting posture is very poor.

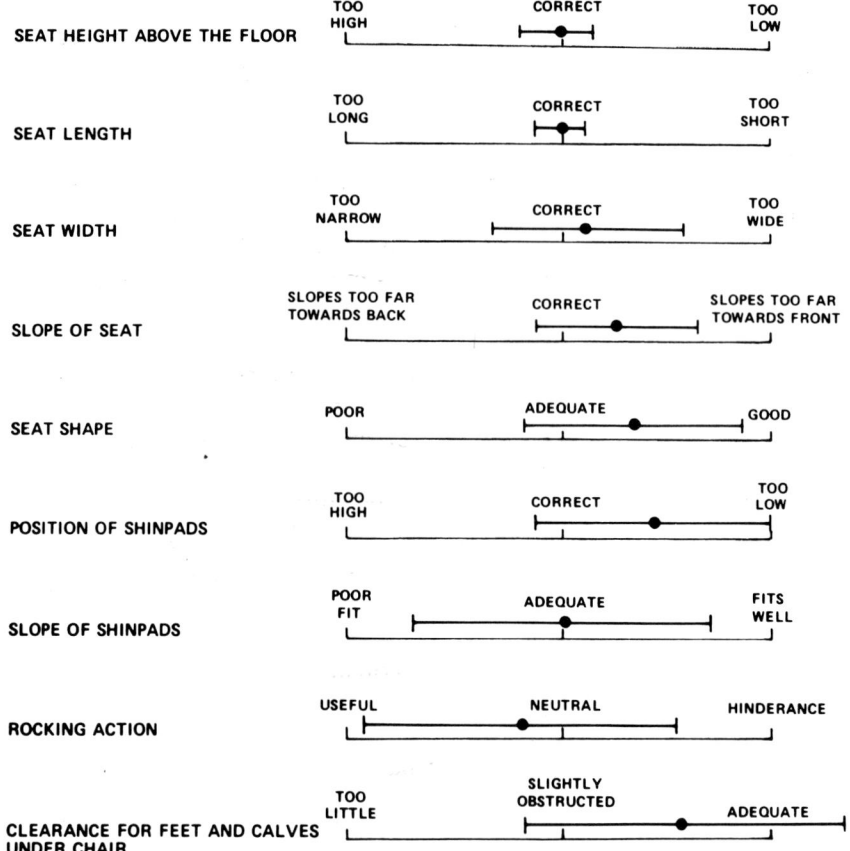

Figure 50. A chair feature checklist used in the evaluation of a kneeling chair. From Drury, C.G., and Francher, M.: Evaluation of a forward-sloping chair. *Applied Ergonomics,* *16*:41–47, March 1985. Reproduced with permission of Butterworth Scientific Limited.

According to Pheasant (1984), "The question of whether or not people know what is good for them is a very difficult one in ergonomics —if they do not, then any data gathered by fitting trials is highly suspect."

Specific Design Considerations Regarding Discomfort

Some seating authorities have specified certain design features as being very critical to reduce discomfort.

Hockenberry (1977) feels that discomfort in long term sitting is mainly due to improper pressure distribution resulting from a poor fit of the seat and backrest shape to the contours of the individual. He lists seven basic pressure factors, several referred to earlier by Hertzberg (1972), which must be considered in chair design to reduce discomfort:

1. Pressures exerted on the ischial tuberosities must be reduced with a slight contouring of the seat in the buttocks region.
2. The backrest must be contoured to provide lumbar support. (As will be discussed in Chapter Five, proper lumbar support will also reduce pressure under the ischial tuberosities.)
3. Pressure under the posterior thighs must be kept to a minimum.
4. A 95 to 105 degree seat-backrest angle will result in an optimal pressure distribution over the seat and backrest.
5. The seat and backrest cushion cover must be elastic enough to pass local pressure loads directly to the cushion.
6. The sides of the seat should not exert pressure on the greater trochanters of the femurs.
7. The backrest and seat should both be contoured with a side to side radius that fits the 5th through 95th percentile of the adult population.

Major chair design faults discussed by Travell (1955) that would result in discomfort are as follows:

1. No lower back support. This is considered by Travell to be the commonest fault in chair design.
2. Improper armrest height.
3. Excessive concavity in the upper backrest resulting in a rounding of the upper back, and pushing the shoulders forward.
4. A backrest that is too vertical.
5. A backrest that is too low in height. One will eventually slump down in the seat to support the upper back.
6. Design factors that will cause the sitter to assume a thigh to trunk angle of less than 90 degrees, resulting in kyphosis of the lumbar spine. Included here would be an insufficient seat to backrest angle, and the effect of sitting on very soft upholstery.
7. A hard front edge to the seat, especially on a chair or sofa with soft seat upholstery. The resulting pressure will interfere with the venous blood flow from the lower legs.
8. The extreme contouring of a bucket seat which will result in excessive pressure over the greater trochanters of the femurs.

The three design factors considered by Keegan (1953, 1962, 1964) to be the most important to minimize discomfort, and applicable to all chairs, are:

1. Proper lumbar support.
2. A minimum angle of 105 degrees between the thighs and trunk to help preserve the lumbar curve.
3. An open or recessed space below the lumbar support for the sacrum

and buttocks. This space is necessary for proper contact with the lumbar support.

Surveys on Comfort

Recent surveys have emphasized the importance of comfort for individuals who spend much of their working day seated.

A survey based on responses from 1,967 FAA air traffic controllers revealed that the chair's fit/comfort was ranked as its most important attribute by the controllers (Kleeman, 1980; Kleeman and Prunier, 1982).

Seat height adjustment was ranked first as the most wanted chair adjustment by the controllers. A chair with upholstered armrests was also preferred by the controllers.

Slightly more than 50 percent of the air traffic controllers were found to be uncomfortable in their chairs. Regarding specific areas of discomfort from sitting:

29.7 percent felt discomfort in the lower back.
16.3 percent felt discomfort in the buttocks.
14.6 percent felt discomfort in the upper back.
9.9 percent felt discomfort in the back of the neck.
6.4 percent felt discomfort in the upper legs.
5.4 percent felt discomfort in the lower legs and feet.

The Steelcase National Study of Office Environments, No. II (1980) conducted by Louis Harris and Associates, Inc. involved interviews with a national cross section of 1,004 office workers and 203 executives.

Forty-six percent of the office workers interviewed felt that a more comfortable chair with good back support would contribute to their productivity either a great deal (26%) or somewhat (20%).

The most important features of a chair to office workers were:

Correct seat height—90 percent
Overall comfort—89 percent
Good lower back support—86 percent
Wheels for moving around easily—83 percent
A swivel base—79 percent
Adjustable seat height—78 percent

By far, the two most important features of an office chair that office workers felt helped them do their job well were good lower back support (66%) and wheels for moving around easily (53%). These were also the main two items that office workers would personally be willing to spend more money for, if purchasing an office chair.

Good lower back support was also ranked first by executives as the chair feature they felt would contribute most to improved comfort for office workers, as the chair feature they felt might best help increase the productivity of the office workers, and as the office chair feature they would be willing to spend more money for.

As stated in the study, "A final indication of the importance of the chair to office workers is the fact that 26% of office workers would take a smaller raise with a specially designed chair for back support and comfort, rather than a large raise with no comfort improvements."

Chapter Five

PRESSURE DISTRIBUTION

EFFECT OF CHAIR FEATURES
ON PRESSURE DISTRIBUTION

As mentioned in Chapter Four, the buttocks are one of the first body parts to experience discomfort from sitting (Slechta et al., 1959a, 1959b; Wotzka et al., 1969; Oshima, 1970). With improper chair design and improper sitting posture, extremely high pressure points and tissue distortion will develop over and posterior to the ischial tuberosities. Ischemia and pain will soon follow, making further sitting intolerable.

Pelvic-Sacral Support

The main support provided by a seat should be over and anterior to the ischial tuberosities (Babbs, 1979). Therefore, the major weight bearing areas on the seat will be the ischial tuberosities and upper half of the posterior thighs (Bennett, 1928; Floyd and Ward, 1967) (Figure 51).

If the only weight bearing areas on the seat are over the ischial tuberosities, extremely high, localized pressure areas will result. If the major weight bearing area on the seat is posterior to the ischial tuberosities, there will be distortion and compression of the gluteus maximus muscles, along with a localized high pressure point over the coccyx (Bennett, 1928; Howorth, 1978; Babbs, 1979) (Figure 52).

The key to proper pressure distribution on the seat, with the weight bearing over and anterior to the ischial tuberosities, is *proper sacral and pelvic support,* thereby preventing or reducing backward rotation of the pelvis and the subsequent lumbar kyphosis. (As used throughout this text, "proper lumbar support" refers to support being given to the upper sacrum and posterior iliac crests.)

DuToit and Gillespie (1979) commented upon how a lumbar lordosis would help transfer some weight bearing to the upper posterior thighs. Watkin (1983) mentioned that when sitting with a lumbar lordosis, about

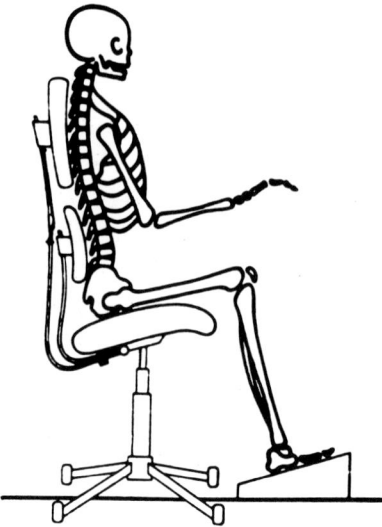

Figure 51. The major weight bearing areas on the seat should be the ischial tuberosities and the upper half of the posterior thighs. From Rizzi, M.: Entwicklung eines verschiebbaren rückenprofils für auto-und ruhesitze. In Grandjean, E. (Ed.): *Proceedings of the Symposium on Sitting Posture.* London, Taylor and Francis, 1969, pp. 112–119. Reproduced with permission of Taylor and Francis Ltd.

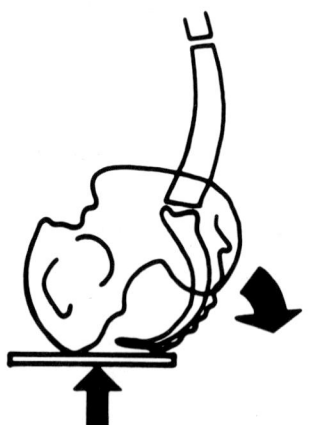

Figure 52. If the major weight bearing area on the seat is posterior to the ischial tuberosities, there will be a localized high pressure point over the coccyx. From Andersson, B.J.G., Örtengren, R., Nachemson, A.L., Elfström, G., and Broman, H.: The sitting posture: an electromyographic and discometric study. *Orthopedic Clinics of North America,* 6:105–120, 1975. Reproduced with permission of W. B. Saunders Company.

25 percent of the body weight that would usually be localized over the ischial tuberosities will be re-distributed over the posterior thighs.

Drummond et al. (1982a, 1982b, 1983, 1985) were able to give specific data regarding the change in pressure distribution when sitting with and without a lumbar lordosis. Pressure distributions were calculated as individuals sat on a pressure scanner with their hands lightly supported in front of the chest or abdomen, and with the feet hanging freely. In able-bodied individuals who were able to achieve a lumbar lordosis when sitting, 18 percent of the sitting pressure was distributed over each ischial tuberosity, and 21 percent of the sitting pressure was distributed over each posterior thigh. In disabled individuals unable to achieve a lumbar lordosis when sitting, there was a shift of pressure distribution posteriorly, with greater point pressures over the ischial tuberosities and sacrococcygeal region. The inability to sit with a lumbar lordosis resulted in 60.3 percent of the sitting pressure being distributed over the ischial tuberosities and sacrococcygeal region, compared to only 39 percent in individuals able to sit with a lumbar lordosis.

The following changes in pressure distribution will occur with a lumbar lordosis:

a. The coccyx will not bear weight on the seat with a lordotic sitting posture. With a marked kyphotic sitting posture, however, weight bearing will occur over the coccyx (Bennett, 1928; Howorth, 1978).

A prolonged kyphotic sitting posture, especially in a moving vehicle, has been implicated as a major cause of coccygeal pain or coccygodynia (Johnson, 1981; Frazier, 1985). Stoshak and Mortimer (1985) referred to a "jean seam coccygodynia" which they attributed to "stiff reinforced seams in blue jeans pressing on the coccyx when the patient sits in a semi-reclining, slumped position on hard, wooden school seats for many hours."

b. There will be better support for the upper trunk against the backrest, with a greater percentage of the body weight being taken by the backrest (Strasser, 1913; Diebschlag and Müller-Limmroth, 1980; Majeske and Buchanan, 1983, 1984; Zacharkow, 1984a) (Figure 53).

Even on a chair with a high backrest the individual will not be able to receive proper support for the thoracic spine and scapulae without proper lumbar support. As a result, the individual will sit with a round-shouldered posture (Kamijo et al., 1982).

Scapular support against the backrest will also improve upper extremity functioning for individuals with weak scapular musculature, particu-

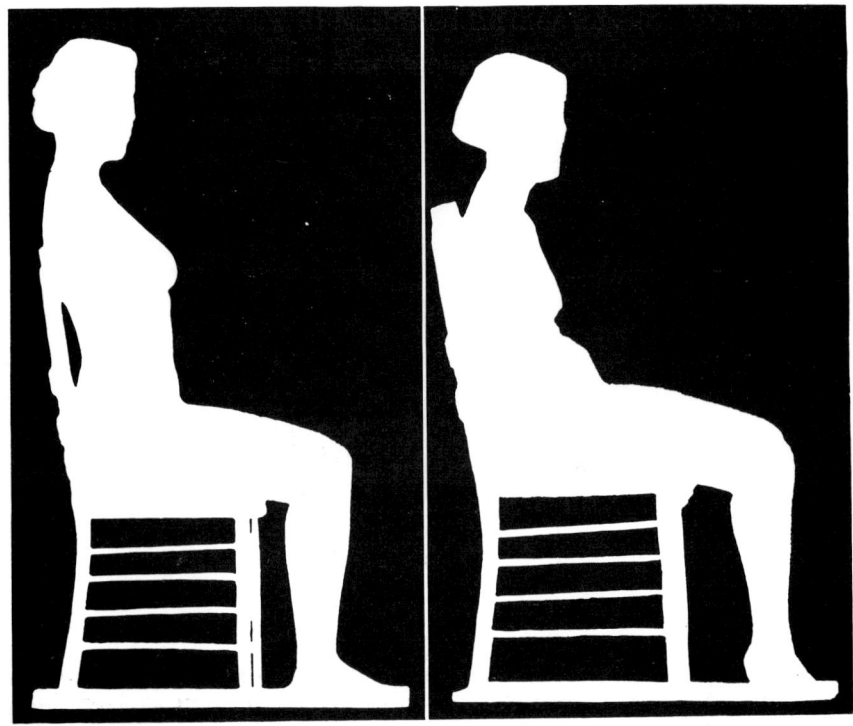

Figure 53.
   Left: With a lordotic sitting posture, there will be better support for the upper trunk against the backrest.
   Right: Kyphotic sitting posture.
   From Kellogg, J.H.: Observations on the relations of posture to health and a new method of studying posture and development. *The Bulletin of the Battle Creek Sanitarium and Hospital Clinic, 22:*193–216, 1927.

larly serratus anterior weakness. The importance of this scapular support is often overlooked in wheelchair sitting posture for disabled individuals.

   c. An important factor in the change in pressure distribution resulting from a lordotic sitting posture is the re-distribution of the trunk body weight. As can be seen in Table II, the largest percentage of the total body weight is in the trunk. In the cadaver studies cited, the segmental weight of the trunk ranges from 44.2 percent to 50.7 percent of total body weight.

   With a lumbar lordosis and the proper thigh to trunk angle, the resting position for the ischials will be further posterior on the seat, the weight line of the trunk will be shifted further forward on the seat, and the upper trunk will receive greater support from the backrest (Strasser,

1913; Bennett, 1928; Kamijo et al., 1982; Watkin, 1983; Zacharkow, 1984a) (Figures 54 and 55).

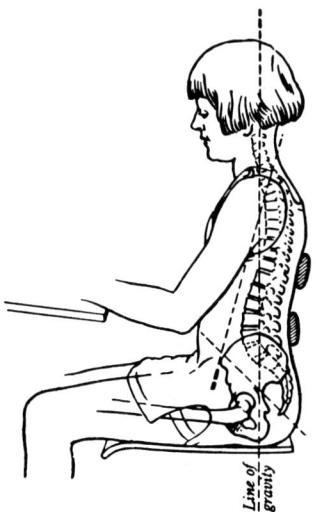

Figure 54. Lordotic sitting posture. Compare the line of gravity and the dotted lines indicating the center lines of the trunk and pelvic cavities to Figure 55. From Bennett, H.E.: *School Posture and Seating.* Boston, Ginn and Company, 1928.

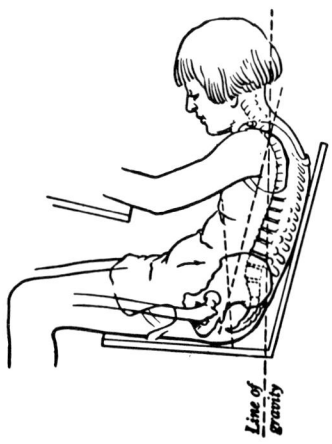

Figure 55. Kyphotic sitting posture. From Bennett, H.E.: *School Posture and Seating.* Boston, Ginn and Company, 1928.

Table II

SEGMENTAL WEIGHT/TOTAL BODY WEIGHT RATIOS FROM VARIOUS CADAVER STUDIES

Investigator	Sample Size	Head	Trunk	Arm x 2	Leg x 2
Harless (1860)	2	7.6%	44.2%	11.4%	36.8%
Braune and Fischer (1889)	3	7.0%	46.1%	12.4%	34.4%
Fischer (1906)	1	8.8%	45.2%	10.8%	35.2%
Dempster (1955)	8	8.1%	49.7%	10.0%	32.2%
Clauser et al. (1969)	13	7.3%	50.7%	9.8%	32.2%

Adapted from Clauser et al. (1969)

Armrests

Proper fitting armrests on a chair are critical to help further reduce body weight on the seat. As can be seen from the cadaver studies in Table II, the weight of both arms ranges from 9.8 percent to 12.4 percent of total body weight. In another study involving twelve living subjects, Drillis and Contini (1966) found both arms to be 11.94 percent of total body weight.

Swearingen et al. (1962), in their study of sitting pressures with eight male subjects, found that the addition of chair arms to a horizontal sitting platform reduced 12.4 percent of the body weight from the seat. Brattgård and Severinsson (1978) found that correct armrest adjustment reduced ischial tuberosity pressures by approximately 25 to 30 percent.

This reduction in ischial tuberosity pressures results from:

    a. Decreased body weight on the seat from proper arm support.
    b. Prevention of a slumped posture of the upper trunk due to proper arm support (Gibson and Wilkins, 1975).
    c. More support of the upper trunk by the backrest due to proper arm support.

The effect of arm support on trunk posture was also recently reported by Nakaseko et al. (1985) in their study involving an experimental keyboard design. One part of the study compared the effect of using a large forearm-wrist support measuring 28 cm from the front edge to the G key, to a small forearm-wrist support measuring 18 cm from the front edge to the G key. Both supports were inclined eight degrees (Figure 56). The individuals using the large forearm-wrist support were found to exert more pressure with their forearms and wrists against the support. In addition, these individuals also sat with a greater trunk inclination. A preliminary study by these authors involving keyboard designs with and without forearm-wrist supports found that the keyboards with forearm-wrist supports "induced the subjects to lean the trunk more backwards."

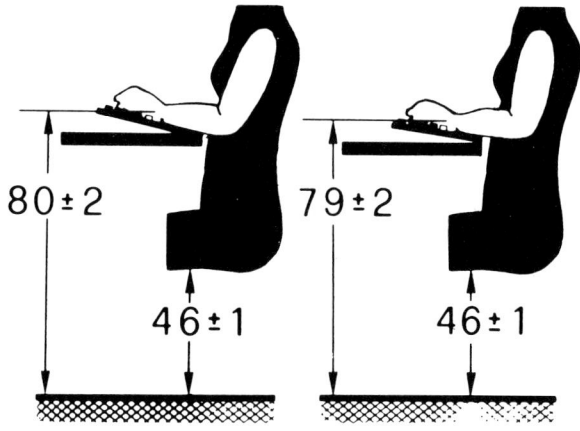

Figure 56.
   Left: An experimental keyboard with a large forearm-wrist support.
   Right: An experimental keyboard with a small forearm-wrist support.
   From Nakaseko, M., Grandjean, E., Hünting, W., and Gierer, R.: Studies on ergonomically designed alphanumeric keyboards. *Human Factors, 27:*175–187, 1985. Copyright 1985, by the Human Factors Society, Inc., and reproduced by permission.

Foot Support and Back Support

The study by Swearingen et al. (1962) first studied individuals sitting on a horizontal platform without back support or foot support. Swearingen et al. (1962) then found that proper foot support would reduce 18.4 percent of the body weight from the seat. They further found that the addition of a backrest inclined 15 degrees from upright, but with the legs dangling and without foot support, reduced 4.4 percent of the body weight from the seat.

The combination of foot support and a backrest with a 15 degree inclination resulted in a 31.3 percent reduction in body weight from the seat. As the arithmetic addition of these two chair features separately would only involve a 22.8 percent reduction in body weight from the seat, the increased reduction in body weight on the seat when these two chair features are combined can be attributed to the greater pressure exerted on the backrest with proper foot support.

Overall, Swearingen et al. (1962) found that the use of armrests, foot support, and a backrest inclined 15 degrees from upright (but without a lumbar support) resulted in a 39.4 percent reduction of body weight from the seat. This percentage of body weight reduction from the seat would be even greater if the backrest had a lumbar support (Diebschlag and Müller-Limmroth, 1980; Zacharkow, 1984a).

A difference in backrest pressure is also observed when comparing a chair with an inclined seat surface to a chair with a declined seat surface. Bendix et al. (1985a) determined the backrest pressure for nine female subjects in chairs with three different seats: a 5 degree seat inclination, a 10 degree declined seat, and a freely tiltable seat. The backrest pressure was found to be twice as high with the inclined seat compared to the other two seats during both desk work (writing, reading, sorting papers) and typing.

An earlier study by Jürgens (1969) also showed a greater backrest pressure with an inclined seat compared to a declined seat. With the declined seat, Jürgens found that a greater percentage of the body weight was shifted to the feet.

The Seat

Seat Contour

Lincoln (1896) observed that "a carved seat, i.e., with a saucer-like hollow of elongated shape to sit in, saves much of the pain which comes from sitting on flat boards."

According to Hertzberg (1972), "A correctly contoured surface, especially one that is also properly cushioned, will permit people to sit longer without discomfort than any flat seat. The contoured surface spreads the load that is normally on the tuberosities to some of the surrounding tissue, greatly reducing the peak pressures."

(The difference between able-bodied seating and wheelchair seating for pressure sore prevention is that with able-bodied seating the objective is not to obtain an overall uniform pressure distribution, but a relatively wider distribution of pressure at pressure points (Asatekin, 1975). Seating for pressure sore prevention, where a uniform pressure distribution is desirable, will be discussed in Chapter Eleven.)

Although Hertzberg's earlier studies (1958) with various seat configurations were never completed, preliminary tests showed that the standard hard, flat seat made test subjects uncomfortable in 1½ to 2 hours. A contoured seat with a one inch thick foam rubber pad did not result in discomfort until after 4 to 5 hours of sitting. Hertzberg (1958) also reported that a group of Air Force pilots sat for 15 consecutive hours without discomfort using a contoured seat pan with a contoured foam rubber cushion.

A study by Kosiak et al. (1958) on sitting pressures under the ischial tuberosities supports some of Hertzberg's findings. Pressures were measured under the ischial tuberosities and at ten other points on the sitting surface, in eleven normal adult subjects while sitting in several different chairs. On a flat, hard surface, Kosiak et al. found most of the pressure to be concentrated under the ischial tuberosities. A hard, contoured seat resulted in a wider pressure distribution than the hard, flat seat. In addition, Kosiak et al. found that a hard, flat seat padded with 2 inches of foam rubber reduced ischial tuberosity pressures over 50 percent compared to the unpadded hard, flat seat.

In a comprehensive study by Ridder (1959), 129 adults selected their preference of basic chair measurements using an experimental chair. The shape of the seat could be individually adjusted since it was made of a series of aluminum plungers capped with slightly concave rubber tips. The final chair measurements revealed that a contoured seat surface was preferred by the subjects. Ridder commented that the addition of one inch of foam rubber to the molded seat surface would increase sitting comfort even more.

Grandjean et al. (1973), in their investigation of multipurpose chairs, found that subjects judged chairs with a slight concavity to the seat surface for the buttocks to be more comfortable than a flat seat surface. For a wider distribution of pressure at pressure points, Grandjean et al. recommended a slightly contoured seat surface padded with 2 to 4 cm of foam rubber.

It is important to realize that excessive contouring or improper location of the seat can have deleterious effects on pressure distribution and sitting comfort:

a. Excessive seat contouring can result in distortion of the gluteus maximus muscles, which are reflected laterally off the ischial tuberosities when sitting (Bennett, 1928).

b. The difference in height between the weight bearing surface of the ischial tuberosities and a potential weight bearing surface of the greater trochanters of the femurs is only approximately 2.5 to 3 cm (Helbig, 1978). Therefore, when the side to side radius of the seat contour is too deep the greater trochanters of the femurs will bear weight (Figure 57). Based on their structure and function, the trochanters are completely unsuited for supporting the body weight in the sitting position, and this will quickly lead to discomfort (Helbig, 1978; Hertzberg, 1972; Hockenberry, 1977).

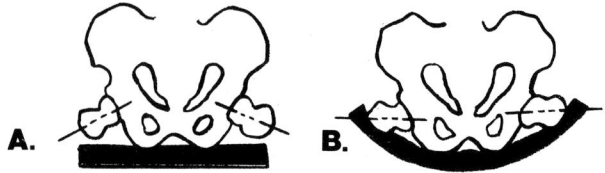

Figure 57.
   A. With a flat seat surface, the weight bearing will be over the ischial tuberosities.
   B. With an excessive seat contour, the greater trochanters will bear weight and the femurs will internally rotate.
   Adapted from Diffrient, N.: The Diffrient difference. *Leading Edge, 1(5):*41–59, June 1984. With permission of Niels Diffrient.

c. As a result of the distortion from an extreme "scoop" to the seat, the femurs will have a tendency to internally rotate (Bennett, 1928; Travell, 1955; Diffrient, 1984). As a result of this internal hip rotation, the greater trochanter will move superiorly (See Figure 57). Therefore, the sciatic nerve will be exposed to pressure just lateral to the ischial tuberosity, where it is normally protected within the depth of the ischio-trochanteric gutter between the greater trochanter and ischial tuberosity (Le Floch, 1981).

d. Any elevation at the posterior aspect of the contoured seat must be located behind the back support. If this elevation is located anterior to the back support, it will tend to slide the sitter forward into a slumped posture (Bennett, 1928) (Figure 58). Actually, the elevation at the posterior aspect of a contoured seat serves no useful purpose, and is best eliminated from the seat design (Bennett, 1928).

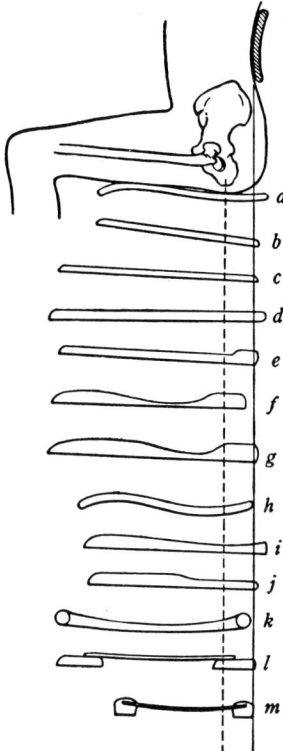

Figure 58. Any elevation at the posterior aspect of a contoured seat that is not located behind the back support will tend to slide the sitter forward into a slumped posture. Note in particular f,g, and h. From Bennett, H.E.: *School Posture and Seating.* Boston, Ginn and Company, 1928.

Upholstery and Upholstery Cover

Regarding optimal pressure distribution on the seat of the chair, latex foams are preferred for cushioning over polyurethane foams. According to Marchant (1972), "If an indenting force is applied to a latex foam, the foam is evenly compressed throughout its thickness. With polyether foam the compression is confined to a localised area near the indentor."

Polyether foam, which is the type of polyurethane foam used most frequently in seating applications, is therefore more prone to shape distortions than latex foam. Evaluations by Cochran and Slater (1973) have substantiated the excellent pressure distributing qualities of latex foam.

Hertzberg (1972) commented that the upholstery cover must be elastic enough to allow local loads to pass directly to the cushion when sitting. With a stiff, non-stretch upholstery cover such as canvas or plastic, the body will be supported by tension in the upholstery cover (Torrance, 1983). This "hammock effect" will adversely affect the pressure distribution of the cushion, resulting in high peak pressures and greater tissue distortion over the buttocks (Chow, 1974; Chow et al., 1976; Barbenel et al., 1978, 1981).

THE OCCUPIED CHAIR AND PRESSURE DISTRIBUTION

An unoccupied chair may give the appearance that it will provide an optimal pressure distribution for the sitter. However, a factor of major importance in pressure distribution is the actual contour and fit of the chair when it is occupied by the individual (Branton, 1970; Asatekin, 1975). For example, the angle between the seat and backrest of a chair may appear satisfactory for proper pressure distribution when the chair is unoccupied, but due to the effect of different degrees of upholstery firmness, when the seat is compressed by the weight of the sitter, the seat to backrest angle may become too acute.

A brief summary will now be given on how a poor chair fit can adversely affect pressure distribution.

Seat Depth

Excessive seat depth will cause the individual to slide forward on the seat in order to avoid pressure at the upper posterior calf. The individual will then be unable to obtain proper support from the backrest and lumbar support, and will usually assume a slumped, kyphotic posture resulting in excessive pressure over and posterior to the ischial tuberosities.

Seat Height

If the seat height is too low for the individual there will be a lack of pressure distribution over the upper posterior thighs. As a result, all the sitting pressure will be localized over and posterior to the ischial tuberosities.

If the seat is too high, the individual will usually move his buttocks forward on the chair seat, to avoid a high cut-off pressure at the distal posterior thigh, and to allow a more stable sitting posture with the feet firmly on the floor. As a result of this position change, the individual will usually end up in a slumped, kyphotic posture since he will not have proper use of the back support. High pressure areas will then result over and posterior to the ischial tuberosities.

Seat to Backrest Angle

If the angle between the compressed seat and backrest results in a thigh to trunk angle less than 90 degrees or greater than approximately 105 degrees, there will be a tendency to sit in a slumped, kyphotic posture with increased pressure over and posterior to the ischial tuberosities, and distortion of the gluteal muscles.

A horizontal seat with an inclined backrest will result in a tendency to be ejected from the seat when one leans against the backrest (Strasser, 1913; Åkerblom, 1948; Ayoub, 1972). Once again, a slumped, kyphotic posture with its characteristic poor pressure distribution will result. An inclined seat, and/or a seat with a slight concavity for the buttocks will counteract this tendency to slide forward on the seat (Bradford and Stone, 1899; Shaw, 1902; Dresslar, 1917; Bennett, 1928; Kroemer and Robinette, 1968, 1969; Rosemeyer, 1972, 1974).

Without a slight concavity or depression for the buttocks, a very firm horizontal seat may actually function as a declined seat. This is due to the difference in height between the weight bearing surface of the ischial tuberosities and the posterior thighs (Carlson et al., 1986) (Figure 59).

Upholstery and Upholstery Cover

With very thick, soft cushioning, or cushioning over a sagging, non-firm base, there will be a tendency for the hips to internally rotate. The effect will be similar to a seat with an extreme "scoop" or concavity for the buttocks, with excessive pressure and tissue distortion over the greater trochanters (Diffrient, 1984). Of course, with the loaded chair, thick, soft cushioning will also result in an acute thigh to trunk angle.

With a smooth-surfaced, low-friction seat cover, as opposed to a woven fabric, there will be a tendency to slide into a slumped sitting posture. Once again, high pressure regions and tissue distortion will result over and posterior to the ischial tuberosities (Branton, 1970; Oxford, 1973; Schaedel, 1977).

Posture

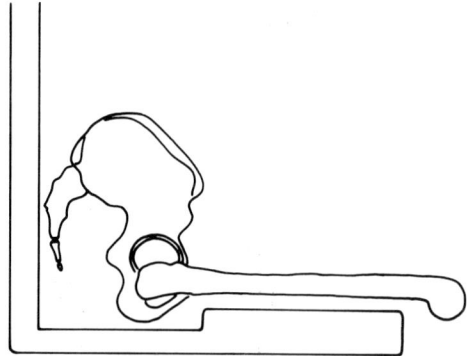

Figure 59. With a very firm horizontal seat, a slight concavity or depression for the buttocks is necessary in order to bring the femurs to a horizontal position. Otherwise, a very firm horizontal seat may actually function as a declined seat. From Carlson, J.M., Lonstein, J., Beck, K.O., and Wilkie, D.C.: Seating for children and young adults with cerebral palsy. *Clinical Prosthetics and Orthotics, 10:*137–158, 1986. Reproduced with permission of the American Academy of Orthotists and Prosthetists.

Backrest

The following features of the backrest can also cause the individual to slide into a slumped, kyphotic posture, resulting in a poor pressure distribution on the seat:

a. The lumbar support is located either too low or too high on the backrest. If the lumbar support is too low, it will push the individual's buttocks forward on the seat. If the lumbar support is too high, it will not support the pelvis and upper sacrum.

b. There is inadequate space or recess for the buttocks below the lumbar support (Oxford, 1973; Asatekin, 1975).

c. The backrest is too vertical (Meyer, 1873). The individual will move his buttocks forward on the seat in order to obtain back support and stabilization in the chair.

d. The height of the backrest is too low for prolonged sitting. The individual will move the buttocks forward and slump in the chair in order to obtain more upper back support (Travell, 1955).

e. There is a forward inclination of the upper part of the backrest, or an excessive concavity (horizontal curvature) to the upper backrest (Bennett, 1928; Travell, 1955; Diffrient, 1970; Andersson et al., 1974d; Asatekin, 1975). This will push the upper trunk and shoulders forward and away from the backrest, and result in difficulty maintaining contact with the lumbar support. Kyphosis of the lumbar spine will result, with poor pressure distribution over the seat and backrest (Figure 60).

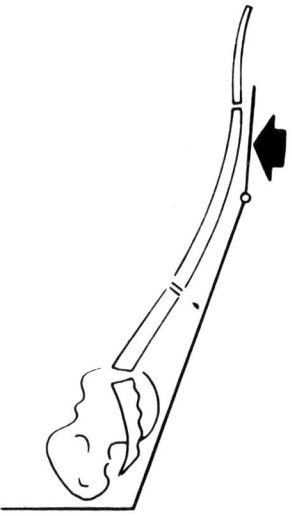

Figure 60. Forward inclination of the upper backrest. From Andersson, B.J.G., Örtengren, R., Nachemson, A., and Elfström, G.: Lumbar disc pressure and myoelectric back muscle activity during sitting. 1. Studies on an experimental chair. *Scandinavian Journal of Rehabilitation Medicine, 6:*104–114, 1974. Reproduced with permission.

ANATOMY OF THE BUTTOCKS

With the hip in an extended position the ischial tuberosity is covered by the gluteus maximus muscle. However, with the hip flexed in the sitting position, the gluteus maximus muscle slides superolaterally off the ischial tuberosity (Minami et al., 1977; Helbig, 1978; Daniel and Faibisoff, 1982). The ischial tuberosity is therefore only covered by skin and subcutaneous fat when sitting (Figure 61).

In a study involving the dissection of six cadavers, Daniel and Faibisoff (1982) gave specifics regarding the soft tissue coverage of the ischial tuberosities (Table III).

Without the proper lumbar support when sitting, pressure will be localized over and posterior to the ischial tuberosities. Compression and distortion of the gluteus maximus will result, leading to discomfort and pain. Nola and Vistnes (1980) and Daniel et al. (1981) have confirmed the

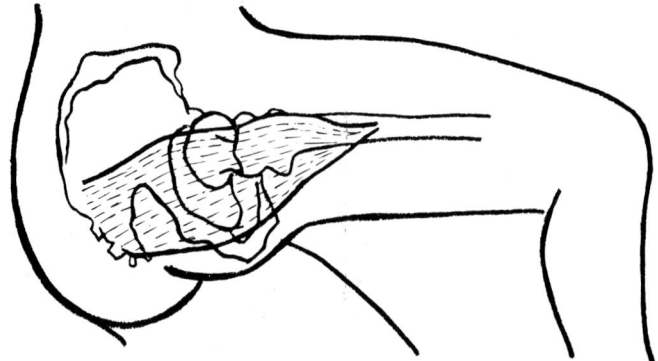

Figure 61: With the hip flexed in the sitting position, the gluteus maximus muscle slides superolaterally off the ischial tuberosity. Adapted from Daniel, R.K., and Faibisoff, B.: Muscle coverage of pressure points—the role of myocutaneous flaps. *Annals of Plastic Surgery, 8:*446–452, 1982. With permission of Little, Brown and Company.

Table III

SOFT TISSUE COVERAGE OF PRESSURE POINTS

	Site		
Tissue	Sacrum (mm)	Trochanter (mm)	Ischium (mm)
Skin	1.0-3.5	0.5-1.5	0.5-2.0
Subcutaneous tissue	5.0-30	5.0-60	5.0-60
Muscle	None	None	Extension: 5.0-45 Flexion: none

From Daniel, R.K., and Faibisoff, B.: Muscle coverage of pressure points—the role of myocutaneous flaps. Annals of Plastic Surgery, 8:446-452, 1982. Reproduced with permission of Little, Brown and Company.

greater sensitivity of muscle to localized pressure, compared to skin and subcutaneous tissue.

Prolonged, localized pressure over the ischial tuberosity may also contribute to the development of ischiogluteal bursitis. Inflammation of the ischiogluteal bursa, which overlies the sciatic nerve and the posterior femoral cutaneous nerve, can produce pain over the center of the buttock and down the posterior thigh. Ischiogluteal bursitis can be easily confused with a herniated disk or acute thrombophlebitis (Swartout and Compere, 1974).

Bi-Ischial Distance

An important measurement for proper seat design and pressure distribution is the distance between the mid-points of the supporting surfaces of the ischial tuberosities in the seated position.

Åkerblom (1948) mentioned that the ischial tuberosities are approximately 3 cm in width and 4 to 5 cm in length. He gave the average distance between the mid-points of the supporting surfaces of the ischial tuberosities as approximately 12 cm in men and 13 cm in women. More specific measurements given by Åkerblom (1948) for the bi-ischial distance, based on 39 female and 51 male skeletons, are given in Table IV.

Table IV
DISTANCE BETWEEN MIDPOINTS OF ISCHIAL TUBEROSITIES

Bi-ischial Distance (cm)	% of 51 Male Skeletons	% of 39 Female Skeletons
9	5.9%	0
10	5.9%	0
11	19.6%	7.7%
12	51.0%	23.1%
13	13.7%	41.0%
14	3.9%	20.5%
15	0	7.7%

Adapted from Åkerblom (1948) with permission.

Diffrient et al. (1981) list the following bi-ischial measurements for various percentile males:

2.5 percentile male—4.8 inches
50.0 percentile male—5.2 inches
97.5 percentile male—5.6 inches

Kira (1976) gives the following dimensions for the ischial tuberosities:

Width —approximately 1 inch
Length —approximately 1½ to 2 inches
Bi-ischial distance —from 4¾ to 6¼ inches

Reynolds and Hooten (1936) and Kira (1976) also mention that females tend to have a larger bi-ischial distance than males. Figure 62 illustrates this characteristic of the female pelvis compared to the male pelvis, with the ischial tuberosities being wider apart and more everted in the female (Ellis, 1977).

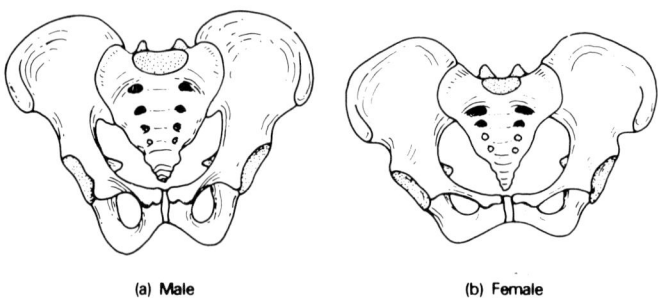

(a) Male   (b) Female

Figure 62. Comparison of the male and female pelvis. Note that the ischial tuberosities are wider apart and more everted in the female pelvis. From Ellis, H.: *Clinical Anatomy,* 6th ed. Oxford, Blackwell Scientific, 1977. Reproduced with permission of Blackwell Scientific Publications Ltd.

Another critical dimension of the buttocks is the length of the perineum, measured from the buttocks cleavage to the front of the genital region. The length of the perineum in the seated position may range from six to twelve inches, with the mean length in the upright seated position being approximately nine inches in males, and seven inches in females (Kira, 1976; McClelland and Ward, 1976).

Figure 63, pertaining to toilet seat design, gives the anatomical relationships and dimensions of the ischial tuberosities and perineum.

BODY BUILD AND PRESSURE DISTRIBUTION

Due to differences in body build and musculoskeletal features, certain individuals may develop discomfort and pain from sitting sooner than others.

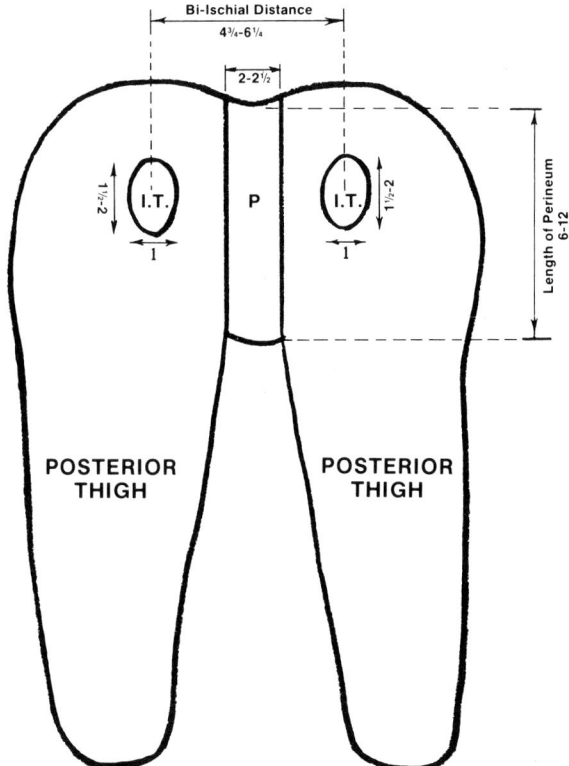

Figure 63. Dimensions of the ischial tuberosities and perineum, pertaining to toilet seat design. I. T. = ischial tuberosity. P = perineum. Measurements are from Kira (1976) and McClelland and Ward (1976). All measurements are in inches.

According to Hertzberg (1972), individuals with small ischial tuberosities or thin gluteal musculature, or both, will usually develop pain sooner than individuals with thick gluteal musculature and ischial tuberosities with broad rocker surfaces.

Dillon (1981) compared the indentation profiles of individuals with different body builds, obtained through strips of wax placed on top of a foam cushion. A lighter individual with a low endomorphy rating produced a narrower and deeper indentation than a heavy individual with a high endomorphy rating.

Using an ultrasonic pulse echo system, Kadaba et al. (1984) were able to map the contour of the buttock-cushion interface for individuals of different body weight. They also found that the individual with the lowest body weight indented deeper into the cushion than the heavier individuals.

With the use of a pressure mat, Reswick et al. (1964) found that the tall and thin male subjects produced the highest pressure concentrations under the ischial tuberosities, with the steepest pressure gradients. The female subjects with ample gluteal padding showed fairly low maximum ischial pressures, without steep pressure gradients.

Using a pneumatic cell pressure transducer, Yang et al. (1984) measured the pressures under the ischial tuberosities of 39 adults while sitting on an unpadded wooden chair. The mean ischial tuberosity pressure was 79.9 mm Hg for females, and 109.0 mm Hg for males.

These studies imply the following regarding body build and pressure distribution:

a. A thin individual with minimal subcutaneous fat in the buttocks region and thin gluteal musculature will produce higher peak pressures under the ischial tuberosities than a heavier individual, who will transmit his body weight over a larger area of the buttocks.

b. Most females will show a better seated pressure distribution under the buttocks compared to males. This is due to the female's greater degree of subcutaneous tissue in the buttocks region. In addition, the ischial tuberosities of the female are wider apart and more everted compared to the male, which will favor a wider indentation profile and a wider pressure distribution. (The relationship between body build and pressure sore development will be discussed in Chapter Eleven.)

Various musculoskeletal factors will also affect the ultimate pressure distribution of the seated individual. As discussed in earlier chapters, these factors include hamstring tightness and the loss of back extension mobility (Stokes and Abery, 1980; Milne and Lauder, 1974).

ASYMMETRICAL SITTING AND PRESSURE DISTRIBUTION

An asymmetrical sitting posture will result in an uneven pressure distribution on the buttocks. As a result, the individual will be unable to tolerate sitting for prolonged time periods.

Cross-Legged Sitting

Whether an individual crosses his knees when sitting for greater stability or due to cultural factors, it will have an adverse effect on pressure distribution:

a. Crossing the knees when sitting will result in kyphosis of the lumbar spine (Schoberth, 1962; Finneson, 1973; Andersson et al., 1974a). As a result, the sitting pressure will be localized over and posterior to the ischial tuberosities.

b. There will be a large increase in weight bearing over one buttock, usually on the uncrossed side (Schoberth, 1962; Dempsey, 1962).

Writing Position

The typical writing position over a horizontal desk is also very asymmetrical. There is a tendency to tilt the body axis to the side of the non-writing arm. This will result in an increased buttock pressure on the side of the non-writing arm (Schoberth, 1962; Alexander, 1966).

Workstation Design

The sitting pressure distribution will also be affected by the specific occupational workstation design (Feiss, 1905; Hünting et al., 1980; Marek et al., 1984). For example, an individual doing data entry work at a computer terminal may have documents positioned off center, to the left. This may result in a trunk shift and rotation to the left, with higher pressures over the left buttock and increased discomfort when sitting due to the asymmetrical pressure distribution.

WEIGHT SHIFTING

Postural changes are necessary when sitting for several reasons:

a. To shift pressure off the ischial tuberosities, thereby preventing or reducing the ischemia, discomfort, and pain from prolonged sitting.
b. To prevent the build-up of temperature and humidity under the buttocks (Clark, 1974).
c. To prevent the static overloading of various muscle groups.
d. To promote proper disc nutrition (Krämer, 1977, 1981; Krämer et al., 1985).

According to Krämer (1981):

"The load-related fluid transport in the human intervertebral disk can be compared to a pump, the function of which is to transport water and metabolites to and fro at the intervertebral disk border. Thus, the nutrition of the disk cells is improved and the removal of metabolites facilitated. All changes in the position of the spine accompanied by intradiskal pressure changes result in either an acceleration or slowing down of the fluid transport with or without an altered direction."

Krämer (1981) also stresses that the

" . . . continuous maintenance of one position leads to an arrest of the load-related fluid transport and is thus disadvantageous for the metabolism of the intervertebral disk. This is in particular true in body postures by which the intradiskal pressure becomes continuously maintained at high levels."

Most current chair and workstation designs, along with common sitting habits, usually result in the individual assuming a kyphotic, anterior sitting posture for prolonged time periods at work. When back pain or discomfort develops, seating authorities usually advocate that the individual lean back and make use of the backrest. This position change will unload the spine and promote fluid transport to the disc.

According to the author, a healthier sitting posture physiologically would be to maintain support with the backrest and arm supports for prolonged periods, only assuming a forward inclined posture (anterior sitting posture) when a position change is needed due to discomfort.

Side to Side Weight Shifting

The problem with shifting weight from side to side is that one buttock must carry a double load while the other buttock carries none (Hertzberg, 1958). As a result, discomfort will eventually result with prolonged sitting since both buttocks never receive adequate pressure relief.

According to Dempsey (1962), "The popular habit of shifting from one buttock to the other, or crossing the legs to relieve one buttock, does not reduce or delay compression fatigue, but instead accelerates the rate of onset of fatigue because greater pressure for each unit area is imposed on one buttock while the other is trying to restore blood circulation."

Forward Weight Shifting

A more optimal way to totally relieve the ischial tuberosities of pressure is a forward weight shift. As with all weight shifting techniques, this will be more easily accomplished with firm rather than very soft seat upholstery (Kroemer and Robinette, 1968, 1969; Ayoub, 1972; Brand, 1980; Black et al., 1981).

A forward weight shift will also be facilitated with a slightly contoured seat surface for the buttocks. In addition, sitting with a lumbar lordosis greatly facilitates the ability to relieve the ischial tuberosities of pressure with a minimal forward weight shift (Zacharkow, 1984a).

It is extremely critical, however, when assuming a forward leaning position in a chair (when shifting weight or leaning over a horizontal desk) that the motion occurs at the hips and not by spinal flexion (Goldthwait, 1909; Mosher, 1919; Schurmeier, 1927; Garner, 1936) (Figures 64 and 65).

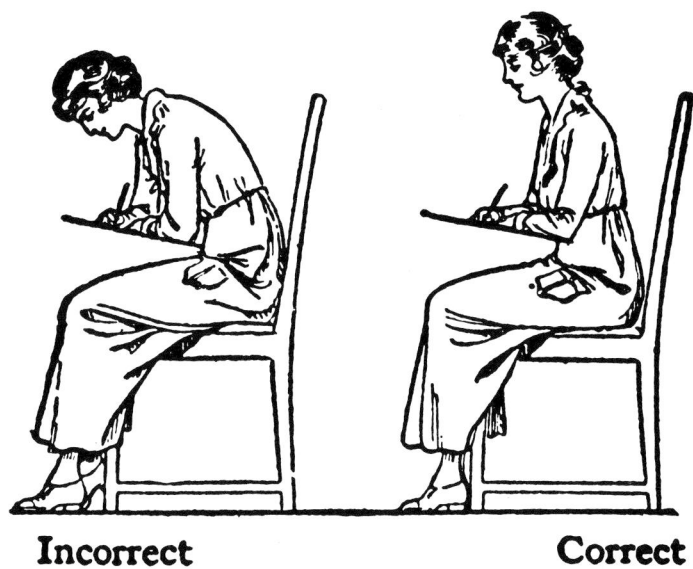

Incorrect   **Correct**

Figure 64. Leaning forward in a chair. From Wood, T.D., and Rowell, H.G.: *Health Supervision and Medical Inspection of Schools.* Philadelphia, Saunders, 1927. Reproduced with permission of the National Board of the Y.W.C.A. and W.B. Saunders Company.

Leg Position

A change in one's leg position when sitting is another way to shift weight temporarily away from the ischial tuberosities. Proper leg position changes require the proper seat height to the chair, along with a rounded front edge to the seat in order to avoid a high cut-off pressure at the distal posterior thigh (Bennett, 1928; Keegan, 1953, 1964; Kroemer and

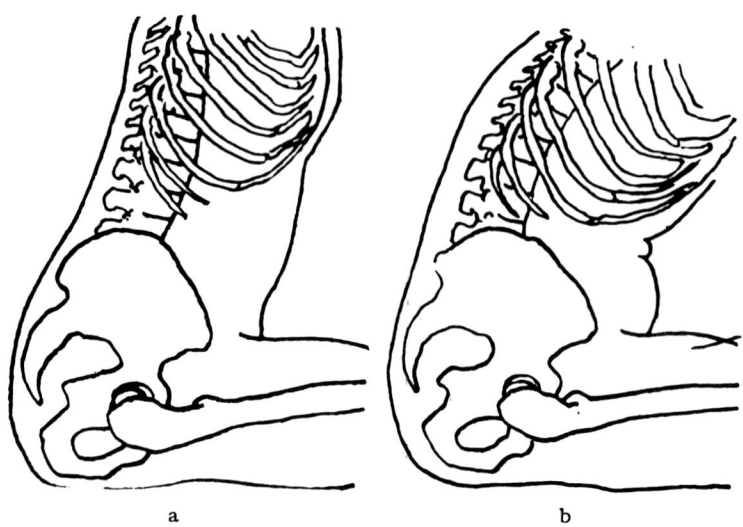

a   b

Figure 65.
    a. When leaning forward from the hips, the abdominal cavity is long and the chest
       and diaphragm are raised.
    b. When leaning forward by spinal flexion, the abdominal cavity is short and
       compressed. The ribs are lowered and the chest cavity is diminished. The upper
       trunk is partly resting on the viscera.
    From Knudsen, K.A.: *A Textbook of Gymnastics,* 2nd ed. *Volume One. Form — Giving*
    Exercises. London, J. & A. Churchill Ltd., 1947. Reproduced with permission of Longman
    Group Ltd.

Robinette, 1968, 1969; Andersson and Örtengren, 1974f; Asatekin, 1975;
Coe, 1979).

    As can be seen in Figure 66, as the knees are lowered, weight bearing
can be temporarily shifted to the lower thighs. The change in joint
angles and muscle tension with leg position changes will also help
reduce the fatigue from prolonged sitting.

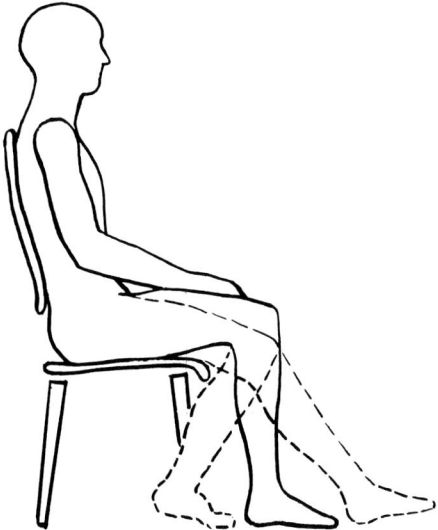

Figure 66. A rounded front edge to the seat, along with the proper seat height, will facilitate leg position changes. Adapted from Keegan, J.J.: Alterations of the lumbar curve related to posture and seating. *The Journal of Bone and Joint Surgery, 35-A:*589–603, 1953. With permission of The Journal of Bone and Joint Surgery.

Chapter Six

SCHOOL SEATING

Bennett (1928) considered going to school to be among the most sedentary of occupations, and the place where permanent habits of sitting are formed. Shaw (1902) commented that "the desks and chairs used in the greater number of our schools are constructed with but the slightest regard for hygienic principles." As a result of improper school seating, Shaw (1902) referred to the "injurious effects as to posture, and wrong habits of carriage, which are borne through life, and sadly enough become more pronounced as the years of life increase." Shaw's comments on school furniture are unfortunately just as pertinent today as they were eighty-five years ago.

The deleterious effects of improper school furniture on the spine have been realized for a long time. As early as 1737, the school regulations of the German princedom of Braunschweig-Lüneburg denounced the bending of the spine in sitting as "unwholesome and injurious" (Kotelmann, 1899; Bennett, 1928). According to Krämer (1981), with the start of prolonged periods of sitting in kindergarten, "very often faulty positions are assumed which jeopardize the further fate of the intervertebral disks."

Among an elementary school population (students ages 6 to 12 years), Mierau et al. (1984) reported the prevalence of low back pain to be 22.8 percent. The prevalence of low back pain was found to increase to 33.3 percent among a secondary school population (students ages 12 to 17 years). In Salminen's study (1984) of 370 Finnish schoolchildren ages 11 to 17 years, 19.7 percent of the students reported present neck and/or back symptoms. Of these students with present neck and/or back symptoms, 58.9 percent reported having symptoms while sitting. Wagenhäuser (1978) also found sitting to be a major exacerbating factor among secondary school students complaining of backache.

Besides the adverse physical effects of improper school seating, a study by Riskind and Gotay (1982) indicated that an individual's physical posture can have carry-over effects on motivated behavior. In a laboratory setting at two different universities, twenty undergraduate students were placed in either a slumped, kyphotic sitting posture or an upright, erect sitting posture. The individuals who were previously placed in a slumped, kyphotic sitting posture later showed significantly lower persistence in a standard learned helplessness task, an insoluble geometric puzzle. These results suggested that "the self perception of being in a more slumped-over physical posture predisposes a person to more speedily develop self-perceptions of helplessness later, following exposure to problems that the person finds to be insoluble."

OBSERVATIONS ON SCHOOL POSTURES

Samuel A. Eliot, a member of the Boston School Committee, in 1833 stressed that "It is the duty of parents and those who act for them to take care that the school-room shall be a place where the children may acquire the use of their intellectual faculties without having their physical organization disturbed or their vital powers debilitated by a constrained position."

Hartwell (1895) determined that elementary school could be classified as a sedentary occupation for 84 to 88 percent of the school period. An English investigation by the Department of Education and Science (1976) found that students tended to use the desk and chair more for both working and listening as they got older. Among the sixteen to eighteen year olds, for example, 73.3 percent of the school period involved using both the chair and the desk.

In regards to prolonged periods of sitting, Bradford and Stone (1899) commented:

> "If children seated, but tired of sitting, are watched, it will be noticed that they either lean upon the desk or slide down upon the chair, to take as far as possible a reclining position, in order to relieve the downward pressure upon the pelvis and the strain upon the lumbar ligaments and fascia, or by twisting the trunk and sitting to one side or the other, relieve one set of muscles temporarily at the expense of a strain of the others."

Detailed Studies on School Postures

Bennett's study (1925, 1928) on school postures was based on 4,637 individual observations in Chicago elementary schools and high schools.

Observations were based on the spinal profile, which was classified as either erect or slumped. The slumped posture involved either a forward slump, a reclined slump, or a slumped spinal profile with the student's sitting position being fairly vertical. Results showed that 59 percent of all observations involved a slumped posture. The worst postures (65 percent slumped) were in reading and writing activities.

An investigation reported by Floyd and Ward (1969) involved observations on forty-two girls and forty-two boys, of mean age 17.2 years, who were in the upper classes of two secondary schools in England. Three aspects of postural behavior most frequently observed involved:

1. Sitting without support from the backrest.
2. The trunk slumped forwards.
3. Both arms leaning on the desk.

This combined posture appeared to be imposed by the necessity to write on a horizontal desk. However, writing was observed to occupy approximately 30 percent of the total time, whereas this desk-supported posture was observed to occur from between 65 to 80 percent of the total time.

Wotzka et al. (1969) made 2,798 observations on students' sitting behavior in auditorium seats during lectures. Approximately 60 percent of the observed time was spent writing by the students, and approximately 28 percent of the time was spent listening. The students were observed to lean against the backrest only approximately 32 percent of the time. Over 80 percent of the time, students rested their lower arms on the writing surface.

MUSCULOSKELETAL STRESS

Both at school and while doing their homework, schoolchildren frequently assume slumped, kyphotic sitting postures.

Lovett (1916) considered prolonged flexion of the spine to be induced by school furniture which failed to properly support the back.

According to Schwatt (1910):

"In the absence of a back rest affording an efficient support to the spinal column there is a collapse of posture, a doubled-up attitude as far as the soft parts permit or the pupil slides downward and forward to nearly a reclining position or rests the arms upon the desk. As a result there occurs a displacement of the spinal column forward in its upper portion and backward in the lumbar region."

Knudsen (1947) found the thoracic spine to suffer the earliest and most frequently from slumped sitting postures, resulting in a decreased mobility. Besides this increased stress on the thoracic spine from improper sitting, gravity will also tend to increase the thoracic kyphosis when standing (Hollinshead, 1969). Not surprisingly, Fon et al. (1980) reported a continuous increase of the thoracic kyphosis with age, through both childhood and adulthood.

In addition to the spinal stress from slumped sitting postures, some of the upper trunk weight will be thrown upon the abdominal viscera. Bennett (1928) considered the forward stoop of a slumped sitting posture to be essentially a "leaning on the viscera."

Kellogg (1896) described the individual breathing in a stooped sitting position as being "constantly in a state of air-starvation, a fact which is evidenced by the disposition to straighten up and draw a long, deep breath every now and then, which is constantly noticed in persons who habitually sit at study or work in a stooped attitude."

Stress to the Upper Thoracic and Lower Cervical Spine

A major part of the school posture problem is that all the sitting activities "encourage the forward position of the head and arms, which in turn tends to draw the head and shoulders forward" (Drew, 1926).

The greatest burden for the neck and upper thoracic spine involves supporting the shoulders and arms. The shoulders are suspended from those trapezius muscle fibers arising from the lower three to four cervical vertebrae and inserting on the acromion process. Therefore, the heavy pull of the shoulders and arms on the lower half of the cervical spine must be counteracted by the back extensors in the region of the lower cervical and upper thoracic spine (Knudsen, 1947).

According to Knudsen (1947):

"It is on the carriage of the upper 5–6 dorsal vertebrae that the shape of the neck and the position of the head depend first and foremost.

This upper half of the dorsal spine is the stiffest part of the whole spinal column, and it is consequently in that part that round back generally begins to develop. By the increased curving of the dorsal spine caused by round back, the upper 5–6 dorsal vertebrae will be inclined forward, the cervical column will move with it into the same inclined position, but in order to be able to look forward one will raise one's head by bending the upper part of the cervical spine backwards;

its slight physiological curve will therefore be increased, i.e., lordosis of the neck will set in as seen in all people with round backs."

Knudsen (1947) also considered round shoulders to be interrelated to round back:

"When the back is rounded, the upper 4 or 5 dorsal vertebrae and the lower 3 or 4 cervical vertebrae are moved forward. From the cervical vertebrae mentioned, the shoulders and arms are suspended as these vertebrae form the origin of portator scapulae, the strongest part of trapezius. The shoulders will tend to place themselves vertically under their points of suspension. If these are moved forward, as in round back, the shoulders will move forward, too. This will affect the pectoral muscles. Their origin and insertion are now brought nearer together. They will adjust their lengths accordingly; they become shorter and they will fix the shoulders in the forward position. In this case round back is the cause of round shoulders, as commonly found in young people with sedentary work during which they sit with rounded backs. In young people employed in bodily work during which the arms and shoulders are brought forward, the stooping shoulders will gradually pull forward their points of suspension, the lower cervical vertebrae, and in this way cause round back. Most likely cause and effect may be sought in both directions in a vicious circle."

Knudsen's comments point to the need for proper arm posture and support with desk work. This is critical for proper shoulder girdle stabilization, and also to help keep the upper thoracic and lower cervical spine as extended as possible (Mills, 1919).

Head-Neck Posture

In 1846, Warren stressed that schoolchildren "should be frequently warned against the practice of maintaining the head and neck long in a stooping position." He emphasized the importance of proper desk height and inclination in counteracting this stooped position.

Recently, in a study involving ten healthy females (mean age 24 years), Harms-Ringdahl and Ekholm (1986) reported that sitting with the head and neck in a relaxed, forward-flexed position, a posture resulting in extreme flexion of the lower cervical and upper thoracic spine, resulted in the experience of pain or discomfort for all ten individuals within 15 minutes. The primary location of pain was the lower cervical and upper thoracic spinal region, although pain referred to the arm or head was also reported. During pain provocation, seven of the subjects reported experiencing one or more of the following vegetative sensations: sweating, nausea, tiredness, dizziness, general coldness or warmth.

With prolonged sitting at a horizontal desk, the slumped head posture will easily become habitual. As a result, when the eyes are raised to look ahead, it has to be accomplished by an increased lordosis in the upper cervical spine. Eventually, "the back of the skull begins to be held contracted back into the upper neck in order to look straight ahead" (Barlow, 1980) (Figure 67).

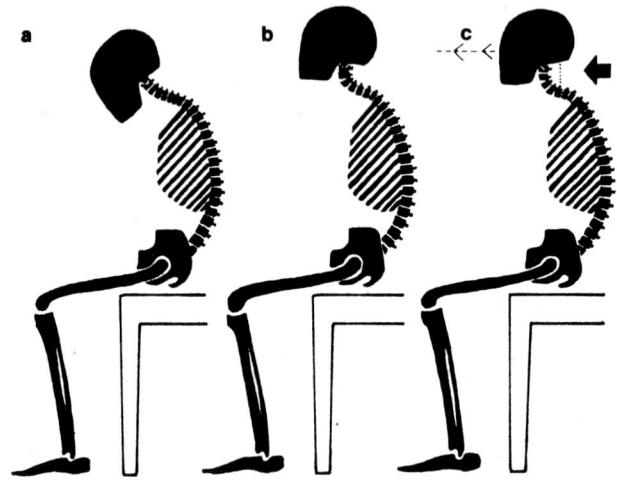

Figure 67.
  a. With prolonged sitting at a horizontal desk, the slumped head posture will easily become habitual.
  b. & c. When the eyes are raised to look ahead, it has to be accomplished by an increased lordosis in the upper cervical spine.
  From Barlow, W.: *The Alexander Technique.* New York, Warner Books Edition, 1980. Reproduced with permission of Alfred A. Knopf, Inc.

Alexander (1918) implicated the prolonged "crouching positions" at the school desk as contributing to the development of a defective kinaesthetic system by altering the proper head-neck relationship. Head posture is considered critical in providing "the reference for the parts of the body in relation to one another and in relation to the surrounding space" (Laville, 1985).

Cohen's (1961) work showed that the neck proprioceptors are critical for orienting the head in relation to the body, and that neck proprioception "plays a very important role in maintaining proper orientation, balance and therefore motor coordination of the body." According to Wyke

(1972), the mechanoreceptors in the spine that primarily provide postural and kinaesthetic perception are more numerous in the apophyseal joints of the cervical region.

Stress to the Lower Thoracic Spine (Thoracolumbar Junction)

The approximation of the thorax to the pelvis in slumped, kyphotic sitting postures will localize stress to the lower thoracic spine and thoracolumbar junction (See Figure 67).

Humphry (1858) considered the lower two thoracic vertebrae along with the upper one or two lumbar vertebrae to be the weakest part of the spine. Although this region of the spine has to bear nearly as much superincumbent weight as the lower spine, Humphry (1858) considered these vertebrae to be disproportionately small for this great weight bearing function. Humphry (1858) also felt that this mid-region of the spinal column would be exposed to maximum leverage from both above and below.

Lovett (1916) considered the thoracolumbar junction to be an area of both weakness and increased mobility. According to White and Panjabi (1978), the lower thoracic spine, unlike the upper thoracic spine, is capable of ample motion in flexion and extension.

In erect standing and sitting, the line of gravity for the upper body is already anterior to the thoracic spine, resulting in a force tending to increase the thoracic kyphosis (Figure 68). In a slumped, kyphotic sitting posture with backward rotation of the pelvis, the gravity line will be even further anterior to the thoracic spine (Figure 69).

The site of maximum spinal stress will be in the region maximally offset from the gravity line, the lower thoracic spine (Chandler, 1933; Alexander, 1977).

Two other points are of importance in regards to the lower thoracic spine. Davis (1955) described a change in the structure of the articular processes, usually found between the eleventh and twelfth thoracic vertebrae, where the articular processes could interlock in a tenon and mortice fashion with compression, thus preventing all movement other than flexion. Macnab (1977) reported how the experimental injection of hypertonic saline into the supraspinous ligament between T12 and L1 may give rise to pain referred to the low back and buttocks.

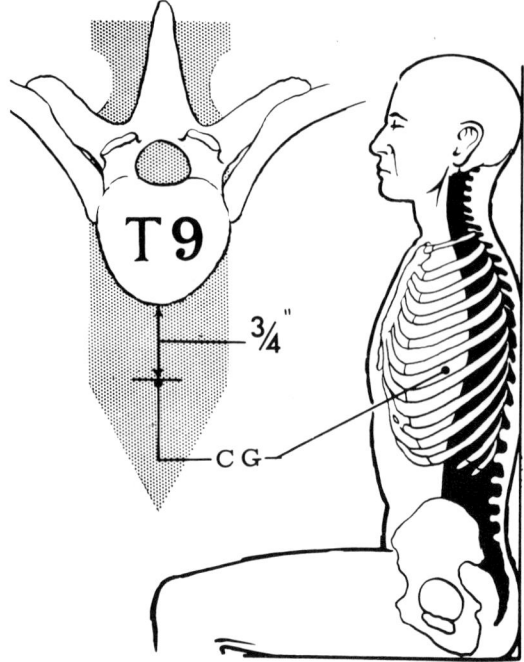

Figure 68. Location of the center of gravity of the upper torso. From Vulcan, A.P., King, A.I., and Nakamura, G.S.: Effects of bending on the vertebral column during +Gz acceleration. *Aerospace Medicine, 41:*294–300, 1970. Reproduced with permission of the Aerospace Medical Association.

Musculature Affecting the Lower Thoracic Spine

Back Musculature. In regards to spinal extension, Rathbone (1934) considered the back to need the most strengthening in the region of the tenth thoracic vertebra to the second lumbar vertebra. According to Wiles (1937), "The weakest part of the extensor mechanism of the spine is in the upper lumbar and lower dorsal regions. The mass of the erector spinae is getting smaller, and the change of curve from concave to convex is taking place, making it mechanically a vulnerable spot."

Knudsen (1947) observed that:

"The joints of the dorsal spine are those that first lose mobility in civilised life because of lack of use. Bendings backward are unconsciously performed in the more mobile loin. Even little children may have dorsal curves that cannot be stretched to the normal. On the other hand, it is seldom the dorsal spine has lost its ability to bend forward, to increase the curve. When the dorsal spine has lost its mobility, one is

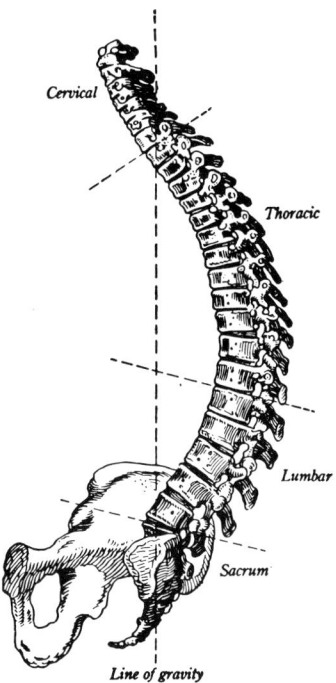

Figure 69. The gravity line in a slumped, kyphotic sitting posture. From Bennett, H.E.: *School Posture and Seating.* Boston, Ginn and Company, 1928.

not able to control its extensors as one controls the extensors of the neck, or the loin, or all other muscles of the skeleton."

Abdominal Musculature. The rectus abdominis muscle, as it connects the rigid thorax to the pelvis, will have its greatest effect on spinal flexion at the more mobile lower thoracic spine and thoracolumbar junction (Lovett, 1916).

In a study involving adolescent females, Toppenberg and Bullock (1986) found the length of the rectus abdominis to be significantly negatively correlated with the thoracic kyphosis. In other words, the greater degrees of thoracic kyphosis were associated with the shorter abdominal lengths.

In a study involving young males (mean age 22 to 23 years old), Klausen (1986) found that individuals who maintained standing posture with predominant activity in the rectus abdominis had a more pronounced thoracic kyphosis than individuals who maintained standing posture with predominant activity in the erector spinae.

It is of interest to note that many of the faulty sitting attitudes of schoolchildren, such as in Figure 70, bear a marked resemblance to the spinal posture assumed during the performance of sit-up exercises.

Figure 70. A faulty sitting attitude of schoolchildren, similar to the spinal posture assumed with sit-up exercises. From Bradford, E.H., and Stone, J.S.: The seating of school children. *Transactions of the American Orthopaedic Association, 12:*170–183, 1899.

Hamstrings. Hamstring tightness can also result in an increased flexion stress at the lower thoracic and upper lumbar spine. Due to the marked backward rotation of the pelvis and backward displacement of the lower lumbar spine, along with the limited hip mobility, the flexion stress will first be localized to the thoracolumbar junction as the student brings his chest forward over the desk.

Milne and Mierau (1979) found a marked decline in hamstring extensibility between the preschool ages of three to five, and schoolchildren age six. The authors related this decrease in hamstring extensibility at age six to the prolonged sitting in school. A further decrease in hamstring extensibility was noted by Milne and Mierau (1979) at the secondary school level.

Lambrinudi (1934) also found that hamstring tightness rarely existed before the age of six. With hamstring tightness, Lambrinudi (1934) found the point of greatest convexity of the spine in flexion to involve the lower thoracic and upper lumbar regions.

In regards to toe-touching exercises, Lambrinudi (1934) commented:

"I have seen children who are unable to touch their toes being exhorted to do so by a series of sudden jerks. The practice of using the back of a growing child as a lever (and a flexible one at that) to stretch the comparatively unstretchable hamstring muscles is a pernicious one. Such movements must cause considerable pressure on the front of the bodies of the vertebrae, and, as judged from the contour, it is upon the lower dorsal and upper lumbar regions that the brunt of the force falls."

Scheuermann's Disease (Adolescent Kyphosis)

Alexander (1977) advanced the hypothesis that Scheuermann's Disease is due to the static spinal load from prolonged flexed anterior sitting postures. "The wedging, kyphosis and irregular ossification best fit the evidence as a subsequent static load effect, resulting mainly from the chair sitting posture" (Alexander, 1977).

Fisk et al. (1984) also mentioned the likeliest stress factor involved in Scheuermann's Disease to be a prolonged static loading, possibly from prolonged sitting periods. These same authors also commented that the hamstring tightness noted by Lambrinudi (1934) and Milne and Mierau (1979) around age six "in all probability reflects the change to prolonged sitting, with the hamstrings resetting to a shorter postural length. Such shortness could alter the programmed pattern of bending so that the spinal joints are used more than the hips."

In adults, an increased incidence of low back problems and disc degeneration of the lower lumbar spine has been found in individuals with clinical and x-ray evidence of previous Scheuermann's Disease (Butler, 1955; Stoddard and Osborn, 1979; Fisk and Baigent, 1981; Krämer, 1981).

DESK-CHAIR RELATIONSHIP

The proper desk-chair relationship is critical in school seating. As Hartwell (1895) expressed, a properly constructed school chair is not enough to assure a healthy sitting posture:

"But a properly constructed chair may be rendered nugatory or even positively harmful, if its occupant is forced to work at a desk whose upper surface is too high or too low, or insufficiently sloped, or whose under surface exercises a cramping influence upon his knees or thighs, or if he is given a desk which, though correct in its proportions, is so placed with relation to his chair that the backward sitting posture is beyond his reach, and the forward sitting posture can only be maintained at the expense of prolonged and wearisome muscular exertion."

Distance

The term Distance, as used by earlier seating authorities, referred to the horizontal distance between the edge of the desk top and the front edge of the seat. A positive distance results when there is a space between a vertical line dropped from the desk edge and the front edge of the seat, a zero distance results when these two lines coincide, and a negative distance results when the edge of the desk overlaps the front of the seat (Kotelmann, 1899; Shaw, 1902; Cotton, 1904) (Figures 71–73).

Cohn (1886) stressed that:

"In right arrangement of Distance lies the kernel of school desk reform. The greater the Distance the more the body will have to fall forward of the form in order that the arms may reach the paper; and the more will the head be obliged to drop and to get near the writing. Thus, whenever we intend to sit upright at a table for a considerable time, we instinctively push the chair so far under the table that the table's edge is vertically over the chair's edge or, if possible, overhangs it by an inch. For the upright position of the head, therefore, the Distance must be nil or, still better, negative."

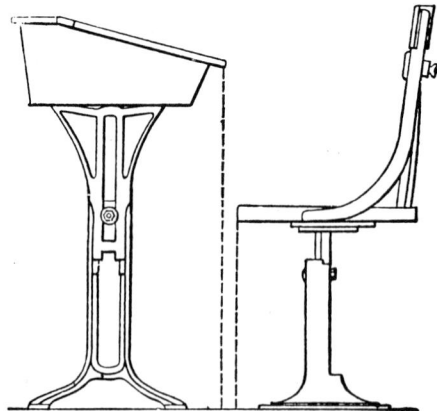

Figure 71. Positive distance. From Shaw, E.R.: *School Hygiene.* New York, Macmillan, 1902.

The Strassburg Commission in 1882 strongly criticized any positive distance:

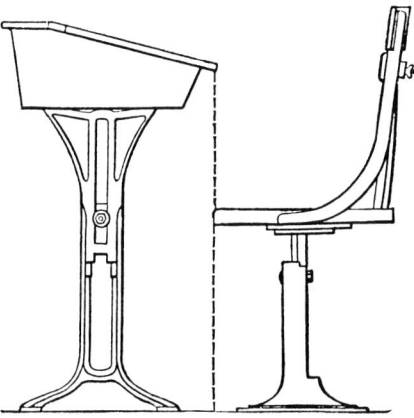

Figure 72. Zero distance. From Shaw, E.R.: *School Hygiene.* New York, Macmillan, 1902.

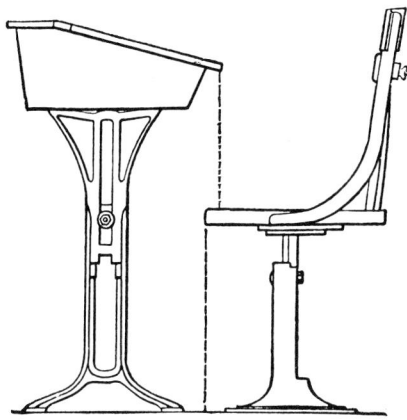

Figure 73. Negative distance. From Shaw, E.R.: *School Hygiene.* New York, Macmillan, 1902.

"This one fault alone is enough to condemn the old school desks, and to make them the more hurtful, the younger the children are who are compelled to sit in them. So much has been said of late years about the evil effects of positive distance that we need not dwell long upon this point. It forces the child, when writing, to support the upper part of his body with his arms, to bring the chest far forward and to bend the head downward too much. In this way it brings the eye improperly near the paper and thus creates short sight artificially. Moreover it gives the child a kind of invitation to twist the spine sideways.

The injurious influences of positive horizontal distance is scarcely less in reading, especially when the desk is not sufficiently sloped. The effect is the most striking in the case of the smaller children. The distance of the form from the desk, and the insufficient support for sitting lead them to prop the head on the left hand, and at the same time to turn themselves to the right about a vertical axis (Figure 74). This brings the left eye nearer to the book than the right eye and makes a good deal more difficult the convergence of the lines of sight, until the right eye is at last left unused altogether. Or the children cross their fore-arms upon the desk's edge and, with heads bent far forward, rest the chin on the back of one hand, bringing the eyes too near the book."

Figure 74. Faulty sitting attitude. Propping the head on the left hand and rotating the torso to the right. From Bobrick, G.A.: *Hygienic Requirements of School Furniture.* New York, Press of Exchange Printing Co., 1892.

Burnham (1892) stressed the importance of a negative distance for writing and reading:

"All authorities condemn the plus distance. The only merit of such a seat is that pupils can easily get out of it. The zero distance is better, but the body cannot easily be held in proper position while writing. The proper conditions for writing can be obtained only by a minus distance. The edge of the desk should overlap the seat by about two inches. The chief objection to this is the difficulty in rising and sitting down. This difficulty is removed by using one of the various kinds of sliding seat or movable chair, or by having the desk-top movable (Figure 75). It is of prime importance during most of the school-work, especially in reading and writing, to have the seat at a minus distance."

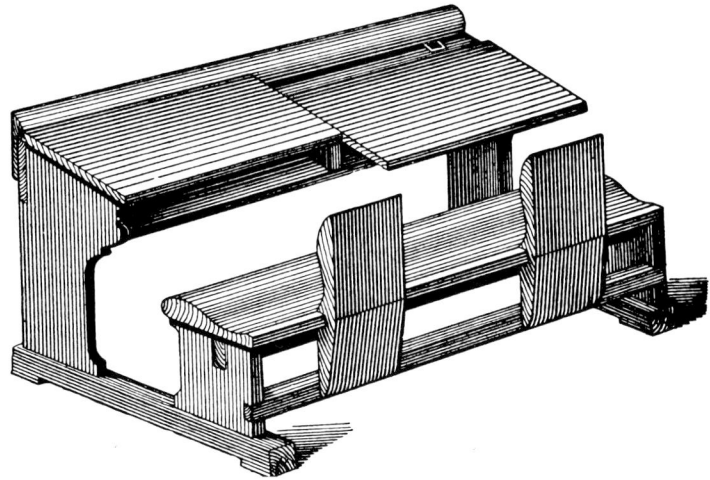

Figure 75. Kunze's school desk with sliding desk top. From Burgerstein, L., and Netolitzky, A.: *Handbuch der Schulhygiene.* Jena, Verlag Von Gustav Fischer, 1895. Reproduced with permission of Gustav Fischer Verlag.

Stone (1900) also discussed the desirability from a hygienic standpoint of a minus distance of several inches. The correct negative distance "should be approximately such that with the elbows against the back of the seat the wrists should reach the edge of the desk."

Cotton (1904) also felt that for a proper writing position the distance from the chair backrest to the edge of the desk should equal the length from the elbow to wrist. Such an arrangement "makes it possible to write freely and easily with the body pretty evenly balanced, or even leaning backward slightly."

With the desk a few inches further forward from this distance, a forward sitting position would be required for writing. Cotton (1904) considered the forward sitting position undesirable for writing because:

1. The back loses its support.
2. The supporting of body weight on the arm tends to rotated postures.
3. The posture tends to round shoulders.
4. The posture tends to bring the eyes too close to the paper.

Schwatt (1910) described the necessity of bending the trunk forward with too large a distance, in order to bring the hands and head above the desk. "The necessity for and the degree of the forward inclination and of the curving of the spine are in direct proportion to the distance." Schwatt (1910) considered the proper minus distance to be "such that the distance between the posterior edge of the desk top and the anterior surface of the seat-back permits of comfortable writing without allowing the back to leave the seat-back and should leave a small distance between the edge of the desk top and the body to prevent pressure upon the chest."

Burgerstein (1915) emphasized that the really significant distance is "that between the slanting back rest and the desk edge. For writing, this space should be a few centimeters greater than the body thickness at the chest; it should correspond to the length of the forearm."

Bennett (1925) noted that since there is no standard seat depth on school chairs, the front edge of the seat is not important in the spacing of the desk from the chair. In agreement with Hartwell (1895), Schwatt (1910), and Burgerstein (1915), Bennett felt the proper distance should be determined from the back support, with the length of the arm and the abdominal diameter being taken into consideration.

Karvonen et al. (1962) observed that if the backrest is far from the desk, the student uses either the desk or the backrest alone as his support (Figure 76). With a closer spacing, the use of both the backrest and desk is possible.

Desk Height

Low Desk Height

A desk height that is too low will result in similar postural faults as an increased distance of the desk from the seat (McKenzie, 1898; Cotton, 1904). The student will be forced to bend forwards, with the body weight being supported by the arms. This will result in a kyphotic spinal posture with round shoulders (Figure 77).

With too low a desk, the student "has to bend his head down to get the proper reading distance. But such a position of the head is impossible for any length of time, since the supporting neck muscles gradually get fatigued. So the head sinks lower and lower, and the spinal column curves out behind" (Kottelmann, 1899).

Pyle (1913) also referred to the stooped-over posture tending to round shoulders, when the desk is too low.

Figure 76. Excessive spacing of the desk from the chair will force the student to use either the desk or the backrest alone for his support. From Bradford, E.H., and Stone, J.S.: The seating of school children. *Transactions of the American Orthopaedic Association, 12:*170–183, 1899.

Figure 77. Posture assumed when the desk height is too low. From Cotton, F.J.: School-furniture for Boston schools. *American Physical Education Review, 9:*267–284, 1904.

High Desk Height

When the desk is too high, the shoulders must be raised (McKenzie, 1898) (Figure 78). As this posture will approximate the origin and insertion of the upper trapezius, increased stress will be placed on the deeper posterior neck musculature in providing stabilization of the head posture (Coe, 1980).

Hartwell observed in 1895 that "the majority of desks in the Boston schools are too high, so much so that in writing the forearm is forced to make a more or less acute angle with the upper arm, while the upper arm is unduly abducted from the body and the right shoulder is unduly raised" (Figure 79).

Kottelmann (1899) explained the writing posture resulting from a high desk as follows:

> "The pupil can not in this case put his elbows on the desk without spreading out the upper arms and raising his shoulders. Since this is uncomfortable, he lets his left arm slip from the desk, keeping only the right one on it in writing." The resulting asymmetrical spinal posture is illustrated in Figure 79.

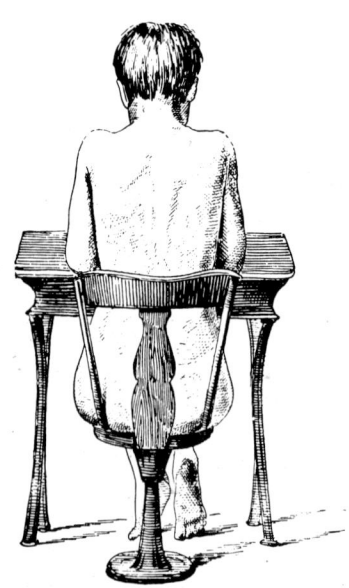

Figure 78. Posture assumed when the desk is too high. From Bradford, E.H., and Stone, J.S.: The seating of school children. *Transactions of the American Orthopaedic Association,* *12:*170–183, 1899.

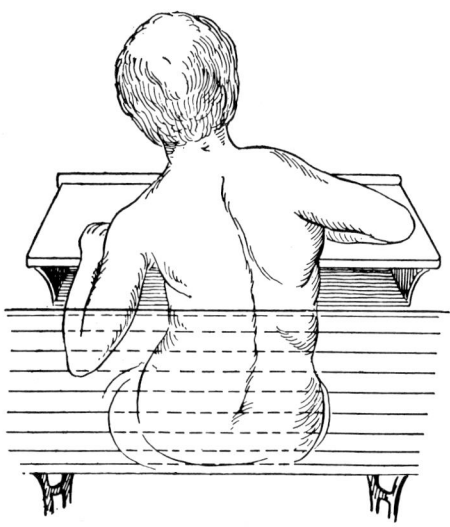

Figure 79. When writing at a desk that is too high, the writing arm will be unduly abducted from the body. From Pyle, W.L. (Ed.): *A Manual of Personal Hygiene,* 5th ed. Philadelphia, Saunders, 1913. Reproduced with permission of W. B. Saunders Co.

Proper Desk Height

According to Schwatt (1910), the proper desk height should allow the student to "place the forearms comfortably upon the desk without raising the shoulders or sinking the head or back."

Cornell (1912) recommended that the desk height should "allow the arm to rest on it naturally when the elbow is at the side."

Bennett (1928) stressed that the desk height, distance (spacing), and slope (inclination) are always interdependent factors in desk position. The following comments from Bennett (1928) pertain to the relationship between desk height and distance (spacing):

> "For any full-sized writing surface in front of the body the elbows are advanced nearly to the desk edge and approximately to the line of the front of the body. With elbows kept close to the sides, this advance involves a minimum of muscular effort, since the upper arm merely swings forward like a pendulum, practically as it does in walking; but as the elbow advances, the pupil remaining erect, it is elevated. Hence the height of a desk in front of the body should be a few inches greater than the elbow-height as measured.
>
> The amount of this increase of height depends on the distance that the elbow is moved forward. It depends also on the length of the upper

arm, since a short pendulum is raised more than a long one in the same amount of forward swing. Therefore the greater the spacing of the desk from the chair the higher the desk must be, and the shorter the forearm the higher the desk must be at the same spacing."

Desk Slope (Inclination)

As early as 1780, Tissot stressed the importance of young people maintaining an erect posture with seated work: "They must bring their work or their book up to their eye level rather than their eyes to their work or book level, for reasons which are obvious."

Bennett (1922) observed that "In requiring a child to sit erect at an ordinary desk while reading or writing, we are demanding a physical impossibility."

Unfortunately, Dresslar's comment in 1917 on school desks is still not outdated: "I believe the chief defect in the desks now on the market is that the desk top is too flat."

Reading

Dresslar (1917) asked teachers to perform a simple experiment in their classrooms to better understand the importance of a properly inclined desk:

"Find a comfortable chair, one in which you may sit erect but not unnaturally so, and then hold a book before you in such a position and at such a distance that you may read the lines most clearly and easily. After finding the position of the book as you prefer it, note the distance that the page is from the eyes (I am assuming that the experimenter has normal vision) and the angle that the page makes with the line of vision, that is, the straight line drawn from the eye to the book. After conscientiously recording this distance and this angle for yourself, try the same experiment with all the pupils of your class. Have them sit in a natural and comfortably erect position at their desks. Then ask them to hold their books in such a position that they can read most readily and easily, and then, while they are so situated, note, again, the two points above mentioned. This may be quickly done by moving down a side aisle and noting the regularity of the demands made on the position of the book. Note especially the relative slants of the books and the desk tops. Assuming that the experiment has been carried out as directed, the result may now be definitely stated: the books will practically be at a right angle to the line of vision, and at an angle of slightly more than forty-five degrees to a line parallel with the floor, and approximately fifteen inches from the eyes of all who have normal vision. And now you ask, who do these facts mean, and what have they to do with the hygiene of school desks?

Unless desk tops are set at the proper angle, children will not, and cannot, sit erect to do their work. Theoretically, one might say that if these are the normal demands for vision, children should be taught to hold their books so. But suppose you try it for ten minutes or a half hour. Of course you will now see where the difficulty lies. Children's arms grow weary and they cannot hold their books so as to get the proper angle of vision for more than a very short time. So the books are put on the desk, and the children's backs are bent, in order to bring the line of vision in the same relative position as it was when they sat erect and held their books in the position demanded.

You may command them to sit erect as often as you think of it, but they will obey, and can obey, only momentarily. The children will bend over their work day after day, unless we devise a practicable desk top that will *necessitate* erect, normal posture for all their work.

How often a boy will slide down into his seat, rest his head on the back of the bench, and stand his book erect on the front edge of the desk. It is easy to call him thoughtless and to command him to sit erect, possibly to reprimand him in some more or less severe way; but the fact is, the boy is really trying to overcome the difficulties and defects in the desk furnished him. The fault lies not with him so much as with the demands that we make upon him."

Bennett (1922) described the child's adjustments of his position for reading at a flat desk as follows:

"We all know how he slides down in his seat until his head is level with the book (Figure 80). When told to sit up, he tries bending his head over until his neck aches; then he leans forward on one elbow and then on both, and then gets his head in his hands and reads with his eyes dangerously near the book, the page in the shadow of his arms and head, and his body cramped in the worst kyphotic position (Figure 81). If he really sits erect, he must hold his book in a position which the arms cannot sustain more than a minute or two, for, if the book lies on the desk, the visual distance is too great and the letters are foreshortened by the angle of about 45 degrees between the position of the page and the line of vision."

In regards to reading posture, Shaw (1902) commented that:

" . . . the line of sight, for the least tax upon the eyes, should fall upon the printed page perpendicular to its plane. With the head in good posture, the page of the book would need to be held at an angle of about sixty degrees from the horizontal and somewhat below the level of the eyes, for easy reading. It is a physiological fact that the eyes are naturally directed a little downward, because such action of the mus-

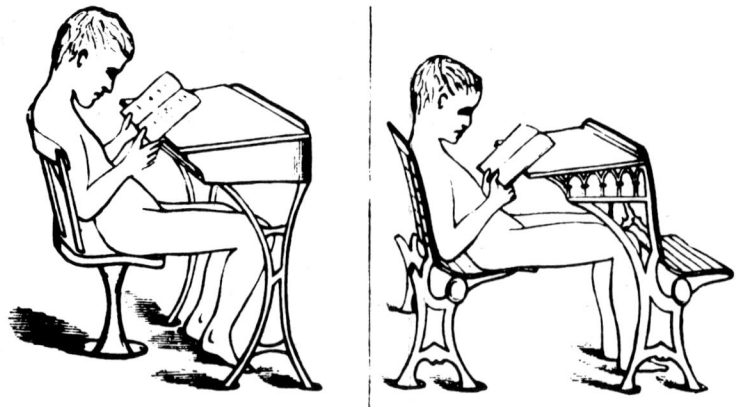

Figure 80. Sliding down in the seat. A position often assumed when reading at a horizontal desk. From Bobrick, G.A.: *Hygienic Requirements of School Furniture.* New York, Press of Exchange Printing Co., 1892.

Figure 81. Kyphotic reading posture with the body weight on the elbows, and the hands supporting the head. From Roth, B.: *The Treatment of Lateral Curvature of the Spine,* 2nd ed. London, H.K. Lewis, 1899. Reproduced with permission of H.K. Lewis and Co. Ltd.

cles of the eyes is not accompanied by fatigue as when the eyes are directed upward, or even on a level, for any length of time."

Writing

Dresslar (1917) considered the problems associated with reading at a flat desk to largely apply to written work:

"If the pupil writes on a flat desk, he will stoop over, for he cannot see his work most easily unless he does so. A teacher may command and exhort her pupils to maintain a healthful posture and they may attempt to obey, but as soon as their attention is withdrawn from their posture to their work, the faulty position will again unconsciously reappear, because the demands of clear vision are more persistently commanding than the advice of any teacher."

Schwatt (1910) also noted the close interrelationship between posture and vision in regards to proper desk inclination:

"The more nearly horizontal the desk top the more strongly must the eyes be turned downward in writing if the head is held vertically. This is accompanied by fatigue and for this reason the head is always rather bent in order to facilitate vision. The desk top should therefore have an inclination which will allow the eyes a free view."

Bennett (1928) considered the flat desk top to have an additional disadvantage for writing purposes, in that the elbow must be at the same level as the writing hand:

"If this level is as low as the elbow in erect posture, the hand is too far from the eyes for clear vision, and one must stoop to attain his visual distance; if it is higher, the forward and outward extension of the elbow involves a spinal twist as well as bend, and the foreshortening and eyestrain cannot be relieved."

Bennett (1928) also commented on the beneficial effects of an inclined desk surface when writing:

"The greater the tilt, the less the foreshortening of the characters or the back and neck bend necessary to correct it, and the more nearly erect one may remain with the same visual clarity. The book is brought nearer to the eyes as well as into better angle, thus lessening the forward leaning or stoop necessary to attain correct visual distance. At any slope the plane of the desk top should coincide with the plane of the underside of the forearm in writing position with elbow down and close to the side. Thus correct posture is maintained in writing, the hand being brought into proper relation to the eye by bending the elbow rather than the back and neck."

With writing, Bennett (1928) stressed that the slope, height, and spacing are always interdependent factors in desk position:

"In general, then, the greater the slope of the desk the higher it should be, and vice versa. The more distant the desk, the more the elbow is advanced and raised, and the higher the desk should be; but the more the desk is raised for the same position of the elbow, the more the forearm is brought toward the shoulder, and hence the more the desk should be sloped and brought nearer."

Other Advantages of a Sloped Desk

When writing or reading on a sloped desk, the upper and lower lines of the paper will be about equally distant from the eyes. This will make changes of accomodation unnecessary when looking from one line to the other (Kotelmann, 1899; Goldthwait, 1923). A sloped desk will also result in better window illumination (Kerr, 1928).

Greater desk slopes will also facilitate leaning against the backrest while reading and writing (Lorenz, 1888; Kotelmann, 1899). With a change in desk inclination from horizontal to forty-five degrees, Bendix and Hagberg (1984) reported that for the ten subjects who were reading, both the cervical and lumbar spine were extended, and the head and trunk changed towards a more upright posture. Eastman and Kamon (1976) also found more erect postures at slanted desks, along with a reduction in fatigue and discomfort.

WRITING POSTURE

Head Posture

In her study of the postural responses of third grade children, Jones (1965) found that the greatest change in posture came with writing at a horizontal desk. The deterioration in posture was characterized by a marked forward and downward displacement of the head.

One hundred years earlier, Dr. Fahrner of Zurich, in his classic book *Das Kind und der Schultisch* (1865), gave a most detailed description of the collapse of the child's posture when writing, due to an improper desk-chair relationship:

"Before the writing begins, the children sit perfectly upright with both shoulder-blades thrown back equally (that is, the shoulders are parallel to the edge of the desk), and the slate or copy-book is so placed before the child that its left margin lies a little to the left of the middle of the body. But as soon as the writing begins all the children move their heads slightly forward and towards the left, without perceptibly altering their attitude in any other way. Soon, however, head after head drops down with a rapid jerk, so that the neck now forms a considerable angle with the rest of the spinal column. In a short time the upper part of the back also collapses, so as to hang from the shoulder-blades, which in their turn are supported by the upper arm. From this moment the scholars are divided into two groups, according to the part of the slate at which they happen to be writing. Those who are writing on the upper half of the slate or at the beginning of a line are able to support themselves on both elbows, and they let their chest sink straight forward against the table. The back in this way becomes curved simply; it becomes what I call a round back. The eyes are from 3 to 4 inches distant from the desk and look straight down upon the writing. For points of support the child uses the front of the chest, the left elbow (which is constantly moved outward till it is a long way from the body) and the right fore-arm anywhere between the elbow and the wrist. But those scholars, who at the critical moment are writing at the end of a line or near the bottom of the slate, cannot any longer support themselves on the right elbow, because it too much overhangs the table and is too far from the body. They are therefore forced to lean on the left elbow alone and, in so doing, not only to bend the spinal column, but to twist it on its axis towards the right. The position is that of the skewed back. The points of support are the left side of the chest and the left elbow, which lies very much to the left of the body and forward from the body; the head is bent towards the left shoulder; the right arm, with its shoulder-blade standing out like a wing, rests on the desk anywhere between the elbow and the wrist; the eyes, now frequently only from 2–3 inches distant from the writing, are rolled considerably towards the right and almost squint over the paper.

In the normal position, the head has its centre of gravity resting upon the bony framework of the spine and is supported by it, so that the muscles of the neck have nothing to do but to balance the head. That slight stoop forward, however, is enough to push this centre of gravity over the front edge of the spinal column. The muscles of the neck must now hold up the head if it is not to drop downward. The work thus laid upon them is considerable. The muscles of the neck are accordingly soon tired out, their tension is relaxed and the work now falls upon the muscles of the back. These in their turn are soon tired out and the child is then forced to lean on other points of support. He tries first one or both elbows. The elbows support the upper arm, the upper arm supports the shoulder blades and the body hangs upon the shoulder blades until they also give way and the chest must needs find a stay and support at the edge of the desk."

Cohn (1886) agreed entirely with Fahrner that "the first slight reaching forward of the head" was the start of the child's postural collapse, and that it must be prevented at any cost. Around this time period, various devices known as straight-holders were invented to help prevent the faulty postures resulting from improperly constructed desks and chairs.

One of the most successful devices was a face-rest designed by Kallmann, a German optician (Figure 82). Kallmann's face-rest consisted of an iron ring enclosed in rubber, which could be attached to any desk at various heights. By supporting the head in an upright posture, Kallmann's face-rest made it impossible to bring the head forward and downward, thereby preventing a stooped forward posture.

Figure 82. Face-rest designed by Kallmann, a German optician. From Burgerstein, L., and Netolitzky, A.: *Handbuch der Schulhygiene.* Jena, Verlag Von Gustav Fischer, 1895. Reproduced with permission of Gustav Fischer Verlag.

Arm Posture

Based on his investigations, Schenk (1894) considered the best writing posture to occur when the abduction of the writing arm is zero degrees; that is, when the upper arm lies lightly against the side of the body. The greater the abduction of the upper arm from the trunk, the more asymmetrical and kyphotic Schenk (1894) found the writing posture to become. In order to prevent abduction of the upper arm when writing, Schenk advocated that the horizontal distance between the desk and the chair backrest should equal the length of the forearm (elbow to wrist) of the person writing (Figure 83).

Floyd and Ward (1964) also noted that with an erect trunk posture, the arms do not have to be abducted in order to rest them on the desk.

It is important that the arms should rest easily upon the desk, rather than the weight of the trunk being supported through the arms and

Figure 83. Schenk's school desk adjusted for writing. From Burgerstein, L., and Netolitzky, A.: Handbuch der Schulhygiene. Jena, Verlag Von Gustav Fischer, 1895. Reproduced with permission of Gustav Fischer Verlag.

shoulders, as when sitting in a collapsed state (Bradford and Stone, 1899; Barlow, 1980).

Whenever an individual leans forward at a desk, supporting the weight of the trunk on his arms, the shoulders will also be raised and drawn forwards. This posture will approximate the origin and insertion of the upper trapezius, and the muscle will become slack and lose its normal resting tone. As a result, increased stress will be placed on the deeper posterior neck musculature in providing stabilization of the head posture (Coe, 1980).

Back Posture

Lovett (1916) considered prolonged flexion of the spine to be induced by school furniture that failed to properly support the child's back. According to Hartwell (1895), a properly designed school chair requires a backrest to "support the sitter's back whether he be quiescent or be actively engaged, e.g., in writing or drawing."

A slightly reclined school sitting posture with an erect trunk was advocated by Lorenz (1888), Schenk (1894), McKenzie (1898), and Schwatt (1910). Bennett (1928) referred to a slightly reclined sitting posture with proper back support as a position of "alert readiness." He considered this to be the proper school sitting posture for the following reasons:

1. There is the least interference with respiration.
2. This posture is best adapted to the various sorts of sedentary work.
3. This posture can be maintained for long periods with the least fatigue.
4. If this posture was made habitual, it would be in the highest degree healthful, comfortable, and graceful (Bennett, 1922).

Dr. Adolf Lorenz of Vienna and Dr. Felix Schenk of Bern were among the first to strongly advocate a reclined sitting posture, with the back properly supported, for writing, reading, and other school activities.

Hartwell (1895) provided a condensed translation of Lorenz's views:

"The forward sitting position, in which the trunk is somewhat bent forward and supported on the desk by the elbows, which we adults prefer, almost without exception, is a dangerous one for the child, since it tends to injure his eyesight through the sinking forward of the head when the neck and trunk muscles become fatigued, and also leads to the production of 'round-back.' The upright (military) sitting position, in seats with perpendicularly placed low-hip or hip and loin supports is too rigorous, calls for an excessive amount of muscular exertion, and does not afford sufficient support to the back of the child either in the writing-periods or in the intervals between them. The reclined-sitting position, in which the back is supported at all times by a properly curved back-support inclined backwards from the inclined seat-surface at an angle of ten degrees to fifteen degrees is to be recommended as the best and simplest means of preventing impaired eyesight and of combating the dangers of rounded back and skewed back. For one who writes in the reclined position, a relatively very large minus distance of 7 to 12 centimeters is demanded, together with an increased desk-slope, to correspond to the inclination of the seat-back" (Figure 84).

Schüldt et al. (1986) recently reported that it is possible to keep the static muscular load of the neck, shoulders, and upper thoracic spine at a low level in a sitting work posture with the thoracolumbar spine inclined slightly backwards and the cervical spine vertical. This posture gave lower neck and shoulder muscle activity levels than a sitting work posture with the entire spine straight and vertical, which in turn gave lower muscle activity levels than a sitting posture with the entire spine flexed forward.

Figure 84. Reclined sitting posture advocated by Lorenz for writing and reading. From Burgerstein, L.: *School Hygiene.* New York, Frederick A. Stokes Co., 1915. Reproduced with permission of Harper and Row, Publishers, Inc.

Specific Backrest Features

Key backrest features for a slightly reclined school sitting posture include the following:

1. A backrest inclination of approximately five to ten degrees (Lorenz, 1888; Schenk, 1894; Schwatt, 1910).
2. A backrest height just below the inferior angles of the scapulae (Cotton, 1904; Schwatt, 1910; Dresslar, 1917). Besides limiting shoulder mobility, higher backrest heights may induce a slumped, kyphotic posture by pushing the shoulders forward.
3. The two most critical regions of back support for holding the trunk in an extended position are:
   a. Pelvic-upper sacral support, to stabilize the sacrum and pelvis, and thereby also support the lower lumbar vertebrae (Cohn, 1886; Branton, 1969; Schoberth, 1969; Lincoln, 1886, 1896).
   b. Support to the lower thoracic spine and thoracolumbar junction (Bradford and Stone, 1899; Cotton, 1904; Rathbone, 1934).
4. Any contouring to the backrest greater than a slight horizontal concavity will tend to push the shoulders forward, resulting in a stooped posture and round shoulders (Bancroft, 1913; Bennett, 1928).

As Bennett (1928) explained:

"It is usually supposed and is probably true that a moderate horizontal concavity increases the comfort of a seat back. Besides conforming better to the shape, it affords support for a certain degree of lateral leaning and turning in the seat. A deep, snug-fitting curve, however, is too confining for comfort. The line across one's back at the shoulders is practically straight, and a curvature in the support here tends to throw the shoulders forward and hinders expansion of the chest and related factors of erect posture. Combination seats, church pews, and benches offer only backs horizontally straight and seem quite satisfactory as far as this factor is concerned. A horizontally straight support below the shoulder blades is not only comfortable but is particularly conducive to expansion of the chest and falling back of the shoulders."

5. The backrest design should allow adequate space for the backward projecting hips.

According to Mosher (1899):

"The shape which the body assumes in sitting, depends upon the position of the pelvis, the head, and the elbows; in standing it depends upon the position of the feet, the pelvis (in relation to the trunk), and the head. A chair which does not permit the pelvis to project behind the shoulder line, unbalances the whole body and makes sitting irksome. A pupil so seated slides the hips forward, and sits with shoulders braced against the back of the chair, thus arching the spine; or he drops the weight of the head and shoulders upon one elbow or both, thereby forcing back the shoulder blade and stretching the muscles which fasten it to the spine, producing the wing-like projecting scapula, so common among school children. At present nearly all the seats in use both in schools and in homes are constructed with backs so shaped that the shoulders, when resting against them, project *behind* the hip line. In my opinion it is this unbalancing of the body which makes children restless and inclined to assume unhygienic postures."

Desk Slope for Writing

Among earlier school seating authorities, Shaw (1902) recommended a desk inclination of 15 degrees for writing, Porter (1906) recommended 15 to 20 degrees, Gould (1905) recommended an inclination of at least 30 degrees, and Dresslar (1917) recommended a desk inclination of at least 45 degrees. More recently, Mandal (1981) renewed interest in desk slopes greater than currently used, advocating a 30 degree inclination for writing.

One problem with steeper desk inclinations is the tendency for books, papers, and pens to slide off the desk (Kotelmann, 1899; Shaw, 1902;

Dresslar, 1917). As a solution to this problem, a horizontal section can be added to the desk, either on one side or at the far edge of the desk.

Another problem with steeper desk inclinations is the tendency for arm fatigue, as the arms will slide down unless pressed against the desk (Schwatt, 1910). However, a compromise must be reached between the potential for arm fatigue at steeper inclinations, and neck stress and fatigue at lesser inclinations (Bennett, 1922). An advantage to steeper desk inclinations is that they will facilitate leaning against the backrest when writing (Kotelmann, 1899). In addition, Lincoln (1896) observed that when leaning against the backrest of a chair in writing, "some support is naturally given to the elbow from behind."

Gould (1905) and McKenzie (1915) both considered that the clearest view of the writing field about the pen point could be obtained if the paper was placed upright, and slightly to the right of the mid-line of the body (for right-handed writing). With a central position of the paper, these authors felt that the writing field would be hidden by the writing hand and pen. As a result, the individual would turn the paper to the left and the pen to the right, along with tilting the head to the left, in order to obtain a clearer view of the writing field about the pen point.

Harmon (1949) considered it important that the child's chair should be able to rotate. With the potential for trunk rotation in uni-manual activities, the ability of the chair to freely rotate with the child could reduce the torque that rotational movements would create in the lower back. Bennett (1928) described the advantages of a swivel seat as permitting the student to "turn so as to face teacher, blackboard, or class or to secure better light on his work, without having to twist his body or sacrifice the symmetrical back support."

READING POSTURE

In addition to writing, a slightly reclined sitting posture with proper back support to keep the trunk erect, with the lumbar spine and lower thoracic spine extended, is also recommended for reading (McKenzie, 1898; Bradford and Stone, 1899; Stone, 1900; Pyle, 1913; Bennett, 1922). Such a position of "alert readiness" will provide proper trunk stabilization along with decreasing the fatigue from prolonged sitting (McKenzie, 1898; Bennett, 1928).

For easy reading with the head in an erect posture, the book needs to be held at right angles to the line of vision, somewhat below the level of the eyes, and at an angle of approximately 60 degrees from horizontal (Shaw, 1902).

According to Bennett (1928):

> "The correct position of a book in reading is at right angles to the line of vision as one sits erect: forty-five to sixty degrees from the horizontal, from fourteen to twenty inches from the eyes (depending on print, light, and the eyes of the reader), and approximately at the height of the chin. With adequate light falling squarely over the shoulder, this gives the ideal visual angle and distance, the most efficient light, and correct posture."

The Problem of Arm Fatigue

However, as Bennett (1928) observed, what schoolchildren will not do is "that theoretically correct thing of sitting erect and holding the book at the ideal height, slope, and distance. The reason is that they cannot so hold it without acutely painful fatigue of the arms."

Adjustable Bookrest

With the inclined writing desk set at the close horizontal distance required for proper back support and arm posture, an adjustable bookrest or reading stand will be required on the desk to fulfill all the visual and postural requirements for reading (Cohn, 1886; Mosher, 1892). Figures 85 to 90 illustrate inclined writing desks with bookrests that were in use approximately one hundred years ago.

For both reading and writing, the most advantageous arm posture would then involve having the upper arm in a vertical position with the elbow close to the side of the body. The forearm should be elevated from the horizontal, coinciding with the plane of the desk top inclination (Figure 91). Such an arm posture would help promote an erect trunk posture, whereas an elevated upper arm posture would increase the flexion force to the upper thoracic spine and promote a round-shouldered posture. A more vertical upper arm posture close to the side of the body will prevent overstretching and help maintain the normal resting tone of the thoracic erector spinae and rhomboidei (Hertel, 1891).

Sitting with the hands in the lap when reading will also promote an erect trunk posture (Bancroft, 1913).

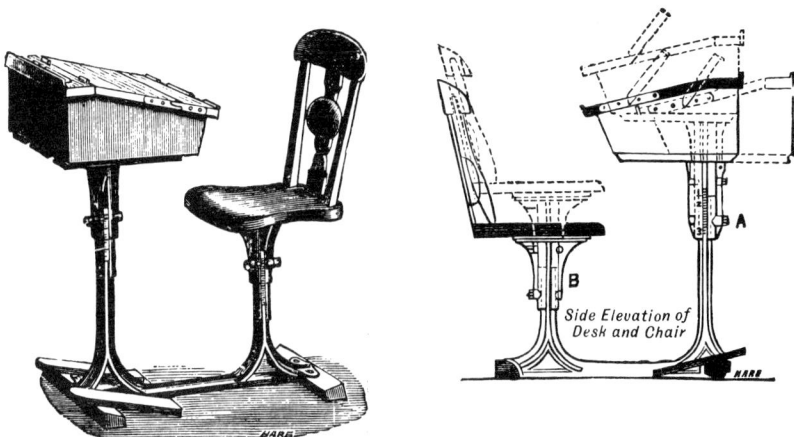

Figure 85. The Glendenning adjustable school desk and seat.

Right: By the application of a key to A and B, the desk and seat may be raised or lowered to any desired height. The desk top also slides horizontally. The writing slope is at 15 degrees, the reading slope is at 40 degrees.

From Roth, B.: *The Treatment of Lateral Curvature of the Spine,* 2nd ed. London, H.K. Lewis, 1899. Reproduced with permission of H. K. Lewis and Co. Ltd.

Figure 86. Faulty reading posture corrected at Glendenning adjustable desk and seat. From Roth, B.: *The Treatment of Lateral Curvature of the Spine,* 2nd ed. London, H.K. Lewis, 1899. Reproduced with permission of H.K. Lewis and Co. Ltd.

Figure 87. The Hygienic Desk, designed by Priestley Smith, an ophthalmic surgeon. The writing slope is at approximately 15 degrees and the reading slope is at approximately 45 degrees. When the book rest is raised for use, its near edge sinks below the surface of the desk, thus creating a groove for the book to rest in. From Cohn, H.: *The Hygiene of the Eye in Schools.* London, Simpkin, Marshall and Co., 1886.

Figure 88. A chair and desk designed by Eliza M. Mosher, M.D. (1899). The desk consists of two parts, a writing table and an adjustable reading stand. The desk is narrow in order to prevent the student from resting his elbows upon it wide apart. It also has a slight slope, and ample depth from front to back for books and papers. The reading stand consists of a moveable headpiece that is attached to the writing table by a pair of brackets having a swivel joint. From Mosher, E.M.: Hygienic desks for school children. *Educational Review, 18:*9–14, June 1899.

Figure 89. Mosher's desk in use for reading. The reading stand has been rotated into position by the student. From Mosher, E.M.: Hygienic desks for school children. *Educational Review, 18:*9–14, June 1899.

Figure 90. When not in use for reading, Mosher's reading stand can also be used as a support for a reference book, such as when doing mathematical work. From Mosher, E.M.: Hygienic desks for school children. *Educational Review, 18:*9–14, June 1899.

Figure 91. Proper arm posture for writing (at Cardot's desk). The upper arm is vertical, and the forearm is elevated from the horizontal, coinciding with the plane of the desk top inclination. From Burgerstein, L., and Netolitzky, A.: *Handbuch der Schulhygiene.* Jena, Verlag Von Gustav Fischer, 1895. Reproduced with permission of Gustav Fischer Verlag.

Chapter Seven

MOTOR VEHICLE SEATING

DRIVING AND BACK PAIN

In an epidemiological study in New Haven, Connecticut, Kelsey (1975a, 1975b) found that driving a motor vehicle, either at work or away from work, was a major risk factor strongly associated with developing a prolapsed lumbar intervertebral disc. Men who spend at least half of the time on their job driving a motor vehicle are approximately three times as likely to develop an acute prolapsed lumbar intervertebral disc as are men who do not hold such jobs (Kelsey and Hardy, 1975). Truck drivers were found by Kelsey and Hardy (1975) to be at very high risk, being approximately five times more likely to develop an acute prolapsed lumbar intervertebral disc than males who are not truck drivers.

In addition, the risk for developing an acute prolapsed lumbar intervertebral disc was approximately twice as great for sitting while driving, compared to prolonged sitting in a chair, regardless of the type of chair (Kelsey and Hardy, 1975).

A further epidemiological study by Kelsey et al. (1984), conducted in Connecticut from 1979–1981, found that the greater the number of hours spent in a motor vehicle, the higher the risk for an acute prolapsed lumbar intervertebral disc. Damkot et al. (1984), in interviews with 303 men, found that the amount of time spent driving an automobile and the number of times getting into and out of a vehicle related to a greater probability of low-back complaints.

An epidemiological survey in Vermont (Pope et al., 1980; Frymoyer et al., 1983) found the driving of cars, motorcycles, buses, tractors, trucks, and heavy construction equipment to be more frequent in patients with low back pain. Their findings suggested "the existence of a complex occupational relationship involving lifting, the use of motor vehicles, and vibrational insult in patients with low back pain" (Frymoyer et al., 1983).

In a study involving the assessment of various back supports for use in

159

car seats (Bulstrode et al., 1983), thirty subjects who were regular drivers and with a three month minimum history of low back pain were asked to list the factors that aggravated their symptoms when driving. The list of factors was as follows:

93 percent—sitting in an incorrect position.
50 percent—movements associated with reversing.
47 percent—sitting in one position for long periods of time.
33 percent—getting in and out of the vehicle.
20 percent—depressing the clutch pedal.
10 percent—inadequate lateral support provided by the car seat.

As a result of extensive history taking and observations, Kendall and Underwood (1968) gave five main causes of backache associated with driving:

1. Entering the car.
2. Leaving the car.
3. Sitting in an incorrect position.
4. Inadequate lateral support.
5. Reaching to the back seat, or reaching for awkwardly placed switches and controls.

Troup (1978) feels that the following interrelated mechanical factors can contribute to the motor vehicle driver's back pain:

1. Muscular effort.
2. Vibratory stress.
3. Shock or impact.
4. Postural stress.

Muscular Effort

According to Troup (1978):

"All driving activities involve the muscles of the trunk. Steering involves the muscles of the upper limb and the extensors and rotators of the spine, rotation adding to the compressive effect of extensor muscle activity because of the added tension in the fibres of the annulus fibrosus of the disc. Pulling on a hand brake has similar effects. Psoas muscle, which originates from the spine, is the principal hip flexor and is active every time the foot is lifted onto the pedal. Accelerating, braking and cornering, and every bump and lurch of the vehicle, tend to move the body in relation to the seat; and all such movements involve a muscular reaction to stabilize the body. The muscular effort of driving therefore, varies with the effort needed to control the vehicle, the road surface, and the number of starts, corners and stops. All involve muscular reactions which add to the temporal pattern of spinal stress."

A poor ergonomic arrangement of the driver's work-space, resulting in the need to adopt a poor posture to improve one's vision or improve one's operation of the controls, will also increase the muscular effort and fatigue of driving (Rebiffé, 1966, 1980; Black, 1966; Hall, 1972).

Vibration and Road Shock

Vibrations are "mechanical oscillations produced by either regular or irregular periodic movements of a body about its resting position" (Grandjean, 1980a). In moving vehicles, the main source of vibration is the interaction of the vehicle with the road surface. With automobiles and trucks, the vertical vibration on the seat is the most critical (Radke, 1964; Parsons and Griffin, 1980).

According to Radke (1964), the source of shock when in a moving vehicle "is most usually the terrain, railroad crossings, chuck holes, and the like, and the effect is typified by its suddenness and violence. Shock may also be experienced when the vibration amplitude reaches the point where it exceeds the travel of any of the suspension elements, including the seat, and 'bottoming' shock occurs."

Improper sitting posture has been implicated in increasing the spinal stress from vibration and road shock. The increased axial loading on the spine from prolonged sitting in a flexed posture and not obtaining proper support from an inclined backrest will result in a loss of height of the intervertebral discs.

As Troup and Edwards (1985) explain:

"The dynamic responses of lumbar vertebrae to static or dynamic loading, or to vibration, are affected by the decrease of stature and the associated loss of disc height. Whenever a compressive load produces a tissue pressure which is greater than the osmotic pressure, fluid is expelled from the disc: this is called the creep effect.

When fluid is expelled from the disc, it narrows and this affects the dynamics of the intervertebral joint. The area of contact between the bearing surfaces of the apophyseal joints changes and the intervertebral joint becomes stiffer. Thus under load the capacity of the tissues to dissipate energy, and their ultimate strength, are reduced. Susceptibility to trauma may also be influenced in that under static compressive load, the transmission of vibration and impact is increased."

Eklund and Corlett (1984) were able to verify the importance of specific chair features based on the reduction in disc height, or shrinkage, with different chairs. Figure 92 illustrates how an easy chair with a high backrest inclined twenty degrees from vertical, with a 4 cm lumbar support, minimizes the creep effect from 1½ hours of sitting compared to a stool without back support.

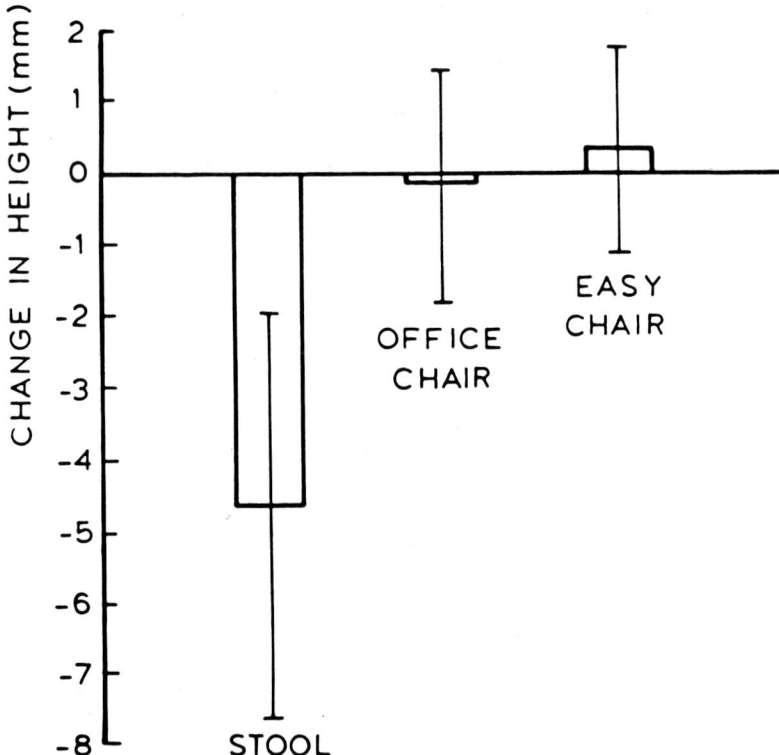

Figure 92. Shrinkage after 1½ hours of sitting in three different chair designs. The means of the three subjects and the standard deviations across the subjects are marked. From Eklund, J.A.E., and Corlett, E.N.: Shrinkage as a measure of the effect of load on the spine. *Spine, 9:* 189–194, 1984. Reproduced with permission of Harper and Row, Publishers, Inc.

Rosegger and Rosegger (1960) emphasized the importance of obtaining proper support from the backrest at all times, in order to minimize the trauma to the intervertebral discs from the shocks and vibrations of tractor driving. If an individual's trunk is upright or leaning forward when accelerated vertically, as opposed to resting against an inclined backrest, there will be a tendency for the head and shoulders to bend further forward. The resulting flexor torque will greatly increase the spinal stress (Vulcan et al., 1970; Troup, 1978). With an inclined backrest, the load will be distributed over both the seat and backrest, thereby reducing the spinal stress from road shock and vibration.

A pelvic-sacral support is also critical to reduce the spinal stress from road shock and vibration. The proper pelvic-sacral support will result in increased support of the trunk against the backrest and it will also prevent the coccyx from weight bearing on the seat (Cooper and Holmstrom, 1963; Howorth, 1978; Diebschlag and Müller-Limmroth, 1980; Kamijo et al., 1982; Zacharkow, 1984a). The spinal stress from road shock and vibration will obviously be increased if the lumbar spine assumes a kyphotic posture with the coccyx weight bearing on the seat (Figure 93).

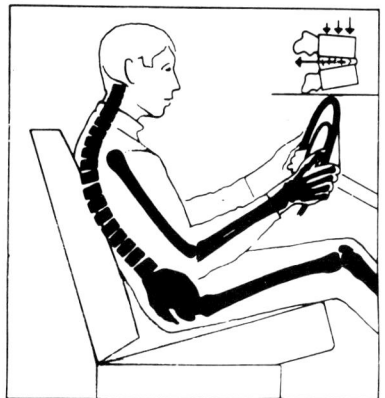

Figure 93. Kyphotic driving posture with the coccyx weight bearing on the seat. From Rizzi, M.: Entwicklung eines verschiebbaren rückenprofils für auto-und ruhesitze. In Grandjean, E. (Ed.): *Proceedings of the Symposium on Sitting Posture.* London, Taylor and Francis, 1969, pp. 112–119. Reproduced with permission of Taylor and Francis Ltd.

A firm, rather than a soft, seat cushion is critical to prevent the coccyx from weight bearing on the seat, and to prevent "bottoming out" with vertical impacts (Hodgson et al., 1963; Cooper and Holmstrom, 1963; Troup, 1978).

Prolonged exposure to vibration when seated has been shown to lead to muscle fatigue, particularly of the erector spinae and oblique abdominal musculature (Wilder et al., 1982). Individuals involved in lifting activities directly after prolonged driving in a flexed sitting posture would therefore be at a very high risk for developing low back pain (Adams and Hutton, 1982).

Recently, animal experiments by Holm and Nachemson (1985) have shown the adverse effect that vibration will have on intervertebral disc nutrition.

Postural Stress

Much of the postural stress with motor vehicle driving results from the particular restraints placed on the driver. The driver's posture is "completely determined by the relative positions of the seat and the controls; his feet are on the pedals, from which they cannot be removed, his hands are on the controls (steering wheel, steering lever, etc.), his eyes are placed where they best satisfy the visual demands of driving." (Rebiffé, 1966).

> "Clearly, of all the relationships which fix the driver's posture, those concerning visual requirements are by far the most restrictive: the direction of the gaze and the position of the eyes strictly determine the direction and position of the head in the driver space; once the head is fixed in a certain position, the freedom of the driver to arrange the body posture is extremely limited because of the small angular differences which can be tolerated for articulation of the head and trunk." (Rebiffé, 1980).

Grandjean (1980 b) differentiates between the car seats for passengers, which he considers "rest chairs" and the driver's seat, which he calls a special "work-chair," the driver being involved in a continuous vigilance task, in a very static sitting posture.

Other causes of postural stress include the following:

a. An awkward, unstable posture adopted by the driver to improve his line of sight, or to improve his ability to reach and operate the controls (Black, 1966; Macaulay, 1969; Kantowitz and Sorkin, 1983).

b. The lumbar flexion and rotation which may occur when entering and leaving the car. This is more prone to occur with a low car seat and a small door. The driver needs to be educated on how to protect his back when getting into and out of the car (Kendall and Underwood, 1968).

c. Individuals with tight hamstrings will usually drive with the lumbar spine in flexion, particularly if the knees are kept extended to reach the foot pedals (Kottke, 1961). The lumbar spinal posture can be improved if these individuals are instructed to move the seat forwards, which will increase the knee flexion and reduce the tension on the hamstrings. However, care must be taken not to increase the degree of hip flexion when moving the seat forwards, as this will also increase the hamstrings tension (Brunswic, 1984 b).

It is also important to note that excessive knee flexion, while being beneficial in relaxing the hamstrings, can be disadvantageous for the application of force to the brake pedal (Pheasant, 1986).

Individuals also need to be instructed in proper use of the backrest when driving. Varterasian and Thompson (1977) calculated the body weight distribution on the driver's seat and backrest of a 1976 two door car, with fifteen automotive industry employees as subjects. They found that the body weight distribution depended a great deal on the individual's driving posture, with some individuals sitting slouched and others more upright. The percent of the individual's body weight on the backrest varied anywhere from 3 percent to 18 percent.

DESIGN OF THE DRIVER'S SEAT

If chair design in the home is faulty, the sitter is made uncomfortable. However, if the design of the driver's seat is faulty, a traffic accident or fatality may result (Black, 1966; Diebschlag and Müller-Limmroth, 1980; Hall, 1983).

Hockenberry (1982) has referred to the potential for traffic accidents when the driver is distracted due to body pressure discomfort. It is not unusual for pressure discomfort on the lower back and buttocks to occur within one to two hours of driving, with a poorly designed driver's seat (Jones, 1969).

Hartnett (1980) describes a particular example of poor design of the driver's work-space:

"In those vehicles equipped with low seat and high steering wheels, it has been found that drivers have to sit too close to the steering wheel. This is necessary in order that they may not only be able to see over it, but gain a proper position for the hands.

The result of this situation is that the driver sits too upright, causing all variations in weight frequency to be borne by the buttocks and spine. This, in turn, causes a "numb bum" effect and the driver tends to shift around on the seat to get relief. In so doing, unnatural forces are exerted on the steering wheel by the driver (he uses it as a means of pulling himself up) causing inaccuracy of steering. At a critical time, such as an emergency, this situation can be disastrous."

Functions of the Driver's Seat

Radke (1964) considers the important functions of the driver's seat to be as follows:

1. Position the driver to give him access to the vehicle controls and optimum visibility.
2. Provide the driver proper support in order to exert the force necessary to operate the controls.
3. Support the body to minimize the muscular effort to maintain the driving posture.
4. Provide the proper contour, padding, and shape to achieve optimum pressure distribution.
5. Minimize the effect of vibration and shock by
   a. Reducing, or at least not amplifying, vibration and shock.
   b. Positioning the driver to minimize the effects of vibration and shock motion.

According to Troup (1972), the driver's seat must serve the following functions:

A. It must resist the following stresses:
   1. The vertical forces due to the effect of gravity on the body.
   2. The forces between the driver and the seat arising from:
      a. Vertical acceleration of the car due to irregularities in the road surface.
      b. Radial acceleration of the car due to change in the direction of travel.
      c. Angular acceleration due to pitching.
      d. Angular acceleration due to rolling.
   3. Reactions arising from external forces applied to the controls.
      a. By the upper limbs to the steering wheel, hand brake, etc.
      b. By the lower limbs to the pedals.
B. It must be adaptable to the individual driver's dimensions and should therefore be adjustable.

Seat Comfort/Discomfort Factors

Kohara and Sugi (1972) considered the following five factors to be important influences on the comfort of automobile seats:

1. The dimensions of the seat.
2. The body pressure distribution on the seat and backrest.
3. The cushion characteristics.
4. The individual's body shape.

5. The final stable posture of the individual. Kohara and Sugi considered this to be the most important factor influencing seat comfort.

Kamijo et al. (1982) found that test subjects considered the sensations of body pressure distribution and driving posture to be the most important factors in the overall evaluation of comfortable driver's seats.

The contribution of different factors to overall seating comfort was as follows:

Sensation of body pressure distribution – 44 percent
Sensation of driving posture – 44 percent
Sensation of having ample room – 30 percent
Sensation of being cushioned – 24 percent
Sensation of being held – 17 percent

Seats rated by the evaluators as comfortable gave good lumbar support, whereas uncomfortable seats had insufficient support to the lumbar region and tended to cause a round-shouldered driving posture (Kamijo et al., 1982).

Automobile seat testing on individuals with and without backache by Kendall and Underwood (1968) implicated the following main causes of discomfort:

1. Insufficient seat length. This would result in poor thigh support and an unstable base for the leg muscles to operate the foot pedals.
2. Soft seats.
3. Lack of lateral support on the seat.
4. Incorrect seat angle.
5. Inadequate seat height.
6. Insufficient backrest height, resulting in inadequate upper back and head support.
7. Lack of lateral back support.

Parsons and Griffin (1980) considered the vertical vibration on the seat to be a major cause of sitting discomfort in cars.

Black (1966) stressed that comfort must not only relate to the seat itself, but must also "relate to ease of vision and ease of control operation – and in fact to the total ergonomics of the driver's work-space."

Specific Features of the Driver's Seat

Seat Cushion

A firm seat cushion, as opposed to a soft cushion, is highly recommended for better protection from road shock and vibration (Cooper and Holmstrom, 1963; Hodgson et al., 1963; Troup, 1978; Woodson, 1981).

Woodson (1981) commented upon the danger of submarining occurring when the occupant sinks deeply into a soft cushion in an automobile accident. Submarining refers to the rotation of the buttocks below the seat belt. This results in slack in the seat belt, and a high probability that the seat belt will end up across the individual's abdomen during the crash impact, resulting in serious or fatal internal injuries (Woodson, 1981).

(In order to avoid abdominal injuries, the proper placement of the horizontal seat belt strap is below the anterior superior iliac spines of the pelvis. However, malposition of the seat belt can still frequently occur from slouching in the seat or crossing the legs [Wells et al., 1986].)

In addition, the lack of support provided by a soft cushion will greatly increase the driver's sitting instability. Increased muscle activity will be required to maintain the desired driving posture, resulting in greater fatigue (Branton, 1966; Kohara and Hoshi, 1966).

When the driver is depressing the foot pedals, the pressure against the distal posterior thigh from the seat will increase. Therefore, it is important that the front edge of the seat be contoured to eliminate a potential high pressure ridge.

Seat Cover

A breathable, woven fabric is preferred over a leather or plastic seat cover for two reasons:

   1. The increased friction from a woven fabric will help keep the individual from sliding forward or sideways on the seat (Branton, 1969, 1970; Hogan, 1982).
   2. A breathable fabric will aid surface ventilation, and will prevent moisture accumulation at the contact areas of the individual's body on the seat and backrest (Oxford, 1973; Babbs, 1979; Woodson, 1981).

In regards to sheepskin seat covers, their most important property is the ability to absorb water vapor, and therefore reduce the humidity at the buttocks-cushion interface. Denne (1979a, 1979b) reported that a resilient natural sheepskin absorbed nearly fifteen times as much water as a pure polyester simulation of sheepskin.

Seat Inclination

An inclined driver's seat is critical for the following reasons:

1. It will help prevent the driver from sliding forwards on the seat (Åkerblom, 1948; Ayoub, 1972). This tendency to slide forward on the seat will always be present if the individual is sitting properly, with his trunk stabilized against an inclined backrest. Vibration will have a similar effect, causing the individual to slide forwards on the seat (Diffrient et al., 1974). Movements of the lower extremity with braking and accelerating will also tend to move the buttocks forward on the seat (Black, 1966).

2. It will facilitate contact of the individual's trunk with the backrest and pelvic-sacral support (Andersson et al., 1974e).

3. It will aid in retention of the driver in the seat during a rapid deceleration, or emergency stop (Jacobs et al., 1980).

Seat inclination angles recommended in the literature are given in Table V.

Table V
THE DRIVER'S SEAT:
RECOMMENDED SEAT AND BACKREST ANGLES

Investigator	Seat Inclination	Backrest Inclination (From Vertical)
Lay and Fisher (1940)	6°-7°	20°-24°
McFarland and Stoudt (1961)	7°	22°
Black (1966)	6 °	21°
Jones (1969)	7°	18°
Diffrient et al. (1974)	10.5°-17°	18°-22°
Babbs (1979)	not given	15°-20°
Grandjean (1980 b)	10°-22°	adjustable to 30° inclination

Backrest Inclination

An inclined backrest, as opposed to a vertical backrest, will reduce the axial loading on the spine, as some of the body weight will be taken by the backrest. With the body weight being distributed over both the seat and backrest, the spinal stress from road shock and vibration will also be reduced (Troup, 1978). The rate of shrinkage of the disc (creep effect) will also be minimized with an inclined backrest (Eklund and Corlett, 1984).

Grandjean (1980b) differentiates between the driver's seat as a "work-chair" and the passenger seats as "rest chairs". Backrest inclination angles reflect this difference, with a slightly greater backrest inclination being recommended for passenger seats. For example, Diffrient et al. (1974) differentiate between an alert posture for the driver and a relaxed posture for the passenger by recommending an 18 to 22 degree backrest inclination for the driver's seat, and a 22 to 28 degree backrest inclination for the passenger's seat.

Backrest inclination angles recommended for the driver's seat are given in Table V. It is important to realize, however, that an excessive backrest inclination will distort the proper axial relationship of the head, thorax, and pelvis. In addition, an increase in shoulder flexion and elbow extension will occur with greater backrest inclinations. This will increase the stress to the shoulder and neck musculature when driving (Hosea et al., 1986).

Backrest Height

Severy et al. (1969) reported that the average backrest height of 21 inches in 1969 model cars offered little or no protection against whiplash. The authors recommended a backrest height of at least 28 inches for protection against whiplash from a rear-end collision.

According to Bogduk (1986):

> "The essential feature of the biomechanics of whiplash is that no direct force is imparted to the head by the vehicle, and the head is not thrown bodily backwards into extension. Rather, the trunk moves forwards from under the relatively stationary head, and thereby imparts a relative extension movement of the head and neck. This has significant implications in the use of head-rests ostensibly to prevent whiplash injury, for in the absence of an actual backward motion of the head, head-rests cannot function by 'cushioning' such a backward motion.
>
> In most vehicles the seat is inclined slightly, or substantially, backwards, and the customary position assumed by most motorists is such that they sit upright or forwards, leaving the head-rest some distance behind. Under these circumstances, a head-rest will not impede the whiplash motion. The back of the seat will strike the trunk sooner than the head-rest will strike the head, if at all, and the trunk will be accelerated

sooner than the head. Consequently, relative motion between the trunk and head, and therefore whiplash, will occur.

Head-rests will be of value only if, at the time of impact, the head is actually resting on the device, or the head-rest comes into contact with the head at the same time as the seat-back strikes the trunk. Under these circumstances, both the head and trunk would be accelerated basically concurrently and to the same extent. Consequently, there would be no relative motion between the head and trunk, and therefore no extension (and injury) of the neck.

Thus, head-rests function not by cushioning a backward movement of the head, but by accelerating the head at the same time and same rate as the trunk. This action is dependent on the head being in contact with the head-rest at the appropriate time, but unfortunately, the position adopted by most drivers is incompatible with these prerequisites, and the potential protective function of head-rests is denied."

An adjustable, padded headrest is critical to provide the proper protection against whiplash. Unfortunately, most headrests (head restraints) are located too far behind the head, and/or they are too low.

In order to provide the proper protection, both a horizontal and vertical adjustment of the headrest are critical. An additional adjustment should allow the headrest to be angled forward from the plane of the backrest. The angle usually recommended is from 5 to 10 degrees (Le Carpentier, 1969; Diffrient et al., 1974; Hiba, 1980).

In addition, the headrest should not be so wide that it obstructs the driver's vision to the rear and sides (Black, 1966; Woodson, 1981; Schmidtke, 1984).

Pelvic-Sacral Support

The proper pelvic-sacral support will prevent the coccyx from weight bearing on the seat, and it will result in greater support for the upper trunk against the backrest (Cooper and Holmstrom, 1963; Howorth, 1978; Kamijo et al., 1982; Majeske and Buchanan, 1983, 1984; Zacharkow, 1984a).

As proper placement of the pelvic-sacral support is extremely critical, height adjustability for the support (vertical adjustment) is important. It should support the upper part of the sacrum, the posterior iliac crests, and the lowest lumbar vertebra (Kottke, 1961; Oxford, 1973; Grandjean, 1980b; Rizzi, 1969). The upper sacral support is critical to help stabilize the pelvis and lumbosacral joint. If only the upper lumbar vertebrae were supported and the pelvis was left unstabilized, fast oscillatory movements of the pelvis rocking over the ischial tuberosities could occur when driving. This would result in increased stress to the lower lumbar spine (Branton, 1969; Panjabi et al., 1986).

The addition of a spinal support to a car seat can result in a marked alteration in the driving posture of an individual who habitually assumes a slumped, kyphotic posture. Health professionals, however, must be cautioned in regards to blindly recommending such supports for drivers' car seats, without examining the relationship of the driver's posture to his work-space. Marked alteration in sitting posture may adversely affect the driver's most ideal line of sight, and his ability to easily reach the steering wheel and other controls (Floyd, 1967).

Certain spinal supports added to car seats may also position the driver further forward on the seat. As a result, the driver may be too far forward from the headrest to receive proper head support in a rear-end collision.

Thoracic Support

In addition to the critical pelvic stabilization provided by proper pelvic-sacral support, a second critical area of support is the lower thoracic spine (thoracolumbar junction). Support in this region is critical in order to promote an erect trunk posture, stabilization of the thorax, and the proper axial relationship of the thorax and pelvis (Rathbone, 1934; Rizzi, 1969).

Lateral Support

Some contouring of the seat and backrest is necessary to provide lateral support for the driver (Figure 94). This will result in better stabilization of the buttocks and thighs on the seat, and of the trunk on the backrest (Roberts, 1963; Thier, 1963).

When rounding a curve without proper lateral support, flexion and rotation of the lower trunk may occur as the driver slides laterally while his shoulders are stabilized by gripping the steering wheel (Kendall and Underwood, 1968; Bulstrode et al., 1983). Proper contouring of the seat and backrest will help assure that the driver's posture remains as symmetrical as possible, thereby minimizing the added spinal stress from asymmetrical sitting. Hall (1972) found that tall drivers stressed the need for pronounced lateral trunk support, whereas shorter drivers emphasized the need for good lateral support from the seat.

Regarding back supports that can be attached to car seats, one particular problem has been noted with inflatable lumbar supports. In a study

Figure 94. Contouring of the seat and backrest to provide lateral support for the driver.

by Bulstrode et al. (1983), the inflatable supports did not provide any lateral support, and the drivers evaluating these supports experienced shearing forces when rounding curves due to the movement of the cushion of air between the trunk and backrest.

Adjustability of the Driver's Seat

In order to prevent many of the problems that can result from a poor driving posture, the driver's seat should be adjustable in as many ways as possible. The most critical adjustments are as follows:

a. Forward-backward adjustment of the entire seat. This will allow the driver to achieve comfortable sitting angles at the hips and knees.

b. Seat height adjustment. This adjustment will enable the driver to obtain the optimal line of sight.

c. Seat inclination adjustment.

d. Backrest inclination adjustment. The angle preferred will be partially determined by the driver's body type and arm length. However, very large backrest angles can reduce the driver's visual field and make access to the controls difficult, along with placing the driver's lower cervical and upper thoracic spine in an extremely flexed posture (Olsen, 1965; Hall, 1983; Pheasant, 1986; Harms-Ringdahl et al., 1986).

e. Pelvic-sacral support adjustment.

f. Thoracic support adjustment (for the lower thoracic spine and thoracolumbar junction).

g. Headrest adjustment, both vertical and horizontal adjustment, along with tilt adjustment. A major postural problem with a non-adjustable headrest is that if it is too low, the driver will be pushed forward at the shoulders. This will result in a kyphotic sitting posture, with additional stress on the neck and thoracolumbar spine.

   h. Steering wheel adjustment, both in horizontal and vertical directions. These adjustments can facilitate a more vertical upper arm posture, thereby reducing the fatigue from keeping the shoulders and arms elevated (Barker, 1985).

Other important features to consider in the design of the driver's seat and work-space include the following:

Adjustable Armrests

Adjustable armrests would help decrease the stress in the neck, upper back, and shoulder muscles when driving. They would result in a more stable and symmetrical sitting posture, with better support of the upper trunk against the backrest (Andersson et al., 1975; Hedberg et al., 1981; Nakaseko et al., 1985).

Left Footrest

A footrest with an adjustable inclination, located to the side of the brake pedal, would result in a more stable and comfortable driving posture. This additional foot support would also help keep the driver from sliding forward on the seat (Jacobs et al., 1980; Darcus and Weddell, 1947).

Automatic Transmission

Andersson et al. (1974e) reported an increase in lumbar disc pressure with both depression of the clutch pedal and gear shifting. For individuals with chronic back pain, these added spinal stresses could be avoided with an automatic transmission.

Chapter Eight

THE OFFICE WORKSTATION

In 1980, it was estimated that there were 5 to 10 million video display terminals (VDTs) in the United States, and more than 7 million VDT operators (Center for Disease Control, 1980). It has also been estimated that by 1990, 40 to 50 percent of all United States workers will use VDTs on a daily basis. At that time, there will likely be approximately 38 million terminal-based workstations of various kinds in offices, factories, and schools (Giuliano, 1982). As a result of the video display terminal, the "paper office" has been transformed into an "electronic office" (Grandjean, 1984a; Giuliano, 1982).

As Grandjean (1984a) explained:

"At the traditional office desk an employee has a great variety of physical activities and a large space for various body postures and movements: he/she may look for some documents, read some texts, exchange information with colleagues, type for a short while and carry out many other activities during the course of the working day.

The situation is entirely different for an operator working continuously at a VDT for several hours or for a whole day. Such an operator is tied to a man-machine system: movements are restricted, attention is directed to the screen or source documents and the hands are linked to the keyboard. These operators are more vulnerable to ergonomic shortcomings, to constrained postures, to unsuitable lighting conditions and to uncomfortable furniture."

Besides the various types of VDT work, fixed postures are also characteristic of other office jobs such as full-time typing and the operating of accounting machines (Hünting et al., 1981; Maeda et al., 1982; Grandjean, 1984c). For example, Maeda et al. (1982) described the typical posture of accounting machine operators as "a continuous sitting posture with the neck and head tilted forward and to the left with some rotation of the head to the left to orient the visual line to the receipts, the left hand being used to turn over the receipts, and the right hand being rapidly used to manipulate the numerical keyboard."

The most frequent musculoskeletal complaints of VDT operators have been found to involve the neck, neck-shoulder region, and the back. The arms, wrists, and hands are also sites for musculoskeletal complaints, but less frequently (Arndt, 1982, 1983; Sauter et al., 1983, 1984; Smith, 1984a; Östberg, 1979; Ong et al., 1981; Cakir, 1980; Elias et al., 1983; Kukkonen et al., 1984). Musculoskeletal complaints have even been reported among VDT operators with minimal job dissatisfaction and minimal psychosocial stress (Smith, 1985).

However, it is important to realize that the musculoskeletal complaints associated with the use of VDTs and other office machines have usually been found in work settings characterized by poor workstation design and poor sitting postures (Hünting et al., 1981; Ong et al., 1981; Starr et al., 1982; Maeda et al., 1982; Sauter et al., 1983). Factors such as non-detachable keyboards, an increased forward inclination of the head, and the lack of proper arm support and back support have been correlated in these studies with an increased incidence of musculoskeletal complaints.

According to the National Research Council's Panel on Impact of Video Viewing on Vision of Workers (1983):

> "Although we cannot draw firm conclusions about the comparative types and incidences of job-related muscular discomfort in VDT and non-VDT workers, the results of studies indicate that many VDT operators do experience significant discomfort. It is likely that this discomfort is largely caused by inappropriate workstation design. Video display terminals are often designed and introduced into offices without the application of relevant human factors design principles. In many instances poorly designed VDTs are simply installed at desks formerly used for traditional office work or placed on whatever furniture happens to be available. Operators are often required to work in cramped spaces that leave them little room to place document holders or manuscripts in positions that allow comfortable working postures. Operators in such situations are likely to experience visual discomfort, muscular discomfort, and fatigue."

The key to reducing the potential for musculoskeletal stress at VDTs and other office machines is a well-designed, adjustable workstation that will provide proper body stabilization for the specific tasks being performed. Without proper body stabilization, the fixed postures associated with VDT and other office machine work will impose sustained static loads on many postural muscles, particularly those of the neck, neck-shoulder region, and back (Maeda, 1977).

Several studies have already demonstrated a reduction in musculo-skeletal complaints or stress, along with an increase in productivity, as a result of properly designed workstations (Rohmert and Luczak, 1978; Dainoff, 1983, 1984c; Secrest and Dainoff, 1984; Pustinger et al., 1985; Ong, 1984). In a study involving telephone operators, VDT workstations incorporating easily made adjustments were found to significantly improve the comfort of the operators (Shute and Starr, 1984). Stewart (1983), however, found the most frequent VDT workplace problem in his evaluations to be a lack of adjustability in the workstation.

The four critical contact points that will determine the individual's workstation posture are:

1. The head and eyes with the screen and source documents.
2. The hands and arms with the keyboard and possibly other writing materials.
3. The back and buttocks with the chair.
4. The feet with the floor or footrests.

(Kroemer and Price, 1982; Kroemer, 1982; Arndt, 1983; Dainoff, 1984d).

Figures 95 and 96 illustrate some important parameters and workstation features at these points of interaction. The remainder of this chapter will explore in detail these four important contact points at the VDT workstation as they determine the individual's workstation posture and his potential for musculoskeletal stress.

HEAD POSTURE

VDT Task Classification

The visual demands at a VDT workstation will depend on the specific task requirements. Dainoff (1982) classified VDT jobs as either:

a. Copy intensive work.
b. Screen intensive work.
c. Interactive work.

Copy Intensive Work

Copy intensive work would include data entry tasks, where the main job is entering information into the computer system. With these tasks, the gaze is directed mainly at the source documents and very infrequently at the display screen. In addition, this work usually involves a very high rate of keying.

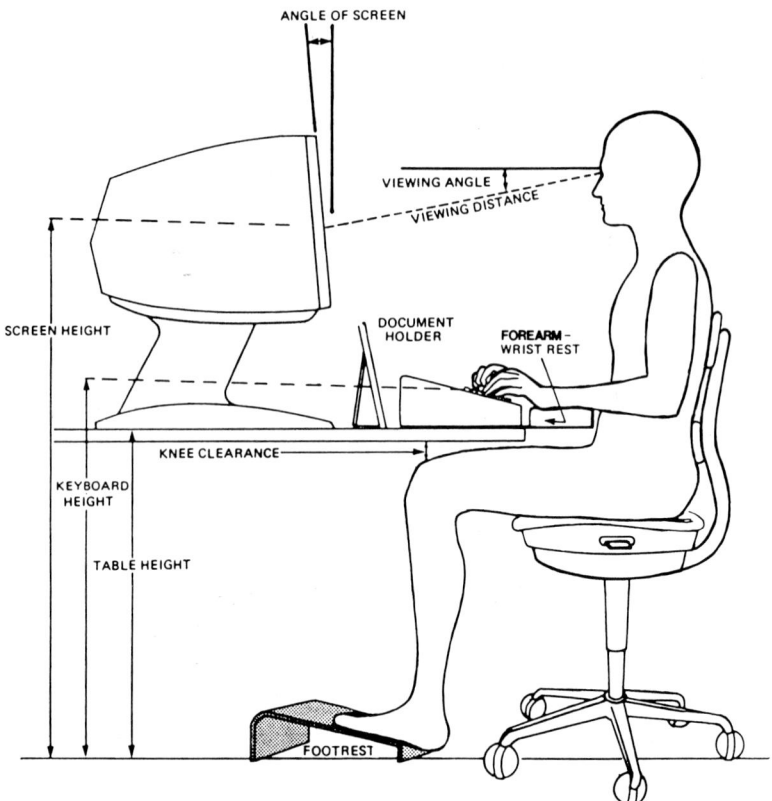

Figure 95. Important VDT workstation terms. Adapted from *Video Display Terminals. Preliminary Guidelines for Selection, Installation, and Use.* Short Hills, Bell Telephone Laboratories, 1983. Copyright 1983, Bell Telephone Laboratories, Incorporated. Adapted with permission.

Full-time typing and accounting machine operating are non-VDT tasks that are copy intensive.

Screen Intensive Work

Screen intensive work would include data acquisition (data inquiry) tasks, where the gaze is directed mainly at the display screen to retrieve information. These tasks usually involve a low to medium rate of keying.

Specific jobs in this category would include directory assistance operators and air traffic controllers.

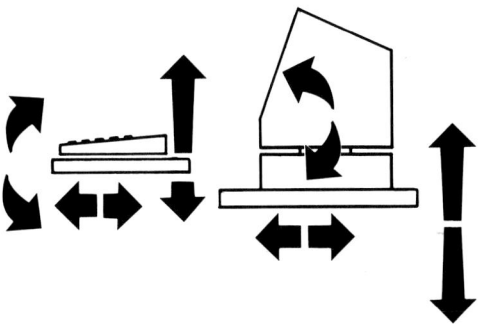

Figure 96. Important adjustability features for (1) the keyboard and its supporting surface, and (2) the display screen and table.

Interactive Work

In interactive work, where the operator engages in a "dialogue" or "conversation" with the computer system, the gaze is alternated between the display screen and source documents. The frequency of alternation of the gaze will depend on the specific VDT task.

Jobs in this category would include airline reservation clerks, computer-aided design engineers, and computer programmers.

Musculoskeletal Stress and Head Posture

Depending on the specific task, the operator's head and neck posture, and even his trunk posture, will be determined by the location of the display screen and source documents. One result of poor ergonomic placement of the screen and source documents will be an excessive forward inclination of the head, which has been associated with an increase in musculoskeletal complaints from the operator (Hünting et al., 1981; Maeda et al., 1982; Sauter et al., 1983).

Neck Pain

An increased forward tilt of the head will result in an increased static loading of the posterior neck muscles, as well as an increase in the cervical spine compression forces (Chaffin, 1973; Less and Eickelberg, 1976). Musculoskeletal complaints involving the posterior neck, shoulders, and upper back have been found to increase with an increase in the forward inclination of the head (Hünting et al., 1981; Maeda et al., 1982; Grandjean et al., 1982b).

Headache

The prolonged static loading of the posterior neck musculature with an exaggerated forward inclination of the head can also be a major cause of headache with VDT and other office machine operators (Travell, 1967; Robinson, 1980; Stewart, 1979; Coe, 1979, 1980; Sauter et al., 1984; Stone, 1984) (Figure 97).

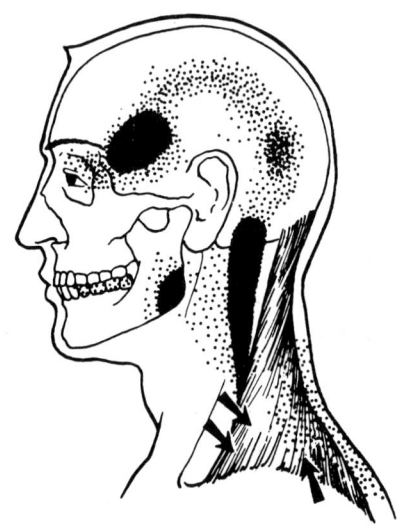

Figure 97. The pain reference pattern of the suprascapular region of the trapezius muscle. Trigger areas are indicated by the arrows, and pain reference zones are indicated by the stippled and black regions. From Travell, J.: Mechanical headache. *Headache*, 7:23–29, April 1967. Reproduced with permission of The American Association for the Study of Headache.

Low Back Pain

An increased forward inclination of the head can also increase the stress on the lower back (Grandjean et al., 1982b).

Sauter et al. (1983), in their study of office workers using VDTs at public agencies in Wisconsin, found a strong significant association of neck posture with back posture:

"The actual data indicate that when neck/cervical postures are poor (inclined or thrust forward), back postures are likewise poor (excessive forward inclination of the back or hunching of the shoulders), and the back tends not to be contiguous with the chair back support."

Major Causes of Poor Head Posture

Non-Detachable Keyboard

With a non-detachable keyboard, the VDT operator will not have the ability to make independent adjustments of the display screen height and the keyboard height (Stewart, 1980; Helander et al., 1984). If the keyboard is used frequently, the table height will usually be adjusted to allow a comfortable keyboard height. This will place the display screen at a less than optimal height, requiring an increased gaze or viewing angle for proper visibility. (The viewing angle is the angle formed by a horizontal line at eye level with the line connecting the operator's eye and the center of the screen. See Figure 95.)

Sauter et al. (1983) reported "a trend toward poor neck posture and lack of back support with increased gaze angle." They also found that the gaze angle was reduced with a detachable keyboard. Therefore, the primary benefit from a detachable keyboard may be an "increased opportunity for improved neck/back/head posture in VDT viewing brought about by a reduced viewing angle" (Sauter et al., 1983).

As most VDT operators seem to prefer a backward leaning trunk posture with full back support, a downward gaze angle of 5 degrees to 20 degrees to the center of the display screen and source documents has been recommended for this sitting posture (Grandjean et al., 1983a; Rühmann, 1984; Sauter et al., 1984). However, recent experimental data from Kroemer and Hill (1986), in a study involving sixteen male and sixteen female subjects, found a preference for somewhat steeper viewing angles.

No Document Holder

Having source documents lying flat on the desk rather than on an adjustable, inclined document holder can be one of the greatest sources of musculoskeletal stress for VDT operators, typists, and other office machine operators (Stewart, 1979, 1980; Cakir et al., 1980; Maeda et al., 1982; Arndt, 1983).

According to Life and Pheasant (1984):

"Laying the copy script flat on the desk beside the keyboard results in the need for increased muscular activity to support the head while it is craned over to read. This would be expected to give rise to discomfort in the neck, shoulders and upper back; and indeed it does."

Their study with twelve skilled female typists also found that the inclined document holder affected the overall posture of the typist, resulting in a tendency to lean the trunk backwards.

Lack of Arm Support

Without proper arm support, the weight of the arms will be transmitted to the posterior neck and shoulder musculature, particularly the trapezius, and the upper trunk will tend to slump forward with an increased forward inclination of the head.

A study of VDT workstations by Hünting et al. (1981) found that the characteristic "hands and arms frequently supported" was correlated with a lower incidence of neck, shoulder, and arm pain among VDT operators. The overall posture of VDT operators may be improved with large forearm-wrist supports, as they will induce a backward leaning trunk posture with better support from the backrest, as well as a decreased loading of the upper trapezius (Nakaseko et al., 1985; Weber et al., 1984).

Improper Chair Design or Chair Fit

As discussed in Chapters Five and Six, a forward inclination in the upper backrest design or an excessive concavity (horizontal curvature) to the upper backrest will tend to push the shoulders and upper trunk forward. As a result, there will be an excessive forward inclination of the head with increased stress on the posterior neck and upper back musculature.

A slumped, kyphotic posture of the trunk with an excessive forward inclination of the head can also occur from the inability to obtain proper back support from the chair. This may result if the seat is too high, and the operator scoots forward to obtain a more stable sitting posture with the feet firmly on the floor.

Poor head posture can also result from too low a backrest height. With inadequate stabilization of the upper trunk, the operator may eventually assume a forward leaning posture due to back fatigue.

Poor Sitting Habits

Of course, poor head posture at the workstation may also just reflect the individual's chronic sitting and standing postural habit patterns.

Neck pain and headache may also result from the awkward head and neck posture observed when VDT operators rest the telephone receiver between their neck and shoulder, while handling customer inquiries as they type at the keyboard (Travers and Stanton, 1984; Travell, 1967).

Workstation Design Based on Visual Demands of Task

In order to prevent a stressful head and neck posture, the position of the display screen and source documents should be based on the amount of time spent viewing each of these sites. (Similarly, the proper location of the keyboard and other writing materials should be based on the amount of time spent keying versus writing [Arndt, 1983].)

For example, in copy intensive work such as data entry tasks, the source documents should be given the most prominent position directly in front of the operator, and the display screen should be positioned to the side (Helander et al., 1984) (Figure 98). With screen intensive work such as data acquisition tasks, the display screen should have the most prominent position in front of the operator (Helander et al., 1984).

In tasks where the gaze is frequently alternated between the display screen and source documents, the best solution is to position the screen and documents side by side, as close together as possible, directly in front of the operator. This arrangement is preferred as there is less potential for neck strain with a frequent mild rotation of the head compared to a flexion-extension movement (Cakir et al., 1980; Wright, 1982; Benz et al., 1983).

The other alternative with frequent switching of the gaze between the screen and source documents is to place the documents immediately in front of and beneath the display screen (Figure 99). The one important advantage of this arrangement is when using multiple documents of various sizes, which may even be stapled together. According to Sauter et al. (1984), "Such documents are difficult to manipulate when placed off to the side of the display, or cannot be readily affixed to a conventional typist's document holder." However, it is still critical to incline the document work surface enough to prevent an excessive forward tilting of the head.

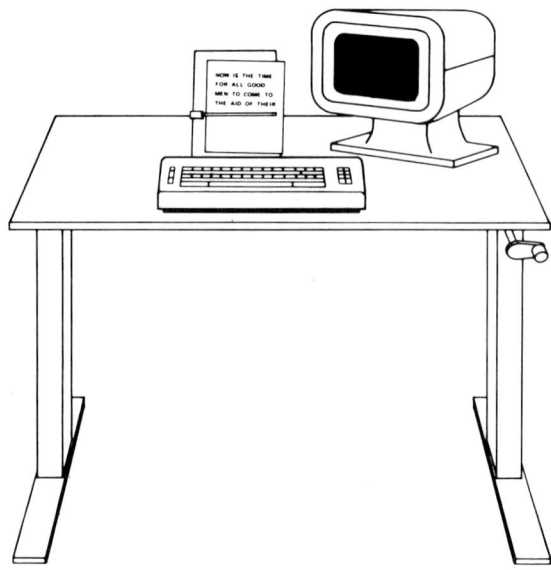

Figure 98. With copy intensive work such as data entry tasks, the source documents should be given the most prominent position directly in front of the operator. The display screen should be positioned to the side. From *Video Display Terminals. Preliminary Guidelines for Selection, Installation, and Use.* Short Hills, Bell Telephone Laboratories, 1983. Copyright 1983, Bell Telephone Laboratories, Incorporated. Reprinted by permission.

Other Visual Factors Affecting Posture

According to Arndt (1983):

"The best efforts to provide a well-designed workstation may be negated by difficulties in seeing the display. The poor quality of the image on the screen (i.e., size, clarity, contrast and brightness) may cause the operator to lean forward in order to see more clearly. It is also not unusual to observe operators leaning forward, sitting off to the side, or slouching down in their chairs in order to avoid reflections on the screen or overhead lights which shine directly in their eyes. The selection of higher quality displays, proper illumination controls, and adjustable screens are thus important considerations for meeting postural as well as visual requirements of VDT work."

Glare

Glare has been referred to as the "single most detrimental environmental factor for VDT operators" (Smith, 1984b). Whether direct glare (from windows or lights) or indirect glare (reflections), there will be a greater

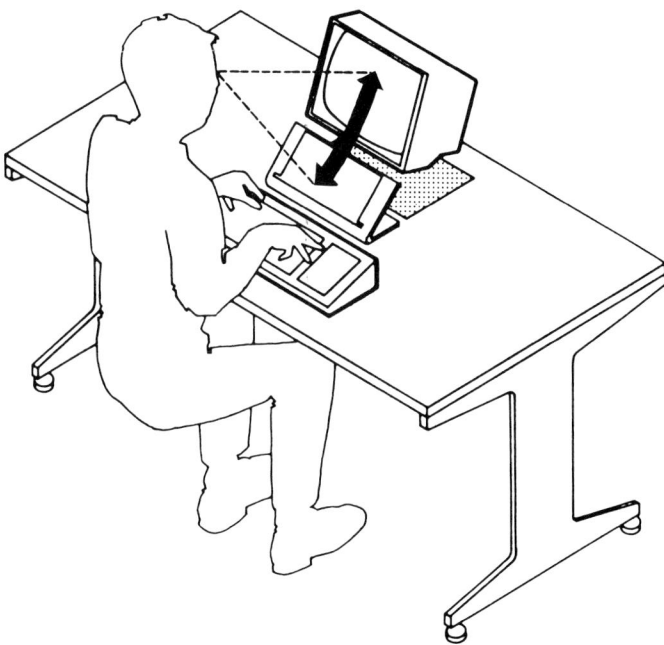

Figure 99. Positioning the documents immediately in front of and beneath the display screen is one alternative with frequent switching of the gaze between the screen and source documents. This arrangement is advantageous when using multiple documents of various sizes. From *Introducing the Visual Display Unit to the Office.* London, BEITA, 1984. Reproduced with permission of the Business Equipment and Information Technology Association.

potential for neck and back pain if the VDT operator adopts a poor sitting posture to reduce the glare problem (Wright, 1982; Shackel, 1983) (Figure 100).

The most important preventive measure against screen reflections is the ability to tilt the display screen (Fellmann et al., 1982; Buti et al., 1983; Shackel, 1983; IBM, 1984). Glare from windows can be alleviated by positioning the VDT away from windows, with the operator's line of sight parallel to the window pane (Stewart, 1980; Benz et al., 1983) (Figure 101). Curtains and blinds can also help control glare from windows. In addition, all light fixtures should be shielded.

To prevent reflections on source documents, plastic jackets and glossy paper should both be avoided (Benz et al., 1983). The document holder should also have a matte finish (Bell Telephone Laboratories, 1983).

Neck strain may also occur due to awkward postures that are adopted to avoid reflections on the keytops. Therefore, all keytop surfaces should

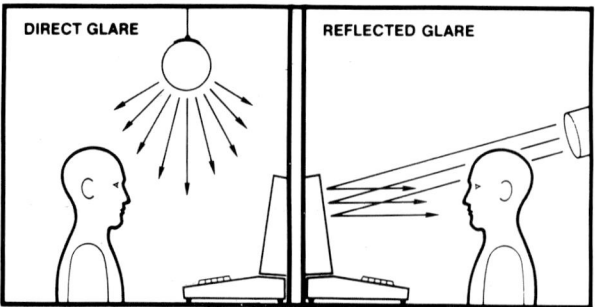

Figure 100. Two sources of visual glare. From Tijerina, L.: *Optimizing the VDT Workstation. Controlling Glare and Postural Problems.* Dublin, OCLC, 1983. Reproduced with permission of OCLC Online Computer Library Center, Inc.

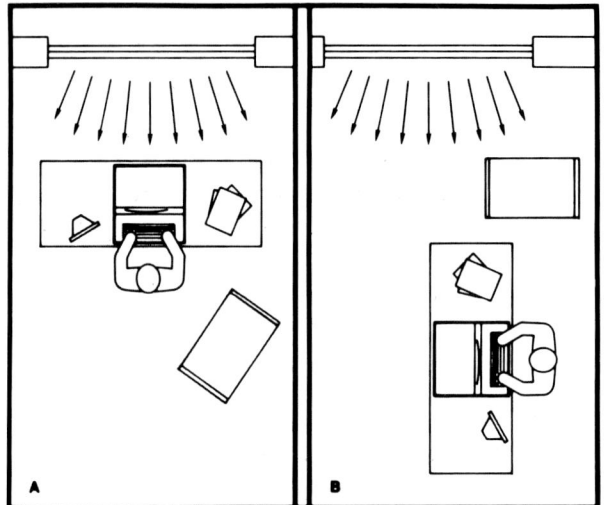

Figure 101. Positioning the VDT workstation to avoid glare from windows.
    A. Incorrect
    B. Correct
From Tijerina, L.: *Optimizing the VDT Workstation. Controlling Glare and Postural Problems.* Dublin, OCLC, 1983. Reproduced with permission of OCLC Online Computer Library Center, Inc.

have a matte rather than a glossy finish (Cakir et al., 1980; Snyder, 1983).

Luminance

Another potential visual problem is a "contrast glare" effect due to an excessive difference in luminance (brightness) between the source document and display screen (Östberg, 1975). As the luminance of a source document is greatest when lying flat on a horizontal desk, another advantage of an inclined document holder will be to reduce the luminance ratio between the screen and the document (Cakir et al., 1980; Stewart, 1980).

The characters on most VDT screens are displayed as a negative image, or light characters on a dark background, which is the opposite of normal printed material. Therefore, a positive image for the VDT screen (dark characters on a light background) has at times been recommended. With a positive image, the contrast in luminance between the screen and source documents will be reduced, thereby improving the adaptation conditions for the eyes, especially with frequent eye movements between the screen and documents (Knave, 1983; Radl, 1983, 1984; Bergqvist, 1984; Sauter et al., 1984). In addition, a positive image can help reduce reflections on the screen, as bright reflections will be less visible on a light background (Sauter et al., 1984).

However, one must be careful in selecting a positive image VDT screen, as there is a greater potential for image instability (flicker) compared to a negative image VDT screen (Bergqvist, 1984; Sauter et al., 1984).

Bifocals

Operators wearing bifocals are very prone to developing neck pain from viewing the display screen. As the bottom lens of bifocal glasses is for reading and viewing nearby objects, the operator wearing bifocals will often end up tilting the head back in order to properly view the screen. This will obviously be a very stressful posture for the neck.

For improved posture and vision, bifocal or trifocal VDT glasses can be prescribed, where one segment of the glasses will be adjusted for comfortable vision at the distance of the display screen (Sauter et al., 1984). For example, with bifocal VDT glasses the upper segment of the glasses can be adjusted for viewing the display screen, and the lower segment can be adjusted for closer work. With trifocal VDT glasses, the middle segment can be adjusted for viewing the display screen, the lower segment for close work, and the upper segment for more distant viewing.

KEYBOARD POSTURE

Proper forearm support at the keyboard is probably the most important factor, and also the most overlooked factor, in regards to musculoskeletal problems involving the neck-shoulder region. First of all, the
gravitational force of unsupported arms will fall upon the neck-shoulder
musculature, particularly the trapezius, resulting in a prolonged static
contraction of these muscles (Andersson and Örtengren, 1974b; Avon
and Schmitt, 1975; Weber et al., 1984; Waris, 1980).

A lack of arm support will also result in a slumped posture of the
upper trunk, as the weight of the unsupported arms will tend to pull the
upper trunk forward and downward, away from the chair's backrest. This
will have an adverse effect on the head and neck posture, resulting in an
increased forward inclination of the head.

Hünting et al. (1981), in their study of VDT workstations, reported a
lower incidence of neck, shoulder and arm pain when the operator's
hands and arms were frequently supported.

Relationship of Keyboard Posture and Operator's Posture

A very confusing issue involves the conflicting recommendations that
are given in various VDT articles, brochures, and books in regards to
proper sitting posture, arm posture, desk height, keyboard slope, etc. at
the VDT workstation. Upon closer examination, it can be realized that a
complex interrelationship exists among the observed VDT workstation
posture and other workstation factors, including: the inclination and
height of the chair backrest, the keyboard slope and height from the
floor, the use of an inclined document holder, the use of forearm-wrist
supports and the size of the supports, and possibly even the arm and
hand lengths of the operator.

One viewpoint in VDT workstation recommendations is based on a
sitting posture with the trunk in a vertical position, the forearms horizontal,
and the upper arms vertical (Figure 102). A chair with a low backrest,
high enough to provide lumbar support, is considered adequate for this
VDT sitting posture (Cakir et al., 1980). To achieve this posture, it is also
stressed that the keyboard should have a minimal thickness, along with
very little slope, usually close to 5 degrees.

After a short time, however, due to the inherent instability of a vertical
trunk posture, the operator may seek further stability from the workstation.

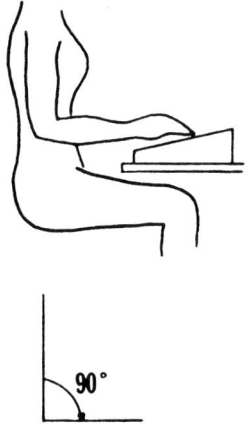

Figure 102. Frequently recommended VDT workstation posture with the trunk and upper arms vertical, and the forearms horizontal. From Arndt, R.: Working posture and musculoskeletal problems of video display terminal operators—review and reappraisal. *American Industrial Hygiene Association Journal, 44:*437–446, 1983. Reprinted with permission by American Industrial Hygiene Association Journal.

A forward leaning posture may be adopted, where the operator can obtain better trunk support by resting the arms on the desk. A very low backrest height and the lack of an inclined document holder will both facilitate a forward leaning posture. There are also certain VDT tasks, such as customer service counter work, where the operator needs to lean forward towards the customer (Launis, 1984; Grandjean, 1984c).

In the vast majority of VDT tasks, however, the operator does not have to lean forward towards a customer. In these job settings, a completely different trunk and arm posture has been frequently observed. Even on chairs not designed for reclined postures, the VDT operators have been observed to lean backwards in their chairs (Grandjean et al., 1983a). In addition to leaning the trunk backwards, commonly observed arm postures do not involve having the upper arms vertical and the forearms horizontal. Instead, the shoulders are usually flexed from 0 to 30 degrees, and the forearms are elevated from 5 to 30 degrees (Arndt, 1982, 1983) (Figure 103).

In field studies by Grandjean et al. (1983a) with an adjustable VDT workstation, the majority of VDT operators at data entry and conversational terminals preferred a backward leaning trunk posture of 10 to 20 degrees from vertical. The "mean body posture" of the VDT operator included a 14 degree trunk inclination, with the forearms elevated 14

Figure 103. Commonly observed VDT arm posture. The shoulders are flexed up to 30 degrees and the forearms are elevated from 5 to 30 degrees. From Arndt, R.: Working posture and musculoskeletal problems of video display terminal operators—review and reappraisal. *American Industrial Hygiene Association Journal, 44:*437–446, 1983. Reprinted with permission by American Industrial Hygiene Association Journal.

degrees, and the shoulders flexed 23 degrees (Figure 104). As Grandjean (1984c) commented, "Many VDT operators in offices disclose postures very similar to those of car drivers. This is understandable: who would like to adopt an upright trunk posture when driving a car for hours?"

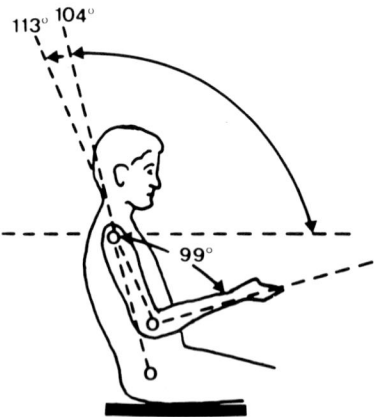

Figure 104. The "mean body posture" at the VDT operators' preferred workstation settings. Based on 236 observations of 59 VDT operators. From Grandjean, E.: Postural problems at office machine work stations. In Grandjean, E. (Ed.): *Ergonomics and Health in Modern Offices.* London, Taylor and Francis, 1984, pp. 445–455. Reproduced with permission of Taylor and Francis Ltd.

According to Grandjean (1984c), the backward leaning trunk posture is the basis for all the other adopted postural elements of the VDT operators, including the flexed shoulders with inclined forearms, and the slightly opened elbow angles beyond 90 degrees. It is therefore important to detail several interrelated workstation features that will help facilitate this backward leaning posture with elevated arms:

a. A backward leaning trunk posture will be facilitated by a chair with a high backrest providing upper back support, along with an adjustable backrest inclination that can be fixed at any angle by the operator. Pressure should be avoided, however, over the outer part of the scapulae and shoulders (Taylor, 1917).

b. A major reason for the elevated forearm posture probably relates to the slope of the keyboard. There will be an optimum efficiency of arm and wrist movements with the forearm angle matching the keyboard angle (Arndt, 1983). Therefore, with greater keyboard slopes, one will probably observe a greater elevation of the forearms.

c. With an inclined document holder, Life and Pheasant (1984) reported a tendency for typists to lean the trunk backwards and to flex the shoulders.

d. Keyboards with forearm-wrist supports will also help facilitate a backward leaning trunk posture (Figure 105). Compared to small supports, large forearm-wrist supports have been found to result in a greater trunk inclination and elevation of the arms, along with greater pressure being exerted on the supports (Nakaseko et al., 1985).

Proper forearm support is essential for maintaining an elevated arm posture. Keeping the upper arms elevated forwards without proper forearm support will produce a high torque about the shoulder joints. The resulting musculoskeletal stress could then only be reduced by leaning the trunk forward from the backrest, thereby reducing the postural torque about the shoulder joints.

Keyboard Slope and Height Relationship

A complex interrelationship probably also involves the preferred keyboard to floor height with the keyboard slope, the distance of the operator from the keyboard, and the length of the operator's arm and hand.

Cushman (1984) and Grandjean et al. (1983a) found that VDT operators preferred a higher keyboard to floor height than the heights often recommended in VDT brochures, which are based on keyboard postures with the upper arms vertical and the forearms horizontal.

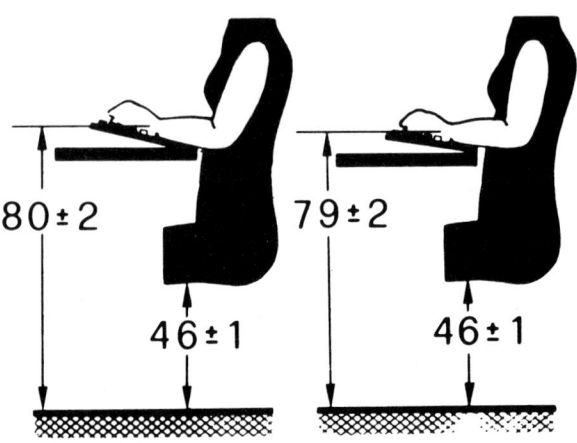

Figure 105.
  Left: An experimental keyboard with a large forearm-wrist support.
  Right: An experimental keyboard with a small forearm-wrist support.
  From Nakaseko, M., Grandjean, E., Hünting, W., and Gierer, R.: Studies on ergonomically designed alphanumeric keyboards. *Human Factors, 27:*175–187, 1985. Copyright 1985, by the Human Factors Society, Inc. and reproduced by permission.

This preference for higher keyboard to floor heights can be explained by the elevated arm postures observed by Grandjean et al. (1983a) and Arndt (1982, 1983) at VDT workstations. Based on the shoulder's arc of motion, as the upper arm is flexed, the elbow will advance forward and be elevated (Figure 106). This will obviously require a greater keyboard to floor height compared to a vertical upper arm posture.

Similar observations were made by Bennett (1928) regarding the writing posture at a desk. He mentioned that at any desk inclination, the plane of the underside of the forearm should coincide with the plane of the desk top.

According to Bennett (1928):

"For any full-sized writing surface in front of the body the elbows are advanced nearly to the desk edge and approximately to the line of the

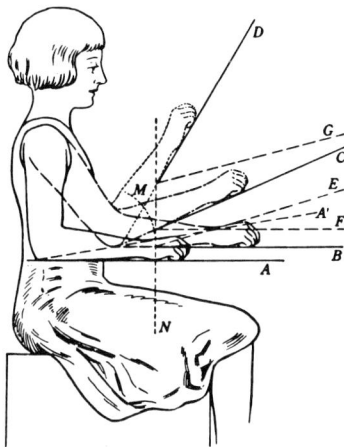

Figure 106. As the upper arm is flexed forward, the elbow will be raised. This will require a higher desk or keyboard to floor height. From Bennett, H.E.: *School Posture and Seating.* Boston, Ginn and Company, 1928.

front of the body. With elbows kept close to the sides, this advance involves a minimum of muscular effort, since the upper arm merely swings forward like a pendulum, practically as it does in walking; but as the elbow advances, the pupil remaining erect, it is elevated. Hence the height of a desk in front of the body should be a few inches greater than the elbow height as measured.

The amount of this increase of height depends on the distance that the elbow is moved forward. It depends also on the length of the upper arm, since a short pendulum is raised more than a long one in the same amount of forward swing."

Bennett (1928) considered the height, slope and horizontal distance of the desk to always be interdependent factors in desk position:

"In general, then, the greater the slope of the desk the higher it should be, and vice versa. The more distant the desk, the more the elbow is advanced and raised, and the higher the desk should be."

A similar relationship probably exists among the keyboard slope, height from the floor, and distance from the operator. For example, a greater slope to the keyboard may require a greater shoulder flexion, as the operator attempts to match the inclination of his forearms to the keyboard slope. The greater corresponding elevation of the elbow would then require a greater keyboard height.

Nakaseko et al. (1985) found that individuals leaned the trunk further backwards with large forearm-wrist supports. This resulted in a greater distance from the shoulder to the keyboard. Therefore, there was an increased shoulder flexion and raising of the elbow, which necessitated a greater keyboard to floor height.

Advantages of an Adjustable Keyboard Slope

Some comfort and productivity studies have shown a preference among VDT operators for keyboard slopes greater than the low sloped keyboards of 5 to 15 degrees that are often recommended in VDT articles and brochures. In a study involving 37 VDT operators, Miller and Suther (1983) found that the overall mean keyboard slope selected was 18 degrees, with the greatest preferred slope being 25 degrees.

Among keyboard slopes of 5 degrees, 12 degrees, and 18 degrees, Emmons and Hirsch (1982) reported that a group of twelve skilled typists showed a preference for the 18 degree keyboard slopes. The keying rates were also higher on the 12 degree and 18 degree slopes compared to the 5 degree slopes.

With the hands in the home row position on a flat keyboard, the distance from the metacarpophalangeal joint to the center of a top row key will be somewhat greater compared to a moderately sloped keyboard (Rupp, 1981). Individuals with short hand lengths may therefore prefer a greater slope to the keyboard, as this may shorten the finger travel distance between keyboard rows (Miller and Suther, 1981, 1983; Suther and McTyre, 1982).

A moderately sloped keyboard will also improve the visibility of the keyboard, especially with a backward leaning trunk posture.

Due to the apparent interrelationship of preferred keyboard slope with sitting posture, forearm-wrist support, keyboard height, and arm and hand length, an adjustable keyboard slope up to approximately 30 degrees would probably be the best solution (Alden et al., 1972). An adjustable inclination to the keyboard supporting surface would help obtain these higher keyboard angles.

Advantages of Forearm-Wrist Support

Neck Stress

A sustained static loading of the trapezius muscle has been found in the keyboard operation of VDTs, typewriters, and calculating machines when the arms are unsupported. This unsupported arm posture is considered to be a major cause of neck pain and headache (Travell, 1967; Waris, 1980; Hünting et al., 1981; Onishi et al., 1982).

Several studies have shown how the upper trapezius muscle load can be greatly reduced with proper arm support (Andersson and Örtengren, 1974b; Avon and Schmitt, 1975; Weber et al., 1984; Mahlamäki et al., 1986).

Lumbar Spinal Stress

Proper arm support can also reduce the loading of the lumbar spine and the lumbar disc pressure, as the weight of the arms will be taken by the arm supports (Occhipinti et al., 1985; Andersson and Örtengren, 1974b).

In addition, with a large forearm-wrist support one will be able to exert greater pressure against the support, which will help extend the upper trunk. This will facilitate a backward leaning trunk posture, with greater support being obtained from the backrest of the chair (Nakaseko et al., 1985). The resulting posture will further help reduce the lumbar disc pressure.

Wrist Stress

Using a thick keyboard without a wrist support can result in a position of extreme wrist extension (Smith, 1984 b; Sauter et al., 1984). This wrist posture will increase the risk of median nerve compression, as the pressure within the carpal tunnel is three times greater with extreme wrist extension (Cailliet, 1975).

A proper forearm-wrist support will help assure that the inclination of the forearm and hand will both follow the inclination of the keyboard.

Alternative Keyboard Designs

The typical keyboard with parallel rows of keys results in an unnatural position of the hands, wrists, and arms when typing. In order to align the fingers at the keys, the keyboard posture is characterized by a marked pronation of the forearms and a marked ulnar abduction of the wrists (Figure 107).

The incidence of hand and forearm impairments (tiredness, cramps, pain) among accounting machine operators was found to be correlated with the degree of ulnar abduction at the wrist (Hünting et al., 1980).

Figure 107. Marked forearm pronation and ulnar abduction of the wrists at the traditional keyboard with parallel rows of keys.

Similar findings were reported by Hünting et al. (1981) regarding the incidence of hand and forearm impairments and the degree of ulnar wrist abduction among VDT operators.

As early as 1926, however, Klockenberg proposed a split keyboard design, with a lateral inclination and horizontal rotation of each keyboard half (Kroemer, 1972) (Figure 108). The advantage of the split keyboard design would be a more natural posture of the hands and forearms at the keyboard, reducing both ulnar abduction at the wrist and forearm pronation. Also reduced would be the excessive shoulder abduction and elevation which the keyboard operator would use to achieve the fully pronated forearm position.

In a more recent study, Zipp et al. (1983) found a significant reduction in EMG activity from the forearm pronators, trapezius, and deltoids with a lateral inclination of the keyboard, which reduced the forearm pronation needed for keying. The horizontal rotation of each keyboard half, which would minimize the ulnar abduction at the wrists, reduced the EMG activity not only from the extensor carpi ulnaris muscle, but also from the deltoids and trapezius.

Therefore, to reduce the marked ulnar abduction at the wrist with traditional keyboard design, Zipp et al. (1983) recommended the split keyboard with the right keyboard half rotated in the horizontal plane counter-clockwise, and the left keyboard half clockwise, each rotation being from 10 degrees to 20 degrees. To reduce forearm pronation, but at the same time not impair visibility of the keyboard, Zipp et al. (1983) recommended a lateral inclination of each keyboard half of 10 degrees to 20 degrees.

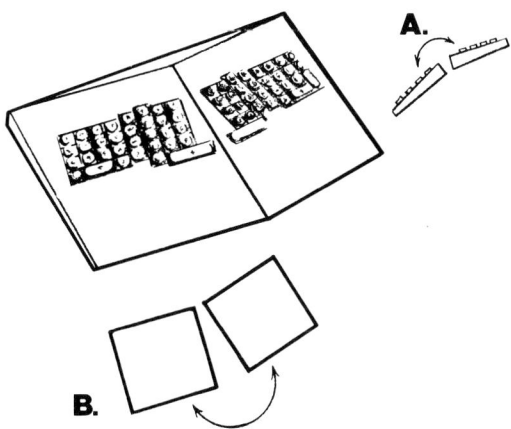

Figure 108. Split keyboard design with a lateral inclination (A) and horizontal rotation (B) of each keyboard half.

In another recent study, Nakaseko et al. (1985) found that 31 trained typists reported less hand and arm pains, along with increased feelings of relaxation, with a split keyboard design, as compared to a traditional keyboard design.

Grandjean et al. (1983b) have suggested that for office jobs involving one-handed operation of traditional keyboards, such as certain VDT data entry tasks and accounting machine operation, the keyboard should be rotated in the horizontal plane to reduce the ulnar abduction at the wrist.

CHAIR STABILIZATION: BACK, BUTTOCKS, FEET

A forward leaning posture at an office workstation with the back unsupported has been correlated with an increased incidence of neck and back pain (Maeda et al., 1980a; Ong et al., 1981; Sauter et al., 1983). With ergonomic improvements at the workstation, however, kyphotic forward bending postures have been found to be almost totally eliminated, as the VDT operators were observed to spend most of the sitting time with the trunk supported by the backrest of the chair (Cantoni et al., 1984).

Proper back support, arm support, and foot support are critical for optimal body stabilization (Darcus and Weddell, 1947; Branton, 1969; Kroemer, 1982; Nakaseko et al., 1985). An increase in operator performance has been observed in several studies after various ergonomic improvements contributing to operator stability, such as non-flexible backrests with lumbar supports, footrests, and the use of inclined arm supports (Rohmert and Luczak, 1978; Ong, 1984; Secrest and Dainoff, 1984).

Proper Backrest Design

As mentioned earlier in this chapter, VDT field studies have revealed that the vast majority of VDT operators prefer a backward leaning trunk posture of 10 degrees to 20 degrees from upright (Grandjean et al., 1983 a). This backward leaning trunk posture is apparently a major factor in the other observed postural adaptations of the operator including the shoulder flexion and elevated forearms.

Therefore, to allow a backward leaning trunk posture with proper spinal support, the operator's chair should not only provide proper pelvic-sacral support, but it should also have a backrest high enough to provide thoracic support. The backrest should also have an adjustable inclination, that can be fixed at any angle preferred by the operator (Grandjean, 1984b; Sauter et al., 1984).

The following factors will all help the operator obtain proper back support and trunk stabilization from a high, inclined backrest:

    a. An inclined seat surface with a non-slip upholstery cover.
    b. Large forearm-wrist supports (Nakaseko et al., 1985).
    c. An adjustable, inclined document holder (Ferguson and Duncan, 1974; Life and Pheasant, 1984).
    d. A detachable keyboard (Sauter et al., 1983).
    e. Proper foot support, with either the feet placed firmly on the floor, or else on a footrest (Coe, 1984).

Footrests

Footrests may be required by smaller VDT operators, either when using chairs with an inadequate seat height adjustment, or when at a VDT table that is too high and non-adjustable.

Important considerations in footrest design for proper body stabilization and comfort are the following (Benz et al., 1983; Cakir et al., 1980; Bell Telephone Laboratories, 1983; Schmidtke, 1984; Rühmann, 1984; Marriott and Stuchly, 1986):

    a. Non-slip upper and lower surfaces are necessary to properly stabilize the feet on the footrest, and the footrest on the floor.

b. The footrest should be adjustable in height and inclination. Fixed footrests also need to be adjustable in horizontal distance from the operator.
c. A fixed footrest should be securely attached to the table or floor. With a moveable footrest, a non-slip floor covering is critical.
d. The footrest should have a large enough surface to allow changes in foot position. This is very critical, as the main disadvantage of a footrest is the restriction of leg movement compared to having the feet firmly and comfortably on the floor (Kotelmann, 1899; Shaw, 1902; Kerr, 1928; Sauter et al., 1984).

REST PAUSES

Besides the need for the operator's chair to provide proper body stabilization, thereby reducing the static loading of postural muscles, adequate rest pauses are also essential to the health and productivity of the operator.

Investigations regarding the effect of rest pauses on productivity have indicated that they tend to increase, rather than decrease productivity (Grandjean, 1980a; Cakir et al., 1980; Floru and Cail, 1985). Rest pauses, however, should not be taken as a recovery period from excess fatigue, but rather prior to the onset of a noticeable degree of fatigue. Frequent pauses taken before the onset of high levels of fatigue are considered more effective than longer, less frequent rest pauses taken after longer periods of continuous work (Cakir et al., 1980; Wright, 1982; Health and Safety Executive, 1983).

According to Wright (1982):

"Rest pauses should not be confused with inactivity or waiting time during VDT use since the operator cannot become disengaged from the work task. Interruptions of unpredictable duration at critical stages of the task while waiting for a machine generated response, may add, rather than reduce, stress. Such a period of apparent inactivity may have no recovery value and may even add to the operator's level of fatigue."

The National Institute for Occupational Safety and Health (1981) recommended the following rest breaks to minimize the visual and muscular problems of VDT operators:

a. A fifteen minute rest break should be taken after two hours of continuous VDT work for operators under moderate visual demands and/or moderate workload.

b. A fifteen minute rest break should be taken after one hour of continuous VDT work for operators under high visual demands, high workload and/or those engaged in repetitive work tasks.

A recent study by Zwahlen et al. (1984) found rest breaks to be highly beneficial in reducing musculoskeletal discomfort with VDT work. However, they also found that two 15 minute rest breaks, in addition to a 45 minute lunch break, were not sufficient in a working day to adequately reduce the musculoskeletal and visual stress in continuous VDT work involving high data acquisition and processing demands.

Chapter Nine

ALTERNATIVE CHAIR DESIGNS

EMPHASIS ON THE BACKREST: MULTIPURPOSE CHAIRS

The Backrest and Stability

As the sitting posture is basically unstable without proper external support, the chair must help fulfill the sitter's need for postural stability (Branton, 1966). Therefore, a good chair is one which "helps the sitter to stabilize his body joints so that he can maintain a comfortable posture" (Oborne, 1982). Stabilization of the pelvis and trunk is critical, as the hip joints are in a mid-position when sitting, and therefore stabilization is not possible through ligamentous restraint (Meyer, 1873).

According to Coe (1979), when one is seated, "the effort is much greater to balance the trunk, especially during arm activity. Unless, therefore, a lower back support is used intelligently for sedentary work, the excessive energy expenditure will fatigue the muscles of the hip and spine. Consequently, comfortable erect postures will be difficult to maintain."

Branton (1966), whose theory of postural instability was discussed in Chapter Three, measured with strain gauges the movement or sway of an individual's center of gravity when sitting with and without back support. In order to prevent the individuals in the experiment from consciously controlling their posture, they wore opaque goggles and their attention was directed to performing a demanding vigilance task.

The four sitting postures investigated by Branton (1966) were:

a. Unsupported sitting in a lordotic upright posture.
b. Unsupported sitting in a slumped, kyphotic sitting posture.
c. Sitting with a high backrest support. A narrow, horizontal backrest was placed just below the scapulae, at an angle of 110 degrees with the seat, resulting in a slumped, reclined posture.
d. Sitting with a low backrest support. A narrow, horizontal backrest was placed against the pelvis, upper sacrum, and lowest lumbar vertebra.

The greatest sway, and therefore instability, was found with the two unsupported sitting postures. The most stable posture, and also the posture found the most comfortable by the test subjects, was with the low backrest support.

Branton (1969) and Schoberth (1969) have both stressed that in order to properly stabilize the pelvis, vertebral column, and trunk, *the upper sacrum and posterior iliac crests must be supported by the backrest.*

Observations and Surveys on the Use of a Backrest

Office Use

Burandt and Grandjean (1963) observed the sitting postures of 246 office employees whose work involved reading, writing, typing, processing of records, and operating calculating machines. Leaning back against the backrest was found in 42 percent of all the observations. This posture was most frequently observed in office workers who complained of back discomfort (45 percent of all observations), neck and shoulder discomfort (49 percent of all observations), and when using a typewriter (64 percent of all observations).

A survey on chairs used by FAA Air Traffic Controllers (Kleeman, 1980) also revealed the importance of the backrest. Among the 1,967 air traffic controllers responding to the survey:

    15.1 percent always use the backrest.
    36.8 percent use the backrest more than half the time.
    30.4 percent use the backrest approximately half the time.
    17.0 percent use the backrest less than half the time.
    0.7 percent never use the backrest.

In regards to the backrest height desired by air traffic controllers, 26.2 percent preferred a backrest just high enough to support the lower back, whereas 63.2 percent desired a backrest height up to shoulder level. This desire for a high backrest was also expressed by office workers (Hünting and Grandjean, 1976).

An important field study by Grandjean et al. (1982a, 1983a, 1984) found that VDT operators prefer a backward leaning trunk posture. According to Grandjean (1984b), "Traditional office chairs with a relatively small back support are not suitable for VDT jobs, since they do not allow relaxation of the full back."

School Use

Observations by Floyd and Ward (1964) on the postural behavior of fifty-five schoolchildren with two different designs of school furniture found that the backrest was used from 50 to 60 percent of the time by the students. Hira (1980), in his observations of university students in India, found that the students sat approximately 40 percent of the time with the trunk supported against the backrest. Bennett (1928) reported that the backrest was used in over 70 percent of the observations made on 4637 elementary and high school students.

Wotzka et al. (1969), in their investigations on the development of an auditorium seat, found that students preferred a high backrest with a pronounced lumbar support. On auditorium seats with a closely adjacent inclined table projecting over the front of the seat, the backrest was used either frequently or always by 90 to 100 percent of the students.

Designs Emphasizing the Need for Back Support

The importance of a well-designed backrest to enable the sitter to obtain full back support when desired is evident in multipurpose chairs. Multipurpose chairs can be used for many different purposes, as they allow forward, upright, and reclining sitting postures (Grandjean et al., 1973). The possible uses for these chairs include in an office with a desk or table, in homes for dining and work areas, and in waiting rooms and meeting rooms (Ridder, 1959; Shackel et al., 1969; Grandjean et al., 1973).

Åkerblom's Chair Design

Åkerblom (1948, 1954) believed there were three basic rest positions for the back:

1. Sitting slumped forward without back support, in a kyphotic, anterior sitting posture. (Of course, other seating authorities have emphasized the potential harm from this sitting posture.)
2. Sitting with only a lumbar support.
3. Sitting with support for both the lumbar and thoracic regions of the back.

With Åkerblom's chair design, the backrest has a forward convexity for lumbar support, and then it slopes backward approximately 25 degrees from vertical. A seat inclination of 5 to 7 degrees enables the sitter to obtain proper support from the backrest without sliding forward on the seat.

Figure 109 gives the basic outline of Åkerblom's chair design, along with the basic resting positions it allows the sitter to assume.

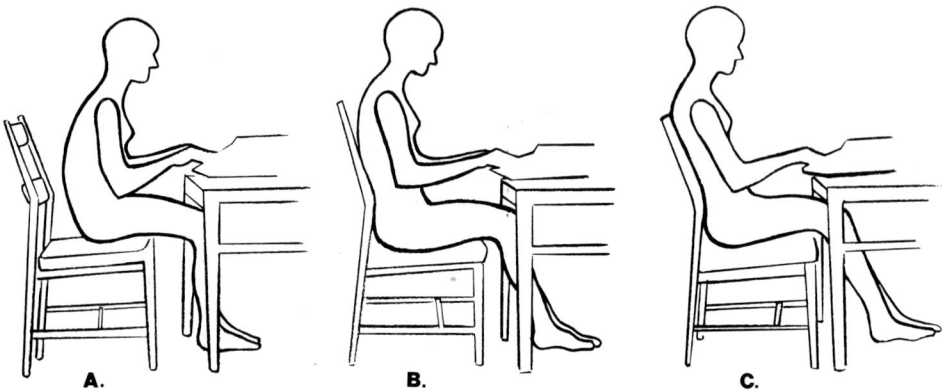

Figure 109. Åkerblom's chair design and three basic resting positions.
   A. Sitting sunken forwards without any external support.
   B. Sitting with a lumbar support.
   C. Sitting with both lumbar and thoracic regions supported.
   Adapted from Åkerblom, B.: *Standing and Sitting Posture.* Stockholm, Nordiska Bok-handeln, 1948. With permission of B. Åkerblom.

Design Recommendations of Grandjean et al.

Design recommendations for a multipurpose chair by Grandjean et al. (1973) were based on the comfort ratings of twelve chairs by fifty men and women. The comfort ratings involved paired comparisons, along with a questionnaire involving body part sensations and ratings of specific chair features.

The multipurpose chair developed, which is considered suitable for both forward and backward leaning postures, has a 7 degree seat inclination along with a slight concavity to the seat surface for the buttocks. The high backrest includes a lumbosacral support, and is inclined 17 degrees from upright (Figure 110).

Grandjean et al. (1973) also recommended an upholstered seat and backrest, with 2 to 4 cm of foam rubber. The upholstery cover should "allow for circulation of air, should not be slippery, and should conserve heat."

A similar design for an office chair, but with an even higher backrest, was more recently recommended by Grandjean (1980a) (Figure 111).

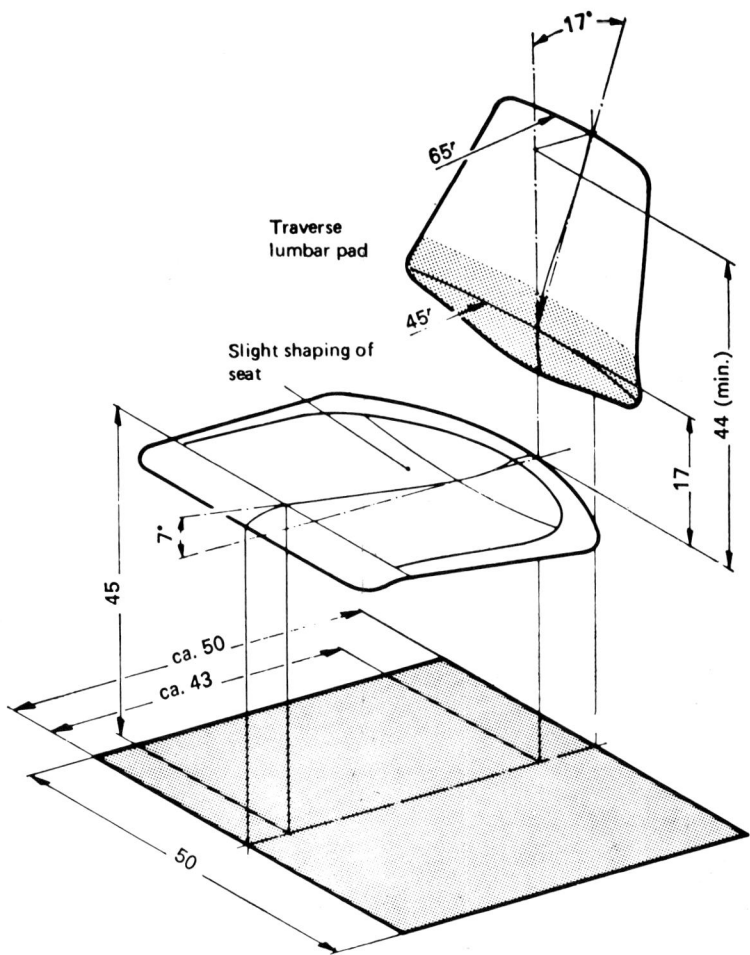

Figure 110. Recommendations for a multipurpose chair by Grandjean et al. (1973). All measurements in cm. From Grandjean, E.: *Ergonomics of the Home.* London, Taylor and Francis, 1973. Reproduced with permission of Taylor and Francis Ltd.

Other Multipurpose Chair Designs and Studies

In Ridder's study (1959) with an adjustable experimental chair, 129 adults selected the design measurements preferred for a chair to be used for dining, writing, and playing table games. Similar to the design recommendations of Grandjean et al. (1973), a high backrest was preferred in order to provide greater back support when leaning back. Also recommended were a contoured seat for the buttocks and a backrest

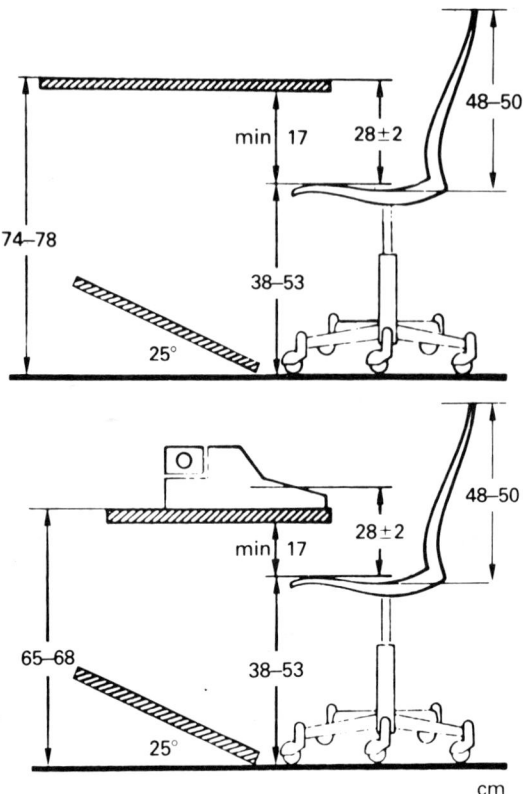

Figure 111. Some important dimensions for an office chair, shown at a standard desk and at a desk for typists. All measurements in cm. From Grandjean, E.: *Fitting the Task to the Man,* 3rd ed. London, Taylor and Francis, 1980. Reproduced with permission of Taylor and Francis Ltd.

inclination of 15 degrees from vertical, along with a horizontal curvature to the backrest (Figure 112).

A study by Shackel et al. (1969) of ten different multipurpose chairs involved the comfort ratings given by three groups of twenty individuals, each involving a different sitting situation. The three different sitting situations were sitting at a desk in an office, eating a meal at a table, and sitting without a desk or table, as in a reception area or meeting room. The chair that received the best comfort ratings, which had a 7 degree seat inclination and a 15 degree backrest inclination from vertical, seemed to be comfortable as it "permits a very restricted range of postures and holds the sitter fairly securely."

Figure 113 gives approximate dimensions for the important features of a multipurpose chair, including seat and backrest inclinations, backrest

Figure 112. Profile of Ridder's (1959) multipurpose seat design. Adapted from Ridder, C.A.: *Basic Design Measurements for Sitting.* Agricultural Experiment Station. University of Arkansas, Fayetteville. Bulletin 616. October 1959.

height, lumbar support placement, and the horizontal contours of the seat and backrest.

As the importance to assume both backward leaning and forward leaning postures has been emphasized with multipurpose chairs, it is of interest to note the Miller chair, described by Bradford and Stone in 1899 (Figure 114). This adjustable chair allowed a change in position from an upright posture (such as for writing) to a reclining position (for reading) with both lumbar and thoracic support.

EMPHASIS ON THE FORWARD LEANING POSTURE

A major criticism of the multipurpose chair design has been its use in work situations where the sitter is forced to maintain a forward leaning posture. Examples of such work situations include assembly work and work at VDT workstations for customer service, where the sitter is required to lean forward towards the customer (Launis, 1984; Grandjean, 1984c; Schmidtke, 1984; Brunswic, 1984b).

The criticism is that with the forward leaning posture, the sitter will lose the spinal support from the backrest of the multipurpose chair. In addition, excessive hip flexion will be required to obtain the forward leaning posture on the inclined seat, and this will also force the lumbar spine into kyphosis.

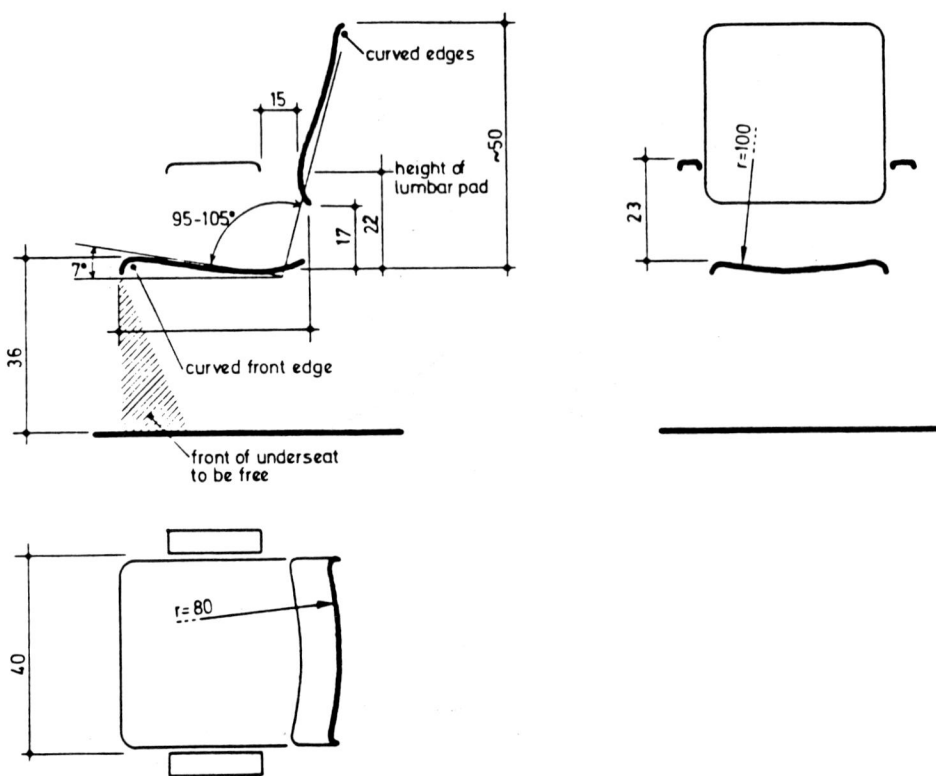

Figure 113. Dimensions for a multipurpose chair. All measurements in cm. Note: These are only approximate dimensions, intended to give a rough idea. From Asatekin, M.: Postural and physiological criteria for seating — a review. *M.E.T.U. Journal of the Faculty of Architecture*, 1:55–83, Spring 1975. Reproduced with permission.

In these situations, where there will be little, if any, use of the backrest, a horizontal seat has been recommended (Kroemer and Robinette, 1968; Hennessey, 1984; Schmidtke, 1984). A horizontal seat will minimize the hip flexion needed to obtain a forward leaning posture, and thereby prevent excessive lumbar kyphosis.

Kroemer and Robinette (1968) felt that the best overall compromise was a horizontal seat with a slightly concave curvature in the center. This design would prevent the buttocks from sliding forward when using the backrest, and it would also minimize the hip flexion compared to an inclined seat when assuming a forward leaning posture. Krjukova (1977) recommended that seats for keyboard operators should be inclined slightly backwards, while the seats for light material assembly lines should be horizontal.

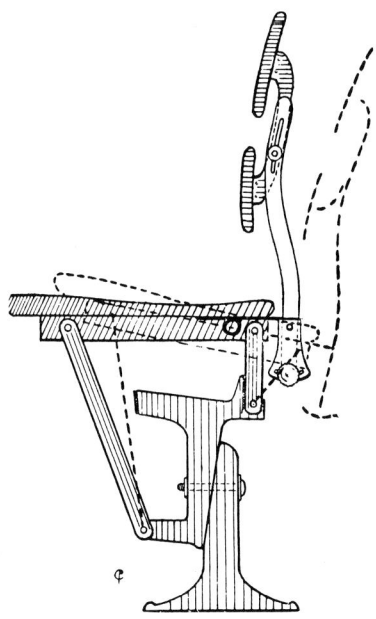

Figure 114. The Miller Chair. As the backrest moves from a reclining to an upright position, the seat loses its slope and moves forward, closer to the desk. From Bradford, E.H., and Stone, J.S.: The seating of school children. *Transactions of the American Orthopaedic Association, 12:*170–183, 1899.

The Forward Sloping Seat

Other individuals have gone beyond the recommendation of a horizontal seat in forward leaning work situations to advocate a forward sloping or declined seat surface (Schlegel, 1956; Mandal, 1976, 1981) (Figure 115). Over the past decade, Mandal (1976, 1981) has been the main advocate of a tilting chair, where the seat can slope forward from 15 to 20 degrees.

According to Mandal, most work takes place in a forward leaning posture without use of a backrest and with excessive flexion of the lumbar spine (reading, writing, assembling, precision work). The forward tilting seat will minimize the hip flexion and lumbar spinal flexion needed to lean over a desk or table, and therefore help preserve the lumbar lordosis. Compared to an inclined seat, Mandal (1981) reported less stretching of the back muscles and a more optimal pressure distribu-

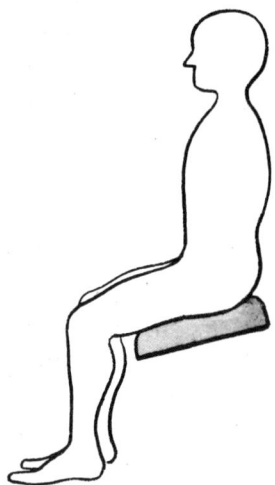

Figure 115. A forward sloping or declined seat.

tion on the seat when a forward sloping seat was used at a horizontal desk.

Criticism of the Forward Sloping Seat

Several individuals have taken issue with the use of forward sloping seats (Kroemer and Robinette, 1968; Branton, 1970; Ward, 1976; Oborne, 1982; Corlett and Eklund, 1984).

Kroemer and Robinette (1968) mentioned that a forward slope to the seat surface would cause the body to slide forward. Increased leg muscle activity would then be necessary to counteract this forward thrust, resulting in increased discomfort and fatigue.

According to Branton (1970), "One cannot be well supported in a seat that slopes forward because then a constant effort is needed to keep from sliding off."

Similar comments were made by Oborne (1982), who referred to the tendency of a forward sloping seat to "destabilize the body and increase its tendency to slip forward. In addition, the supporting advantages of the backrest will be less apparent."

Corlett and Eklund (1984) commented on the forward sloping seat as follows:

> "The negative aspect of this form of seat is that the sitter is intrinsically unstable, the resultant force from the seat having a horizontal component which must be countered by a force at the feet or by friction

between skin, clothes and seat. The former requires muscular activity and the latter introduces an additional source of discomfort, arising from the drag on clothing as a result of the relative movement between the sitter and the seat. The provision of a backrest, particularly one with any degree of backward slope, must add another component of forward thrust, pushing the sitter off the seat and requiring further compensatory forces from the feet."

Among concerns expressed by Ward (1976) regarding the forward tilting seat were that with the lumbar spine unsupported, there would be unnecessary static contractions of the trunk musculature. Ward also questioned the ability to make postural changes when using the forward tilting seat at work: "If the sitter is tilted forward he must brace himself against the floor and/or desk in order to counteract the forward displacement of his centre of gravity, and will have few opportunities to change this posture."

Studies Involving the Forward Sloping Seat

With x-rays, Burandt (1969) analyzed the pelvic posture of twenty-one female switchboard operators on seats without backrests which had either a forward slope (decline) of 6 degrees or a backward slope (incline) of 6 degrees. A more upright posture of the pelvis was found on the inclined seat, whereas the pelvis was tilted backward much more on the declined seat.

(Bennett (1928) had earlier expressed the view that a more erect posture would result with an inclined seat. He felt that the tendency for the ischials to slide backward rather than forward on an inclined seat would help facilitate a flexion at the hips, rather than the lumbar spine, as one leans forward over a desk.)

Jürgens (1969) compared the pressure distribution on the seat, backrest, and feet with two different chair designs. One chair design, modelled on Åkerblom's multipurpose chair design (Figure 109), had an inclined seat. The other chair had a seat surface of which the posterior one-third was horizontal, and the anterior two-thirds sloped forward. In both forward and backward leaning postures, the chair with the forward sloping seat always resulted in greater pressure on the feet and less pressure against the backrest compared to the inclined seat design.

Bendix and Biering-Sørensen (1983) analyzed the posture of the trunk on four different seats without a backrest: a horizontal seat and seats with forward slopes of 5, 10, and 15 degrees. The ten healthy adults in the study sat for one hour with the elbows placed on a horizontal table, reading a horizontally placed text. The spinal curves were measured at six intervals over the hour.

The results showed an increase in the lumbar angle with an increase in the forward slope of the seat. There was an average increase of 1.4 degrees in the lumbar angle for every 5 degree increase in the forward slope of the seat.

Although the lumbar spine moved toward lordosis with an increase in the forward slope of the seat, the major adaptation to an increased forward slope did not take place in the spine. For example, with an increase in forward slope of the seat from 0 degrees to 15 degrees, the lumbar angle showed an increase of only 4 degrees.

In a supplementary study by the authors (Bendix and Biering-Sørensen, 1983), when the seat and the pelvis were both tilted forward 15 degrees, the lumbar angle increased approximately 13 degrees. However, when the subjects assumed their comfortable posture on the 15 degree forward sloping seats, the lumbar angle increased by only 4 degrees, rather than 13 degrees. Therefore, only approximately one-third (4 degrees out of 13 degrees) of the body adaptation to the 15 degree forward sloping seat took place in the lumbar spine. The other two-thirds of the body adaptation to the forward sloping seat involved an extension of the hip joints.

A slight tendency was also found in the main study for an increase to occur in the forward inclination of the trunk over the one hour of sitting. This was particularly noticed with the 15 degree forward sloping seat, but not with the horizontal seat.

A comfort evaluation by the subjects after one hour of sitting showed a preference for the 5 degree forward sloping seat and the horizontal seat. The worst comfort rating was for the 15 degree forward sloping seat.

Bendix et al. (1985a) investigated the effects of three adjustments of an office chair on foot swelling, backrest pressure, and lumbar muscle activity; on healthy females while doing work at a horizontal desk (writing, reading, sorting papers, etc.) and while typing. The three different chair adjustments in the study were: a fixed forward seat slope (decline) of 10 degrees, a fixed seat inclination of 5 degrees, and a freely tiltable seat from an 8 degree incline to a 19.5 degree decline.

The results showed a tendency for less foot swelling with the 5 degree seat inclination compared to the declined and freely tiltable seats. No statistically significant difference for the EMG load levels of the lumbar erector spinae was found with the three different seats.

The pressure against the backrest was found to be twice as high with the inclined seat compared to the other two seats (Figure 116). In addition, backrest pressure was found to be approximately 50 percent greater with typing compared to desk work. This greater backrest pressure with typing could probably be explained by the elevation and inclination of the manuscript which induced a more upright posture.

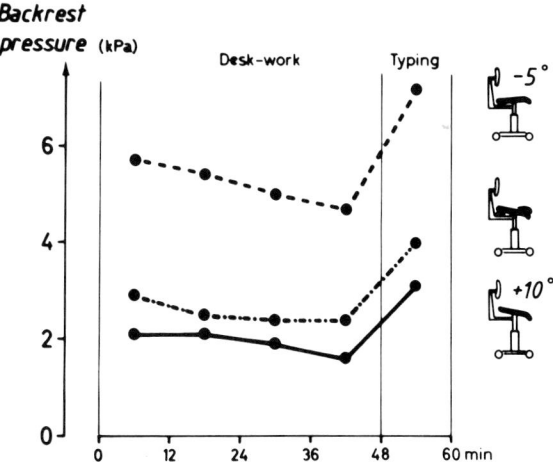

Figure 116. Mean backrest pressure with three different chair designs, for both desk work and typing. (From a laboratory study involving nine healthy females.) From Bendix, T., Winkel, J., and Jessen, F.: Comparison of office chairs with fixed forwards or backwards inclining, or tiltable seats. *European Journal of Applied Physiology,* 54:378–385, 1985. Reproduced with permission of Springer-Verlag.

Although there was a significant preference for the freely tiltable seat over the other two seats in a one hour laboratory study, there was a statistically very insignificant preference for the freely tiltable seat in the eight hour field studies done at the subjects own workplaces. According to Bendix et al. (1985a), this "might reflect that some people get more tired from long-term sitting on a tiltable seat, compared to the other seats, but prefer it for shorter periods."

Kneeling Chairs

A kneeling chair is a unique design of a forward sloping seat without a backrest. In order to keep the buttocks from sliding forward on the seat, the sitter's legs are supported just below the knees by a large padded support. As a result, the sitter's lower legs are tucked under the seat (Figure 117). This sitting posture results in a more open thigh to trunk angle than when using more conventional chairs, along with a greater degree of knee flexion.

Figure 117. Kneeling chair.

Diffrient (1984) commented that the kneeling chair would most benefit those individuals who wanted to work in a forward leaning posture over a desk. Among the disadvantages of this design, Diffrient mentioned that without a backrest one is unable to assume a backward leaning posture and obtain back support. Therefore, the sitter is limited to forward leaning and upright postures.

Diffrient (1984) also mentioned the potential problems of constant pressure on the shins, and the fact that the sitter's feet are in a cramped position. Thompson (1985) also referred to the potential problem with pressure on the knees, along with the awkward positioning of the toes. The possibility of increased discomfort in the knees and lower legs suggested by Diffrient (1984) and Thompson (1985) was also noted in the comfort ratings of university students after two hours of sitting in a kneeling chair (Porter and Davis, 1983).

The only extensive study of the kneeling chair reported in the literature to date is by Drury and Francher (1985). This study involved the comfort ratings for a kneeling chair with a 15 degree forward slope to the seat. The kneeling chair was evaluated over a 2½ hour session by a "typing group" consisting of six secretaries, and a "terminal-using group" consisting of six university students who frequently used computer terminals. Five one-half hour training periods were previously given to all the subjects in order to adjust to the kneeling chair.

The comfort ratings showed that the greatest discomfort with the kneeling chair involved the legs and knees, and to a lesser extent the back and buttocks. Overall greater discomfort was reported by the terminal-using group, whom unlike the typists remained seated for the entire 2½ hour period. Major problems with the kneeling chair encountered by the typing group involved getting into and out of the chair, and problems with swivelling the chair.

According to Drury and Francher (1985):

> "Despite the training given in use of this chair, the overall comfort was not particularly good. Results were worse in overall magnitude than the earlier prototype conventional chair tested, and discomfort increased with time-on-task rather than remaining level.
>
> Body parts affected by the novel chair were primarily the legs, particularly knees and shins. Pressure on the shins, particularly from female subjects wearing high boots or skirts which ended at shin-pad level, was perhaps the most vociferous complaint. Knee discomfort, presumably from the acute knee angle, was noticeable. For this increased leg discomfort there was little or no corresponding decrease in back discomfort. Although subjects had the theory of the chair explained to them during training and tried to sit with a lordotic spine, they often slumped forward to give a kyphotic curve instead.
>
> In summary, the authors have distinct reservations about recommending the chair tested for either prolonged industrial seating or for jobs requiring frequent chair entry and egress."

A more recent study by Lander et al. (1987) compared a kneeling chair to a conventional chair design with a slight seat inclination and a ten degree backrest inclination. The study involved twenty healthy subjects viewing an educational videotape for thirty minutes.

Paraspinous muscle electromyography initially demonstrated lower cervical and lumbar muscle activity when sitting on the kneeling chair. However, a gradual increase in cervical and lumbar muscle activity occurred with the kneeling chair over the thirty minute period, so that the final EMG levels were higher with the kneeling chair compared to the conventional chair.

Comfort ratings revealed a slight preference for the conventional chair regarding both low back comfort and overall comfort. Comments contributing to a lower comfort rating for the kneeling chair included the lack of lower back support, and the inability to easily change position.

According to Lander et al. (1987), it appeared unlikely that the kneeling chair would be of "significant benefit in the prevention of low back or neck pain for the majority of individuals who are forced to sit for prolonged periods."

Chairs for Supported-Standing

Chairs for supported-standing, or sit-stand chairs, are mainly designed for high workplaces with inadequate leg room (Figures 118 & 119). Examples of such work situations would include assembly work on large vertical frames and certain surgical procedures (Bendix et al., 1985b; Corlett et al., 1983). (The idea of a sit-stand chair has been around for some time. Figure 120 illustrates Schindler's (1890) design of a sit-stand chair for the schoolroom.)

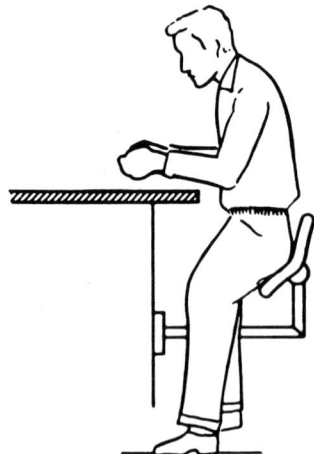

Figure 118. A chair for supported-standing. From Roebuck, J.A., Kroemer, K.H.E., and Thomson, W.G.: *Engineering Anthropometry Methods.* New York, Wiley-Interscience, 1975. Reproduced with permission of J.A. Roebuck, Jr. and the American Industrial Hygiene Association Journal.

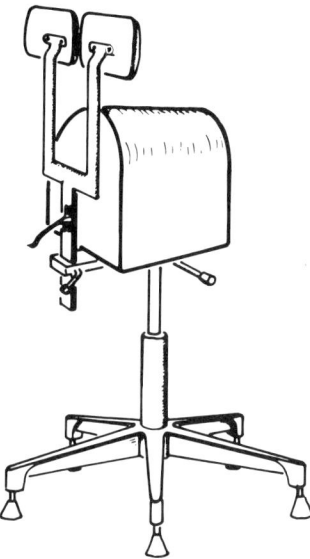

Figure 119. A saddle chair, designed for supported-standing. From Bendix, T., Krohn, L., Jessen, F., and Aarås, A.: Trunk posture and trapezius muscle load while working in standing, supported-standing, and sitting positions. *Spine, 10:*433–439, 1985. Reproduced with permission of Harper and Row, Publishers, Inc.

Bendix et al. (1985b) analyzed the posture of twenty-four healthy subjects performing a simulated repetitive assembly task at a workplace with poor leg room. The postures studied were standing, sitting on a conventional low office chair, and sitting on a saddle chair that allowed a supported-standing posture with the thighs sloped forward 45 degrees (Figure 121).

Results showed that from standing to supported-standing the lumbar spine moved toward kyphosis, although not as much kyphosis as with the low office chair. However, the forward inclination of the entire trunk that occurred with the low office chair did not occur with the change from standing to supported-standing. The authors concluded that "if leg space is poor, variation between supported-standing and standing should be encouraged, and an ordinary office chair avoided."

The major problem with sitting on a low chair where the workplace has poor leg room is that the individual will be further away from his work. This will require greater forward leaning of the trunk with potentially greater spinal flexion. A supported-standing posture will allow closer access to the work, along with the other benefits of sitting over

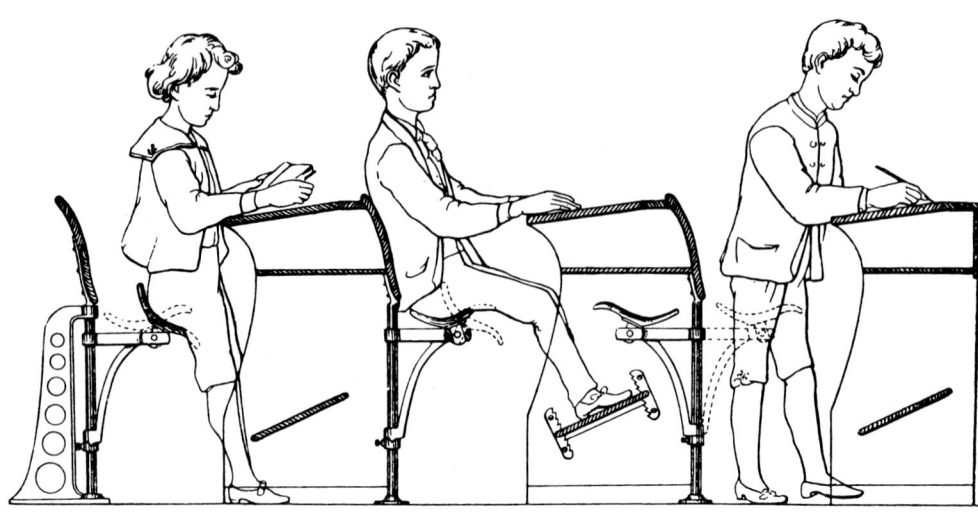

Figure 120. Schindler's (1890) chair design for the schoolroom could be adjusted for supported-standing, sitting, and standing. From Burgerstein, L., and Netolitzky, A.: *Handbuch der Schulhygiene.* Jena, Verlag Von Gustav Fischer, 1895. Reproduced with permission of Gustav Fischer Verlag.

Figure 121. The three postures studied by Bendix et al.: standing, supported-standing on a saddle chair, and sitting on a conventional low office chair. From Bendix, T., Krohn, L., Jessen, F., and Aarås, A.: Trunk posture and trapezius muscle load while working in standing, supported-standing, and sitting positions. *Spine, 10:*433–439, 1985. Reproduced with permission of Harper and Row, Publishers, Inc.

standing such as improved precision and stability, and a decreased energy expenditure (Bendix et al., 1985b). Compared to a chair with a horizontal seat and a low backrest, Eklund and Corlett (1985) found a sit-stand seat to decrease the rate of disc compression (shrinkage) in a light assembly task with restricted knee space.

Recently, a sit-stand posture chair was developed for use by surgeons during microsurgical procedures (Congleton et al., 1985). In addition to allowing a forward sloping of the thighs, the chair also incorporates a saddle seat design to prevent the surgeon from sliding forward on the seat. This sit-stand posture chair was used by thirteen surgeons during actual surgical procedures, and it received much higher comfort ratings compared to a commonly used surgical chair.

OVERLOOKED FACTORS INFLUENCING SEATED WORK POSTURES

As many alternative chair designs such as forward sloping seats and kneeling chairs have been advocated in an attempt to minimize the lumbar flexion with forward leaning postures, it is important to first analyze why people have a tendency to assume a slumped, kyphotic sitting posture with prolonged sitting at a table or desk.

Bancroft (1913) referred to the tendency for the shoulders to be drawn forward in all work situations where the hands and arms are used in front of the body, such as with desk work. In addition, the forward inclination of the head that will occur at a horizontal desk in order to improve one's visual angle, will also have a tendency to throw the body out of erect balance, and into a slumped posture. Therefore, the head, neck, shoulders, and upper back will all have a tendency to contribute to a kyphotic posture of the lumbar spine. This is especially true if the individual cannot obtain support from a properly contoured backrest. The excessive energy expenditure required to balance the trunk in an erect posture without proper back support will also favor a slumped, kyphotic sitting posture (Staffel, 1884; Coe, 1979).

Branton's work (1966) with strain gauges showed that the individual's center of gravity tends to drift slowly forward in a slumped, kyphotic anterior sitting posture. This would relate to Fahrner's comment (1865) that the trunk will drop forward by a series of jerks until its weight is supported by the forearms or elbows on the desk or table.

The greater the forward slope with a declined seat or kneeling chair, the greater will be the forward displacement of the sitter's center of gravity, and the greater will be the trunk instability. Therefore, there will be a greater tendency to slump forward and seek stabilization from the desk or table (a closed-chain position).

The poor sitting habits and incorrect movement patterns of most individuals will also result in excessive thoracolumbar flexion when leaning over a desk (Mosher, 1914, 1919; Zacharkow, 1984b).

The Chair—Desk Relationship

In addition to all these factors predisposing the individual to assume a slumped sitting posture, one of the most important to consider is the relationship of the chair to the desk. This relationship is frequently ignored, and it is often taken for granted that a forward leaning position is the only postural alternative for all desk activities.

As discussed in Chapter Six, critical factors to consider in the chair-desk relationship are the height of the desk, the inclination of the desk, and the horizontal distance from the edge of the desk to the seat (Hartwell, 1895; Scudder, 1892; Shaw, 1902; Bennett, 1928; Karvonen et al., 1962). The sitter will be forced to seek support from the desk and assume a slumped, forward leaning posture for one or several of the following reasons:

   a. The desk height is too low.
   b. The desk is not inclined, but horizontal.
   c. There is too great a distance between the desk and chair.

A proper chair-desk relationship will enable the sitter to obtain proper pelvic-sacral support and lower thoracic support from a well-designed backrest, along with proper arm support from the desk. As discussed in Chapter Six on School Seating, many seating authorities have advocated reading in a slightly reclined sitting posture. Lorenz (1888) also advocated writing in a slightly reclined posture while obtaining back support from a chair with a high backrest and an inclined seat. Schüldt et al. (1986) recommended sitting with the thoracolumbar spine slightly reclined and the cervical spine vertical, as this posture kept the static activity of several neck and shoulder muscles at a low level.

The proper adjustments in height, inclination, and horizontal distance are also critical at VDT workstations for the keyboard and its supporting surface, the source documents, the writing surface, and the display screen, in order to obtain proper back support from the chair (Kroemer, 1982).

Posture Studies with an Inclined Desk

Bendix and Hagberg (1984) studied the sitting posture of ten subjects while reading at desk inclinations of 0 degrees, 22 degrees, and 45 degrees. They found that as the desk inclination increased, both the cervical and lumbar spine became more extended, and the position of the head and trunk changed toward a more upright posture. This positive effect on lumbar spinal posture with an inclined desk will be greater than the effect from using a forward sloping seat (Bendix, 1984). In addition, unlike a forward sloping seat, the inclined desk will also improve the posture of the cervical spine (Bendix, 1984).

Eastman and Kamon (1976) also reported more erect postures at inclined desks as opposed to horizontal desks during reading and writing tasks, along with a reduction in fatigue and discomfort at the inclined desks.

Less and Eickelberg (1976) compared the force changes at the neck for seventeen subjects performing a simulated assembly task at either a horizontal work surface or a work surface with an 18 degree inclination. The greater cervical spinal flexion at the horizontal work surface resulted in greater neck muscle tension forces and cervical compression forces compared to the inclined work surface.

Interestingly, the ergonomic adaptation that was considered the most significant among a group of Norwegian factory workers in reducing the musculoskeletal sick leave at their cable making and wire assembly plant was the ability to adjust the height and inclination of the work table (Westgaard and Aaräs, 1985).

Chapter Ten

SEATING FOR THE ELDERLY: EASY CHAIRS

The percentage of the United States population age sixty-five and over is projected to increase from 11.4 percent in 1981 to 13.1 percent in the year 2000 and 21.7 percent in the year 2050 (Bureau of the Census, 1982). Within the age group sixty-five and over, the proportion of individuals who are sixty-five to sixty-nine years old is getting smaller, while the proportion of individuals age seventy-five and older is increasing. In 1970, the proportion of the elderly who were age seventy-five and older was 38 percent. By the year 2000, it is projected that approximately 44 percent of the population age sixty-five and over will be age seventy-five and older (Kovar, 1977).

In addition, the percentage of aged persons in nursing homes increases with age, with 10 percent of all individuals age seventy-five and over, and 22 percent of all individuals age eighty-five and over being in nursing homes (Pearson and Wetle, 1981).

As the elderly will spend a larger portion of the day sitting compared to younger individuals, and since many nursing home residents are chairfast (Kovar, 1977; Wells, 1980), these statistics reflect the need for well-designed easy chairs for the geriatric population.

CRITIQUES OF GERIATRIC CHAIRS

Easy chairs should be selected so that the individual can "sit in comfort and get in and out with ease. Failure to do this has implications not only for comfort and ease of use, but also for the health, independence and social well-being of the user. Difficulty in getting out of a chair may contribute to general loss of mobility and social isolation" (Institute for Consumer Ergonomics, 1983a).

In a survey of geriatric ward furniture, Wells (1980) inspected 749 chairs. Approximately one-third of the chairs were considered inadequate because of low seat heights, low back heights, instability, and extremely uncomfortable seats. Shipley (1980) observed that many geriat-

223

ric chairs in nursing homes had seat heights and back heights that were too low, along with the backrest being too upright.

Snyder and Ostrander (1972), in their study of six nursing homes, found the following shortcomings in geriatric chairs:

1. Little variety of geriatric chairs was provided. A variety of chair seat heights, seat depths, and arm heights was considered necessary for the diversity of residents.
2. Although plastic and vinyl seat covering serves a function for the incontinent individual, it becomes too hot and is slippery.
3. Chairs with greater backrest inclinations are more difficult to get out of.
4. Cushioning was found to be insufficient to provide comfortable seating when occupying a chair for several hours.
5. Chair arms were often too high or too narrow for the individual to obtain enough leverage to lift himself out of the seat.
6. The seat height was often too high, making it difficult for the individual to place his feet firmly on the floor. Footstools were rarely available.
7. The width of the chair was often too narrow for large or heavy individuals to be comfortable.

Survey of Users

Munton et al. (1981) conducted a survey among the arthritic and elderly regarding the individual's easy chair at home. Most of the 379 people in the survey were in the age group fifty to eighty, and 42 percent of the individuals could not walk unaided.

Some results of the survey are as follows:

a. Forty-two percent of the individuals had some degree of difficulty in rising from their easy chair, with 18 percent of this group experiencing great difficulty in rising or not being able to rise unaided.
b. Thirty-four percent of the individuals felt that the seat height was too low.
c. Seventy-five percent of the individuals did not have room to tuck their feet under the chair when getting up.
d. Thirty-four percent of the individuals had easy chairs that caused pain or discomfort with sitting. The location of the pain caused by the easy chair was as follows:

Neck	20.1 percent
Shoulders	16.4 percent
Back	20.3 percent
Lower Back	23.2 percent

Buttocks   10.3 percent
Thighs   20.3 percent
Other   10.8 percent

e. Sixty-three percent of the individuals used extra cushions in their chair. Sixty percent of these individuals used the cushions to support the back, 17 percent used the extra cushion to sit on, and 23 percent of these individuals used the extra cushions both to sit on and for back support.

The ten factors considered to be most important by easy chair users in this survey were as follows:

Order of Rank	Chair Feature
1	Easy to get out of
2	Comfort
3	High Seat
4	Easy to get into
5	Easy to move
6	Fireproof
7	Angle of backrest
8	Stability
9	Castors fitted
10	Footrest

RISING FROM AN EASY CHAIR

"The less mobile a person is the more important his easy chair becomes to him. Such mobility problems may include difficulties in rising. Thus it is important to have a chair designed for easy rising, so as to encourage the retention of what mobility a person might have.

The ability to get out of the chair is one of the major factors governing independence for the elderly and the arthritic. Difficulties in rising may have a profound impact on their lives—distress caused by resulting pain, and low self-esteem and frustration at their inability to perform this apparently simple act may discourage further attempts. A more inactive life and heavy reliance on others for day-to-day care may result in a diminution in their quality of life" (Munton et al., 1981).

The Home Accident Surveillance System of the United Kingdom recorded 494 accidents for the years 1979 and 1980 that were associated with easy chair use at home. Sixty-six percent of these accidents were associated with getting out of the chair. Approximately ten times as many accidents were associated with rising from an easy chair compared to getting into an easy chair (Institute for Consumer Ergonomics, 1983f).

The breakdown of these 494 accidents involving easy chairs was as follows:

326   (66 percent)   Rising from the chair
 35   (7 percent)   Getting into the chair
 63   (13 percent)   Falling out of the chair
 65   (13 percent)   Involved tripping
  5   (1 percent)   Involved the chair tipping

Ellis et al. (1979) determined the knee joint forces involved in rising from a normal chair without the aid of armrests. They found that the tibiofemoral force could rise to approximately seven times body weight at about the time the individual loses contact with the chair when rising. The patellofemoral force was found to rise to between two and six times body weight.

The obvious hazards for the geriatric population associated with rising from an easy chair, along with the decrease in muscle strength, and increase in joint pain and stiffness associated with this age group, warrant special attention to the specific chair features that can facilitate rising.

Effect of Chair Features on Rising

Open Space Beneath Front of Seat

An open space is critical between the floor and the front of the easy chair so that the feet can be placed underneath the seat when rising. This will facilitate rising since the individual's center of gravity can be more easily brought over the feet (Keegan, 1953; Damon et al., 1966; Norton et al., 1975; Munton et al., 1981; Wheeler et al., 1985) (Figure 122).

Seat Height

The need for a high seat to aid rising conflicts with the need for a seat low enough to permit a wide range of knee angles and leg positions when sitting, and thereby provide more comfort and relaxation (Norton et al., 1975). For example, one of the preferred relaxing postures in an easy chair that a low seat height helps facilitate is sitting with the legs stretched forwards (Oborne, 1982).

If independent rising is given priority, a higher seat height should be chosen for the individual, and a footstool should then be provided if necessary (Institute for Consumer Ergonomics, 1983g). A higher seat

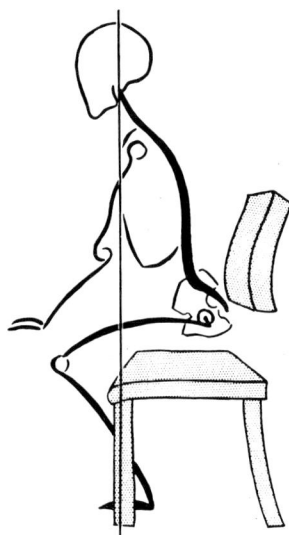

Figure 122. An open space between the floor and the front of the easy chair will facilitate rising. From Keegan, J.J.: Alterations of the lumbar curve related to posture and seating. *The Journal of Bone and Joint Surgery, 35-A:*589–603, 1953. Reproduced with permission of The Journal of Bone and Joint Surgery.

height will not only help decrease the stress in the muscles and joints of the hips and knees, but it will also reduce the range of motion needed at these joints for rising (Burdett et al., 1985).

The critical seat height measurement is the loaded seat height. A seat height may appear adequate to facilitate rising, but with very soft seat upholstery, or with hammocking of the seat suspension, the seat height may actually be too low when loaded with the sitter's body weight (Institute for Consumer Ergonomics, 1983e, 1983g).

Seat Depth

The deeper the seat, the greater the effort required to move the body forward to the front edge of the seat in preparation for rising (Norton et al., 1975; Munton et al., 1981).

Also, excessive seat depth will prevent the individual from receiving proper support from the backrest, especially the pelvic-sacral support.

Backrest Inclination

An excessive backrest inclination can make leaning forward prior to rising from the chair more difficult. In addition, an excessive backrest inclination may result in neck strain, as extreme lower cervical and upper thoracic flexion will be required to bring the head forward and hold the gaze horizontal (Olsen, 1965; Institute for Consumer Ergonomics, 1983a; Harms-Ringdahl et al., 1986).

Seat Inclination

With an inclined backrest being essential on easy chairs, an inclined seat is necessary to help prevent the individual from sliding forward on the seat. However, too large a seat inclination will require greater effort to get to the edge of the seat in preparation for rising (Institute for Consumer Ergonomics, 1983a). An increased forward lean of the trunk may then be necessary to initiate rising (Wheeler et al., 1985).

Seat Cushion

A firm seat cushion is essential to aid in the upward thrust of the body when rising. A thick, soft cushion can make rising difficult, as it will reduce the effective seat height when compressed by the sitter (Laging, 1966; Norton et al., 1975; Finlay, 1981; Munton et al., 1981; Finlay et al., 1983; Clark and Faletti, 1985).

In addition, as Kroemer (1971) emphasized, "the body often 'floats' on soft upholstery, and the posture must be stabilized by muscle contraction."

Armrests

Armrests can cause a considerable reduction in all the knee forces when rising: the tibiofemoral force, the patellofemoral force, and the quadriceps muscle force (Seedhom and Terayama, 1976).

Finlay et al. (1983) have emphasized the critical interaction between seat height and arm height. In regards to facilitating rising, they found that individuals "did not benefit from optimal arm height if seat height was low, and vice versa."

The conflict regarding armrest design for easy chairs is that the most suitable armrest height for comfort when the individual is seated is different from the optimal height for getting into and out of the chair (Loud and Gladwin, 1983). The best compromise is a sloping or shaped armrest, with the armrest height at the front optimal for rising, and the armrest height at the rear optimal for sitting comfort (Laging, 1966; Institute for Consumer Ergonomics, 1983c) (Figure 123).

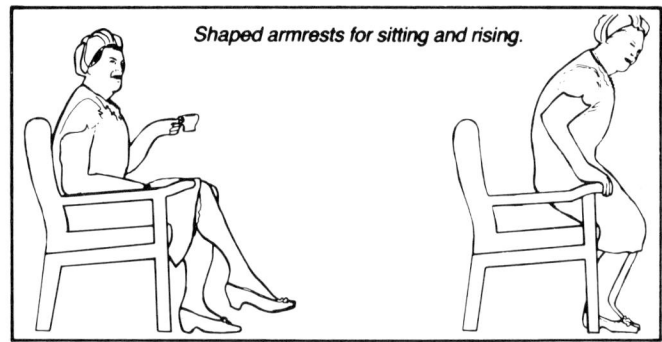

Figure 123. A shaped armrest, with the armrest height at the front optimal for rising, and the armrest height at the rear optimal for sitting comfort. From Institute for Consumer Ergonomics: *Selecting Easy Chairs for Elderly and Disabled People.* Loughborough, University of Technology, June 1983. Reproduced with permission of the Institute for Consumer Ergonomics.

According to the Institute for Consumer Ergonomics (1983a):

1. The armrest height to the floor at the front of the chair "should be high enough to provide support to the sitter in the initial stages of rising from the chair and continue to do so until a stable standing posture has been achieved."
2. The armrest height to the seat at the rear of the chair "should provide comfortable support for the arms when seated. This will mean that the armrest is lower at the back than at the front."
3. "The armrests will need to be sloped or in some way shaped to satisfy the need for the armrests to be higher at the front than at the back. The armrests should provide an unpadded grip area which protrudes slightly beyond the front of the seat. All edges should be rounded to enable a comfortable, firm grip, although the area where the arms will rest when seated should be broad and comfortable."

If the armrests end behind the front edge of the seat, they will give poor support for rising (Institute for Consumer Ergonomics, 1983g). These short armrests will fail to give support to the individual whose balance is being gained as he is nearly erect (Norton et al., 1975). Pushing on these short armrests when rising may also cause the chair to slip backwards (Institute for Consumer Ergonomics, 1983g).

Norton et al. (1975) have also stressed the need for removable armrests on easy chairs used by chairbound individuals. This feature will greatly facilitate transferring the individual from bed to chair and vice versa.

OTHER IMPORTANT EASY CHAIR FEATURES

Backrest

Important features for the backrest of an easy chair are as follows:

1. The backrest should be high enough to support the thoracic spine and scapulae.
2. A headrest will be necessary with a backrest inclination of approximately 25 to 30 degrees (Kohara, 1965; Diffrient et al., 1974).

According to the Institute for Consumer Ergonomics (1983d, 1983g), an adjustable headrest is necessary for optimum comfort on high-backed chairs. However, it is difficult to incorporate a fixed headrest design in the backrest due to the variability in trunk length, back shape, and the individual preference for cushioning behind the head or neck. The need for adjustable headrests was also stressed by Hiba (1980) and Branton (1984).

The headrest should be adjustable not only in height, but also in horizontal distance from the backrest, and in inclination. Le Carpentier (1969) reported a 90 percent range of 2.5 degrees to 12.5 degrees for the angle of the headrest forward from the plane of the backrest.

3. A slight horizontal curvature of the backrest can help stabilize the trunk (Ridder, 1959; Institute for Consumer Ergonomics, 1983a).
4. A pelvic-sacral support is critical to stabilize the pelvis, and support the upper sacrum (and thereby the lower lumbar spine).

Examples of some easy chair backrest profiles based on sitting trials are shown in Figures 124 and 125.

Seat to Backrest Angle

According to Diffrient et al. (1974), the seat to backrest angle of an easy chair should be at least 100 degrees. Seat to backrest angles of 100 to 105 degrees are recommended for conversation, reading, and watching television (Diffrient et al., 1974).

Grandjean et al. (1969) found that healthy individuals preferred an easy chair with a seat to backrest angle of 101 to 104 degrees for reading, whereas a 105 to 108 degree seat to backrest angle was preferred for rest. Le Carpentier (1969) reported that males preferred an easy chair with a

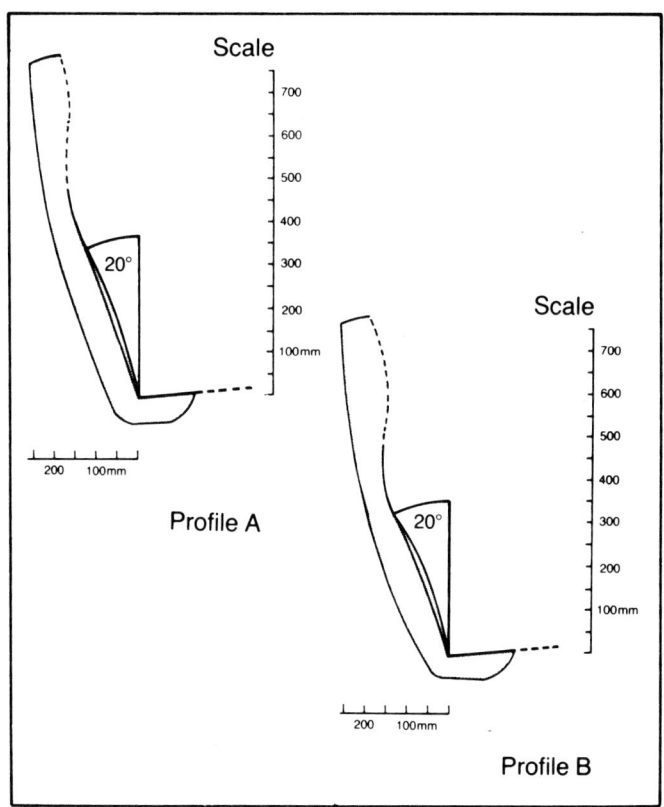

Figure 124. Two backrest profiles, based on the sitting trials of the Institute for Consumer Ergonomics (1983). From Institute for Consumer Ergonomics: *Selecting Easy Chairs for Elderly and Disabled People.* Loughborough, University of Technology, June 1983. Reproduced with permission of the Institute for Consumer Ergonomics.

113 degree seat to backrest angle, but females preferred a seat to backrest angle of 105 degrees.

Individuals with limited hip mobility, such as due to osteoarthritis, will find larger seat to backrest angles more comfortable (Atherton et al., 1980).

Table VI gives the seat inclinations, backrest inclinations, and seat to backrest angles for easy chairs as recommended by various authors and studies.

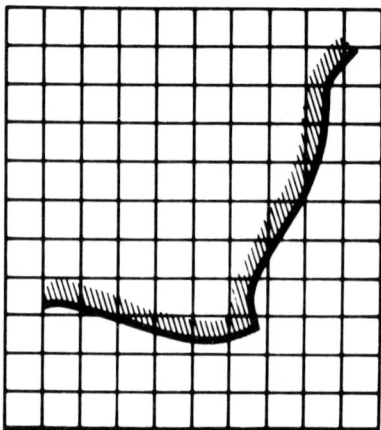

Figure 125. Easy chair profile for individuals with low back pain, based on the sitting trials of Grandjean et al. (1969). Grid 10 cm × 10 cm. From Grandjean, E.: *Fitting the Task to the Man*, 3rd ed. London, Taylor and Francis, 1980. Reproduced with permission of Taylor and Francis Ltd.

Chair-Table Relationship

In their study of six nursing homes, Snyder and Ostrander (1972) emphasized the importance of the chair-table relationship. The seated individual must be able to get in close to the table.

According to Snyder and Ostrander (1972):

1. There will be a greater amount of personalized social interaction during a meal served at tables that have a proper chair-table fit, than at those that do not.
2. There will be more negative affect in a misfit situation than a good fit situation.
3. There will be more food eaten where there is a good fit than where there is a misfit situation.
4. There will be more anxiety and/or stress behavior where chair and table do not fit than where there is a good fit.
5. Different amounts of time devoted to eating and to verbal socializing will be found between fit and misfit seating arrangements.
6. The chair-table relationship is particularly important in dining rooms, recreation rooms, and therapy spaces where a table is essential to the performance of the activity.

Table VI

EASY CHAIRS:
RECOMMENDED SEAT AND BACKREST ANGLES

Investigator	Population	Seat Inclination	Seat to Backrest Angle	Backrest Inclination (From Vertical)
Diffrient et al. (1974)	Average Adult	15°	103°	28°
Le Carpentier (1969)	Sample of British Male Adult Population	9°	113°	32°
Le Carpentier (1969)	Sample of British Female Adult Population	12°	105°	27°
Grandjean et al. (1969)	Healthy Adults (Reading)	23°-24°	101°-104°	34°-38°
Grandjean et al. (1969)	Healthy Adults (Resting)	25°-26°	105°-108°	40°-44°
Grandjean et al. (1969)	Adults with Back Pain (Resting)	19°-21°	105°-108°	34°-39°
Institute for Consumer Ergonomics (1983 a)	Elderly and Disabled Individuals	7°-10°	100°-103°	20°
Fernie et al. (1987)	Ambulatory Elderly with Various Disabilities	9°	102°	21°
Batchelor and Farmelo (1975)	Disabled Individuals (Hemiplegia)	10°	100°	20°

GERIATRIC SITTING COMPLICATIONS

Pressure Sores

As the majority of pressure sores occur in individuals over age seventy, the chairfast geriatric population will be at very high risk (Barbenel et al., 1977; Andersen and Kvorning, 1982). The most common pressure sore in the geriatric population is over the sacrum (Clark et al., 1978; Andersen and Kvorning, 1982). Barton and Barton (1981) attributed over 50 percent of sacral pressure sores to the sitting position. Bailey (1967) referred to the "characteristic geriatric low sacral sore" from sitting in a slumped posture in a reclining chair.

The geriatric population will be very prone to developing high pressure points and high shearing forces over the buttocks when sitting (Swearingen et al., 1962; Bennett et al., 1981). Major factors accounting for this greater potential for tissue deformation include occlusive arterial disease, muscle atrophy and the loss of subcutaneous tissue, and the decrease in skin elasticity and resiliency with age (Kirk and Chieffi, 1962; Daly and Odland, 1979; Torrance, 1981). Braverman and Fonferko (1982a) reported a marked degradation of the elastic fibers of the skin after age seventy. An abnormal thinning of the walls of the dermal microcirculatory vessels, especially the postcapillary venules, is another change reported in the older geriatric population (Braverman and Fonferko, 1982b).

The seat cushion and cushion cover must provide the proper protection against the development of pressure sores in those high risk geriatric individuals. These concerns will be discussed in detail in Chapter Eleven.

Peripheral Nerve Entrapments

Peripheral nerve entrapments are common in the geriatric population, and can result in significant pain, paresthesia, and weakness. They may lead to permanent nerve damage (Reddy, 1986). Various aspects of the individual's prolonged sitting posture, such as crossing the knees or leaning on the elbows, are often implicated in the etiology of several peripheral nerve entrapments.

Peroneal Nerve Entrapment

Crossing the knees is the most common cause of peroneal nerve entrapment in the elderly (Reddy, 1986). With crossing the knees, compression of the common peroneal nerve can occur between the head of the fibula and the opposite patella.

The common peroneal nerve is especially vulnerable to compression in thin individuals, and in those individuals with underlying peripheral neuropathy. Elderly diabetic individuals are particularly prone to develop a peroneal nerve palsy from leg-crossing (Reddy, 1986).

Ulnar Nerve Entrapment

Leaning on the elbows with pronated forearms can compress the ulnar nerve within the ulnar groove (Reddy, 1986). Compression of the ulnar nerve within the cubital tunnel may result from a resting arm position of full forearm pronation, along with extreme elbow flexion (Wadsworth and Williams, 1973).

The armrest design should incorporate an open space directly beneath the elbow. As a result, proper support can be provided for the forearm, avoiding pressure on the elbow (Keegan, 1953).

Radial Nerve Entrapment

Hartigan (1982) has implicated the geriatric wheelchair as a major cause of radial nerve entrapment in the elderly. Compression of the radial nerve can result from pressure of the back of the individual's arm against the hard, tubular vertical frame of the wheelchair backrest.

Chapter Eleven

DISABLED SEATING:
THE WHEELCHAIR AND PRESSURE SORES

According to Nelham (1981):

"The wheelchair occupant cannot achieve his or her full potential at school, place of work or in recreational activities unless a comfortable, functional posture is achieved with the minimum risk of the development of pressure sores."

With the wheelchair dependent population the interrelationship among an individual's sitting posture, spinal posture, pressure distribution, comfort, fatigue, and energy expenditure becomes very critical. It is therefore unfortunate that in regards to wheelchair seating and pressure sores, many important concepts from the literature on able-bodied seating have been ignored.

Zacharkow (1984a, 1984c) feels that several misconceptions still dominate current thinking on proper wheelchair seating and pressure sore prevention:

1. Proper wheelchair sitting posture is with the backrest and trunk vertical, the thighs horizontal, and a 90 degree angle between the thighs and trunk.
2. Whenever possible, a wheelchair should have as low a backrest as possible, and no armrests. There should also be little, if any, body weight on the footrests of the wheelchair.
3. Since many wheelchair dependent individuals lack normal sensation over the buttocks, lower back, and legs, comfort can be ignored regarding wheelchair sitting posture.
4. Pressure sores are inevitable when the external pressure on the sitting surface exceeds the normal capillary blood pressure of 12 to 32 mm Hg.
5. The main reason wheelchair dependent individuals develop pressure sores is because of a failure to do a wheelchair push-up every 5 to 15 minutes.
6. Sacrococcygeal and trochanteric pressure sores are all due to the recumbent position. The supine bed posture is the cause of all

sacrococcygeal sores, and the side lying bed posture is the cause of all trochanteric sores.

As discussed in Chapters Two and Three, the sitting posture for able-bodied individuals is basically unstable without additional external support. This is because the ischial tuberosities, with their rounded shape resembling a pair of very short rockers on a rocking chair, provide only a linear base of support (Meyer, 1873; Hope et al., 1913). In addition, the hip joints are in an intermediate position when sitting, and the trunk cannot be locked relative to the thighs by ligamentous restraint (Åkerblom, 1948; Meyer, 1873; Coe, 1983).

Branton (1966) hypothesized that there is a continual need for postural stability when sitting, so that the seated person "spontaneously takes up such postures as will allow him to sit stably, while relieving his brain and muscles from greater exertion than would be necessary otherwise."

> "If the seat does not allow postures which are both stable and relaxed, the need for stability seems to dominate that for relaxation and postures are adopted which rigidify the body internally in compensation. In other words, if seat features fail to stabilize him, the person must stabilize himself, e.g. by crossing his legs, or by supporting his head on his hand. This must be at some extra cost in muscle work" (Branton, 1966).

For able-bodied seating, the backrest, armrests, foot support, and seat cushion are all critical stabilizers. These chair features will help support the seated individual so that he can maintain a stable posture whether working or relaxing. The proper chair dimensions (backrest height, seat height, seat inclination, seat depth, etc.) are also critical in providing essential stabilization for the sitter.

The aim of proper support for all types of chairs and workstations for able-bodied seating should be optimal stability with minimal restraint (Branton, 1966, 1970; Shipley, 1980; Hedberg et al., 1981; Schaedel, 1977; Dempsey, 1962; Nakaseko et al., 1985). Kohara and Sugi (1972) consider a good, final stable posture to be the most important factor influencing seat comfort. According to Branton (1966), the greater the sitting instability, the greater the discomfort.

This theory on the able-bodied sitter's need for postural stability has important implications for wheelchair seating:

1. If the wheelchair does not provide for proper body stabilization, the wheelchair dependent individual must stabilize himself. However, this will require an additional expenditure of energy due to increased

muscular effort. To achieve this stabilization, the wheelchair sitter will usually adopt slumped, kyphotic postures and/or asymmetrical postures. For example, the buttocks may be slid forward on the seat in order to obtain better trunk stabilization from the backrest; one hip may be brought to rest against the inside of the armrest for better pelvic stabilization; an arm may be hooked around one of the wheelchair push handles.

Goldthwait (1909) stressed that "there is a certain definite amount of energy available for expenditure with each individual; that this energy can be expended in many different ways, but that if expended for one thing it no longer is available for other efforts, except after periods of recuperation." His remarks are particularly relevant to the wheelchair dependent population, considering the various degrees of weakness and paralysis encountered (paraplegia, quadriplegia, hemiplegia, multiple sclerosis) and the limited potential for intrinsic stabilization of body segments when the wheelchair stabilizers are inadequate.

2. The stability of the spine is largely dependent on the extrinsic support provided by the trunk musculature. "The lack of inherent or intrinsic stability of the vertebral column and the importance of the trunk muscles are clearly demonstrated if one tries to hold an unconscious person upright" (Nachemson, 1966).

Similarly, many wheelchair dependent individuals have paralysis or marked weakness of the trunk musculature (erector spinae, lower abdominals), thereby eliminating or greatly reducing any extrinsic spinal stability. This leaves the intrinsic spinal stability, or the support provided by the spinal ligaments.

However, it has been shown that the adult thoracolumbar ligamentous spine fixed at the base and free at the top can support only approximately $4\frac{1}{2}$ lb (2 kg) before buckling occurs. This $4\frac{1}{2}$ lb load approximates one-sixteenth of the average combined weight of the superincumbent torso, head, and upper extremities (Lucas and Bresler, 1961; Lucas, 1970; Morris, 1973).

A paralytic kyphosis and/or scoliosis can easily result from prolonged overstretching of the spinal ligaments, due to the lack of proper trunk stabilization provided by the wheelchair. (Improper back height, back inclination, pelvic-sacral and thoracic support, arm support, etc.) These stability failures of the spine induced by inadequate external support will not only have adverse effects on sitting balance, transfer ability, and respiration, but they will also greatly increase the wheelchair sitter's risk of pressure sores (Zacharkow, 1984a).

3. All wheelchairs, whether manual or electric, and regardless of whether propelled independently or with assistance, are moving vehicles. Therefore, Roberts' (1963) remarks on vehicle seating are very applicable to the disabled population in regards to proper stabilization, both with wheelchair seating and as automobile drivers and passengers:

> "Every time a vehicle sways, rounds a curve, goes over a bump, accelerates or slows, the passenger is thrown slightly off balance. He then expends energy, usually without thinking about it, in using his muscles to regain stability."

4. Branton (1966, 1969) mentioned that even if an individual appears to sit still, his body is continuously moving. The freedom of the pelvis to move, which will be present in all sitting postures where the upper sacrum is not supported by the backrest, will result in "continuous hunting" or relatively fast oscillatory movements of the pelvis rocking over the ischial tuberosities.

The resulting shearing force on the buttocks from sitting postures without proper pelvic and sacral support is one of the most overlooked factors in the etiology of pressure sores (Zacharkow, 1984a). The additional oscillatory stresses due to vibration and road shock in moving vehicles (wheelchairs, automobiles) while in a slumped, kyphotic sitting posture is another major etiological factor in pressure sore formation that merits further attention (Patterson, 1984).

5. Proper wheelchair stabilization is also critical in the execution of manual tasks by wheelchair dependent individuals (Do et al., 1985). With seated able-bodied individuals, voluntary movements of the upper limb are preceded and accompanied by stabilizing contractions of the pelvic and trunk musculature, in order to assure postural stability (Bouisset and Zattara, 1981). Compared to able-bodied individuals, however, paraplegic individuals demonstrate both a stronger and a longer activation of residual scapular and trunk muscles with a manual task (Do et al., 1985). This would reinforce the need for the enhanced postural stability from the wheelchair backrest, footrests, armrests, and seat cushion to help reduce the physiological strain of manual tasks (Andersson and Örtengren, 1974b; Lundervold, 1951a).

PRESSURE SORES: INCIDENCE, COSTS, COMPLICATIONS

Pressure sores continue to be a major problem for many disabled individuals. In the Greater Glasgow Health Board area, the incidence of pressure sores in a one day survey was found to be 8.8 percent among all hospital patients and individuals receiving a home visit from a district nurse (Jordan and Clark, 1977; Barbenel et al., 1977; Clark et al., 1978). In another one day survey involving a Scottish community, the incidence of pressure sores was found to be 9.4 percent in the Borders Health Board area (Jordan, Nicol, and Melrose, 1977). A Danish investigation by Abildgaard and Daugaard (1979) reported a pressure sore incidence of 11 percent among 528 hospitalized patients examined on one day.

Many groups of disabled individuals have a very high incidence of pressure sores, including those individuals with spinal cord injuries, hemiplegia, multiple sclerosis, cancer, and the disabled geriatric population. This particular chapter section will focus on the spinal cord injured population. The other disabled groups will be discussed in the final section of this chapter.

The Spinal Cord Injured Population

Pressure Sore Incidence

In the Greater Glasgow Health Board area, the pressure sore incidence was 21.6 percent for individuals with paraplegia and 23.1 percent for individuals with quadriplegia (Jordan and Clark, 1977). The review by Dinsdale (1974) placed the incidence of pressure sores from 25 percent to 85 percent in paraplegics and quadriplegics. Slightly over 50 percent of admissions to a major spinal cord injury unit have been reported as being due to pressure sores (El-Toraei and Chung, 1977).

Data from the United States Regional Spinal Cord Injury Systems for the years 1975 to 1980 showed that 40 percent of spinal cord injured individuals developed pressure sores during the time period from onset of injury to initial rehabilitation discharge. Approximately 6 percent of these individuals had severe pressure sores, penetrating the dermis to involve subcutaneous adipose tissue and muscle (Young and Burns, 1981a). There was approximately a 30 percent incidence of pressure sores in each of the five follow-up years beyond initial rehabilitation discharge. Five to seven percent of spinal cord injured individuals developed severe pressure sores during each of the follow-up years (Young and Burns, 1981b).

Pressure Sore Complications

Over 4 percent of deaths among the spinal cord injured have been directly attributed to pressure sores (Geisler et al., 1977). In addition, chronic pressure sores are a major etiological factor in renal amyloidosis among the spinal cord injured (Hackler, 1977). Renal failure, the leading cause of death in this population, is often due to renal amyloidosis (Hackler, 1977; Dalton et al., 1965).

Pressure sores are also a potential source of bacteremia (Bryan et al., 1983). Galpin et al. (1976) reported that among twenty-one patients with sepsis attributed solely to pressure sores, bacteremia was documented in sixteen of these patients. Despite appropriate antibiotic therapy, the mortality rate was 48 percent.

Amputation is another potential complication of pressure sores. Data from the National Spinal Injuries Centre of the United Kingdom from 1944–1981 reported that 83 individuals with spinal cord injuries underwent a major limb amputation. In 23 percent of these individuals, the reason for the amputation was a deep pressure sore over a bony prominence, also involving the underlying joint or bone (Grundy and Silver, 1983, 1984).

Narsete et al. (1983) reported that spinal cord injured individuals admitted to a hospital for pressure sore surgery required an average hospitalization of 87 days, compared to an average hospitalization of 14 days for all plastic surgery patients.

Medical Costs

Sather et al. (1977) estimated that the cost of treating an individual with pressure sores is from $15,000 to $30,000, depending on the number and severity of the pressure sores. According to Motloch (1978), the medical costs associated with the healing of a pressure sore can range from $10,000 to $46,000. Young and Burns (1981b) reported that in the first four follow-up years after initial hospital discharge, spinal cord injured individuals with severe pelvic pressure sores averaged approximately $15,000 a year more in hospital costs than spinal cord injured individuals without pressure sores.

It has been estimated that the yearly cost for the healing of pressure sores in the United States is over $2 billion (Motloch, 1978)!

Most Common Pressure Sores

The three most common pressure sores are those over the ischial tuberosity, sacrococcygeal region, and greater trochanter of the femur (Figure 126). Based on 2,500 spinal cord injured individuals treated over a five-year period, El-Toraei and Chung (1977) reported that 21 percent of the pressure sores were over the ischial tuberosities, 19 percent were over the trochanters, and 15 percent were over the sacrum.

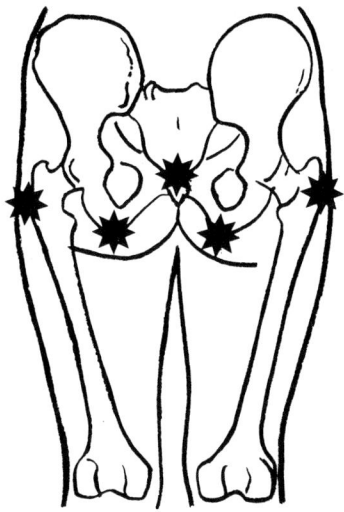

Figure 126. The three most common pressure sores are those over the ischial tuberosity, sacrococcygeal region, and greater trochanter of the femur. Adapted from Spence, W.R., Burk, R.D., and Rae, J.W.: Gel support for prevention of decubitus ulcers. *Archives of Physical Medicine and Rehabilitation, 48:*283–288, 1967. With permission of the Archives of Physical Medicine and Rehabilitation.

A ten year retrospective Scottish survey (January 1971–December 1980) of patients with pressure sores in a regional plastic surgery unit was reported by Wyllie et al. (1984). The majority of these individuals were either paraplegic or quadriplegic, due to spinal cord damage or multiple sclerosis. Of the 270 pressure sores documented, 24 percent of the sores were over the ischial tuberosities, 18 percent were over the sacrum, and 17 percent were over the greater trochanters. Within a year, 12.2 percent of the 270 pressure sores recurred.

The highest recurrence rate was for ischial pressure sores. Recurrences were documented in ten of the forty-one individuals with ischial sores, with eight of these individuals having more than one recurrence (Wyllie et al., 1984; McGregor, 1984).

Eleven percent of admissions to the spinal injury unit of Royal Perth Hospital in Western Australia from 1971–1978 were due to pressure sores (Noble, 1979). Of the 355 total pressure sores among 259 spinal cord injured individuals, 32 percent of the pressure sores were over the ischial tuberosities, 17 percent were over the sacrum, and 9 percent were over the trochanters.

Constantian and Jackson (1980), in a review of 2,319 pressure sores among a spinal cord injured population, also found the ischial pressure sore to be the most prevalent. In addition, out of 432 primary ischial sores, 64 percent recurred. A smaller survey (Berry, 1980) of paraplegic patients who had surgical repair of pressure sores between 1973 and 1977 showed a 47 percent recurrence of ischial pressure sores.

In a review by Hentz (1979) of 115 spinal cord injured individuals who underwent surgical closure of a pressure sore, ischial pressure sores were also the most prevalent. Although ischial pressure sores are the most prevalent among the spinal cord injured population, this is not the case for all disabled populations. As will be discussed in the last section of this chapter, the majority of pressure sores affecting the disabled geriatric population involve the sacrococcygeal region.

The three most common pressure sores are usually associated with completely different predisposing positions:

1. Ischial sores: from sitting.
2. Sacrococcygeal sores: from lying supine.
3. Trochanteric sores: from the side lying bed position.

However, this is a major misconception in the etiology of pressure sores. As will be discussed in this chapter, many sacrococcygeal and trochanteric sores are also caused by poor sitting posture. Barton and Barton (1981) attribute over 50 percent of sacrococcygeal pressure sores to the sitting position. Petersen and Bittmann (1971) reported that 36 percent of individuals with sacral pressure sores were not confined to bed rest, but were up in wheelchairs.

Petersen and Bittmann (1971) also reported that 28 percent of individuals with trochanteric pressure sores were up in wheelchairs, and not confined to bedrest. The trochanteric pressure sore due to sitting is distinctly different from a trochanteric pressure sore due to the side lying bed position. It will occur over the posterior aspect of the greater trochanter, as opposed to the more lateral aspect of the trochanter (Griffith and Schultz, 1961; Sather et al., 1977; Watson, 1983).

Overall, more pressure sores are due to sitting, as opposed to the recumbent position (El-Toraei and Chung, 1977; Porreca and Chagares, 1983; Jordan and Clark, 1977; Jordan, Nicol, and Melrose, 1977).

Comments by Thiyagarajan and Silver (1984) are pertinent to this greater prevalence of pressure sores among chairfast versus bedfast individuals:

"Many patients admitted to hospital with chronic sores might have avoided admission if appropriate advice had been given. Most had developed their severe sores after prolonged sitting because their medical and nursing attendants misguidedly thought that sores were caused by going to bed and would heal if the patient sat in a chair. Other patients had been told that as the sore was on the sacrum there would be no direct pressure on it and they could continue to sit. Medical attendants were often unaware of the risks that shearing strains place on the skin over the sacrum when patients sit in a chair or transfer from bed to chair."

INHERENT PROBLEMS WITH CURRENT WHEELCHAIR DESIGN

Poor sitting posture is the most ignored factor in the etiology of pressure sores. This includes both intrinsic factors that predispose the individual to poor sitting posture and extrinsic factors, specifically the wheelchair. As Professor of Orthopaedic Surgery Robert Roaf (1976) said:

"I am always appalled by the bad sitting posture into which our present design of wheelchairs seems to force patients. I feel we have given inadequate attention to this as a contributory factor in the aetiology of pressure sores."

The Wheelchair Seat Upholstery

Due to the need to make a wheelchair portable, a hammock (sling) seat is standard on the vast majority of wheelchairs. Gibson et al. (1975) and Rang et al. (1981) feel that the unstable base of a hammock seat invites pelvic obliquity (Figure 127). The tendency for the hips to adduct and internally rotate on a hammock seat will result in a narrow, triangular base of support (Pope, 1985b) (Figure 128).

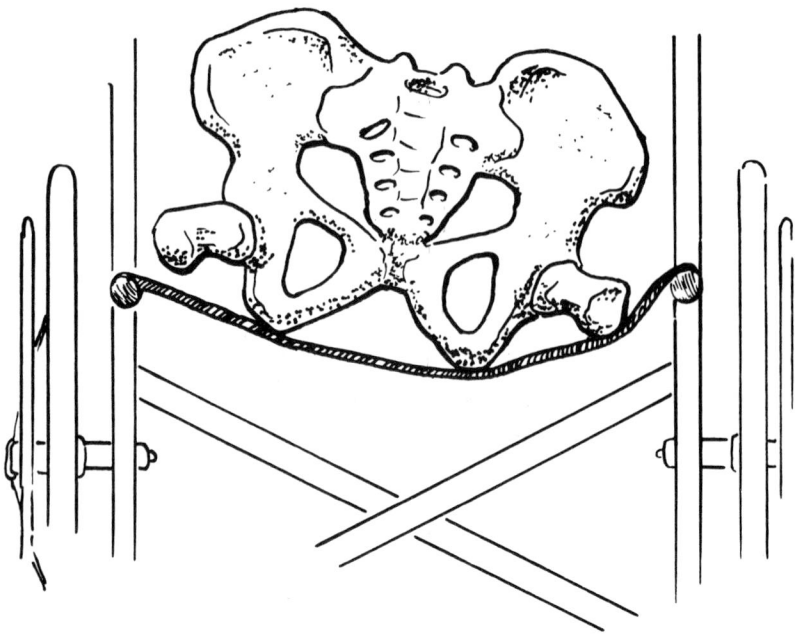

Figure 127. Pelvic obliquity on hammock seat. From Zacharkow, D.: *Wheelchair Posture and Pressure Sores.* Springfield, Thomas, 1984. Reproduced with permission of Charles C Thomas, Publisher.

Sitting with a pelvic obliquity will result in an increased potential for pressure sores over the lower ischial tuberosity and lower greater trochanter (Winter and Pinto, 1986). The hip on the high side of the pelvis may become subluxated or dislocated from being kept in a chronically adducted and flexed position. In addition, proper perineal care can be difficult if the hip on the high side becomes fixed in an adducted position (Stauffer, 1974).

The pelvic instability resulting from the unstable base of a hammock seat will result in insufficient end-support for the spine and trunk (Lucas and Bresler, 1961; Lucas, 1970; Rocky and Nelham, 1984). This can be an initiating factor in the development of scoliosis in individuals with weak or paralytic trunk musculature. Taft (1973) feels that scoliosis is seen more frequently in muscular dystrophy children when sling seats are used, due to the asymmetrical position of the pelvis on the sling seat.

Another problem associated with the hammock seat is the increased shearing forces over the lateral buttocks and posterior aspects of the greater trochanters (Knapp and Bradley, 1970). Griffith and Schultz

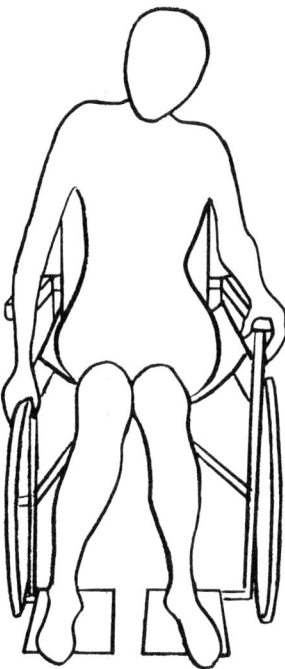

Figure 128. Tendency of the hips to adduct and internally rotate on a hammock seat will result in a narrow, triangular base of support.

(1961) and Guthrie and Goulian (1973) have implicated the sloping sides of the hammock seat as a cause of trochanteric pressure sores.

In two British surveys of wheelchair users, discomfort caused by the hammock seat was mentioned as a problem with current wheelchair design (Platts, 1974; Fenwick, 1977). The discomfort associated with the hammock seat was also considered a major problem with geriatric wheelchairs by Snyder and Ostrander (1972) in their study of six nursing homes in New York.

The Wheelchair Backrest Upholstery

A hammock backrest is standard on the vast majority of wheelchairs and can contribute to several complications. A hammock backrest provides no pelvic-sacral support. The individual is forced to sit in a slumped, posterior sitting posture (Figure 129). This puts the center of gravity of the trunk either over or behind the ischial tuberosities, with less weight being distributed over the posterior thighs.

Figure 129. The hammock backrest will result in a slumped, posterior sitting posture. From Zacharkow, D.: *Wheelchair Posture and Pressure Sores.* Springfield, Thomas, 1984. Reproduced with permission of Charles C Thomas, Publisher.

The coccyx may also become weight bearing when the individual sits with a kyphotic lumbar spine (Howorth, 1978; Johnson, 1981; Frazier, 1985). As the back upholstery continues to stretch, the lower sacrum may also become weight bearing.

So, depending on how kyphotic the sitting posture, the ischial tuberosities, coccyx, and lower sacrum will have an increased risk of pressure sore formation (Zacharkow, 1984a). The additional point pressure from the stiff midline seam of tight jeans against the coccyx and lower sacrum will result in an even greater risk of pressure sore formation (Stoshak and Mortimer, 1985).

The stress put on the posterior ligamentous structures of the lumbar spine in a relaxed, kyphotic sitting posture is considered to be a major cause of low back pain (Keegan, 1953; Kottke, 1961; Cyriax, 1975). The survey of wheelchair users reported by Platts (1974) revealed that many individuals required cushions for their lower backs due to sitting discomfort.

In the survey reported by Fenwick (1977), the most frequent site of pain from wheelchair sitting was in the back. Fenwick (1977) also commented that, "Not surprisingly, we found that discomfort, especially in the back and buttocks, increased with the amount of time spent sitting in the wheelchair."

It is felt that scoliosis can be initiated by a kyphotic sitting posture in individuals with weak or paralyzed erector spinae muscles of the thoracolumbar spine (Koreska et al., 1976; Gibson and Wilkins, 1975; Wilkins and Gibson, 1976; Gibson et al., 1978). According to these authors, a kyphotic sitting posture places all the posterior ligaments of the lower thoracic and lumbar spine under continuous stress. The facet joints, which help resist lateral bending when the spine is extended, become open in the kyphotic sitting position. As a result, lateral bending of the thoracolumbar spine is facilitated, along with pelvic obliquity. Lancourt et al. (1981) reported that many of the spinal cord injured children and adolescents they studied had developed a long thoracolumbar kyphosis. They mentioned that this deformity appeared to develop rather quickly compared to the scoliotic deformities.

Much earlier, Stone (1900) and Lovett (1900, 1916) discussed how extension of the lower thoracic and lumbar spine tends to prevent lateral curvature, while flexion allows it. These authors also stressed the importance of proper backrest support for these spinal regions, warning against the prolonged stretching of the back muscles by the continued maintenance of flexion of the spine.

Effect on Cervical Spine

With a slumped, kyphotic sitting posture, the gravity line of the head will pass further anterior to the cervical spine, and there will be an increased demand placed on the posterior neck musculature (Jones et al., 1961; Gray et al., 1966; Bunch and Keagy, 1976). An increase in neck muscle activity will also be required to keep the head erect and the gaze horizontal.

The greater the slump and thoracolumbar kyphosis, the greater will be the forward thrust of the head, resulting in a marked increase in activity from the upper trapezius and other posterior neck musculature (Gray et al., 1966). A greater than 50 percent increase in muscle tension at the back of the neck has been reported when going from an erect to a slumped sitting posture (Gray et al., 1966).

The alteration in the shape of the cervical spine in a slumped, kyphotic sitting posture would probably resemble the contour described by Ingelmark (1942). From radiological examinations of sixteen patients with pain in the middle and lower neck, and trapezius muscle tenderness, Ingelmark (1942) found an absence of the normal cervical lordosis at the C4 to C7 level, and a hyperlordosis above the C4 level. Pope (1985a) recently referred to this characteristic hyperextension of the upper cervical spine in her study on sitting instability of severe physically disabled, wheelchair dependent individuals.

Sitting in a wheelchair in a marked kyphotic posture will also make diaphragmatic breathing and digestion more difficult (Goldthwait, 1909; Strasser, 1913; Mosher, 1914; Asatekin, 1975; Bunch and Keagy, 1976).

Nonadjustable Armrests

There is a good possibility when a wheelchair has nonadjustable armrests that they will end up being too low (occasionally too high) for the individual in the chair. This is unfortunate because proper fitting armrests can decrease the myoelectric activity in the trapezius muscles when sitting (Andersson and Örtengren, 1974b; Avon and Schmitt, 1975). They can also reduce the disc pressure in the lumbar intervertebral discs (Andersson and Örtengren, 1974b; Occhipinti et al., 1985).

Proper fitting armrests will help facilitate a backward leaning trunk posture, with increased support of the upper trunk against the wheelchair backrest. When the armrests are either removed or too low, the trunk will end up in a slumped kyphotic position (Brattgård, 1969; Gibson and Wilkins, 1975; Brattgård et al., 1983).

Armrests that are too low can also result in the individual's leaning to one side in his wheelchair in order to obtain some arm support (Diffrient et al., 1974). This will result in an uneven pressure distribution on the buttocks. In individuals with weak or paralyzed shoulder musculature, subluxated shoulders may result from armrests that are too low.

Armrests that are too high can cause pain in the shoulder, neck, and upper back muscles by keeping the shoulders constantly elevated (Diffrient et al., 1974).

Another problem with current wheelchair design is the narrow tubular construction and insufficient padding of the armrests (Snyder and Ostrander, 1972). This may not only cause discomfort, but can also lead to compression of the ulnar nerve and pressure sores over the elbow.

The Wheelchair Frame (Seat and Backrest Angles)

Most standard wheelchairs have the entire frame inclined 3 to 5 degrees, along with a 7 to 10 degree bend in the back uprights, just above the armrests. However, it must be remembered that with a hammock backrest and seat, and with the tendency for wheelchair upholstery to stretch out quickly, the seat and back inclinations will easily vary from chair to chair.

A problem that will always be present is the tendency to slide forward on the wheelchair seat and assume a slumped sitting posture (Zacharkow, 1984a; Crewe, 1983) (See Figure 129). This will occur both when sitting relaxed and when propelling the wheelchair. With a horizontal or near horizontal seat and an inclined backrest, there will always be a force tending to push the buttocks forward (Strasser, 1913; Åkerblom, 1948; Ayoub, 1972). This will result in an increased upward rotation of the pelvis, an increased kyphosis of the lumbar spine, and more weight bearing concentrated over the ischial tuberosities, coccyx, and lower sacrum, along with a dangerous shearing force over the same areas. With movement of the wheelchair, vibration and road shock will increase both the tendency to slump, and the shearing forces on the buttocks (Diffrient et al., 1974).

For many wheelchair dependent individuals with poor trunk balance, the typical slightly reclined backrest will be the equivalent of a vertical backrest for an able-bodied individual (See Chapter Three). As a result, adequate sitting balance will only be possible by moving the buttocks forward on the seat, thereby obtaining proper trunk stabilization from the wheelchair backrest (Meyer, 1873; Staffel, 1884; Strasser, 1913; Bunch and Keagy, 1976). Although the resulting posture will be more stable, kyphosis of the lumbar spine will result, with an increase in pressure and shearing forces over the buttocks.

In the wheelchair survey reported by Fenwick (1977), "far more patients considered the backrest too upright than at too much of an angle." More wheelchair users were also found to complain of back pain if they considered the backrest to be too upright (Fenwick, 1977).

The trunk stabilization provided by the proper backrest inclination is also critical for the wheelchair dependent individual to execute manual tasks (Andersson and Örtengren, 1974b; Do et al., 1985).

In reclining wheelchairs with a greater backrest inclination, there will be an even greater sliding force present on the seat than with standard wheelchairs. When the backrest is lowered from a 10 degree inclination to a 30 degree inclination, a 36 percent increase in shearing force on the buttocks has been reported (Brattgård et al., 1983). Increased shearing forces on the buttocks are also a problem with powered reclining wheelchairs (Warren et al., 1982).

Seat Height

The standard seat height of most wheelchairs is from 19½ to 21 inches (49.5 to 53.3 cm). Most studies on sitting for able-bodied individuals recommend seat heights from 16 to 20 inches (40.6 to 50.8 cm) (Kroemer, 1971; Diffrient et al., 1974). Considering that the recommended lower seat heights are for chairs where the feet rest comfortably on the floor, a 19½ to 21 inch (49.5 to 53.3 cm) wheelchair seat height seems acceptable when taking the 2 inch (5.1 cm) minimum floor clearance needed for the footrests into consideration.

The problem arises when one realizes that all wheelchair users need seat cushions that may add from 1 to 3 inches (2.5 to 7.6 cm) to the seat height when they are compressed. As a result, many wheelchair-dependent individuals will be sitting too high to make use of the desks and tables in the able-bodied environment that have heights based on 16 to 20 inch (40.6 to 50.8 cm) chair seat heights (Zacharkow, 1984a; Brattgård, 1969).

In regards to geriatric wheelchairs in nursing homes, Snyder and Ostrander (1972) commented that the geriatric wheelchairs "literally tower over the conventional living room furniture." Some of the important implications mentioned by Snyder and Ostrander (1972) regarding the fit between the geriatric wheelchair and tables in dining and recreational areas are as follows:

1. There will be a greater amount of personalized social interaction during a meal served at tables that have a proper chair-table fit, than those that do not.
2. There will be more negative affect in a misfit situation than a good fit situation.
3. There will be more foot eaten where there is a good fit than where there is a misfit situation.
4. There will be more anxiety and/or stress behavior where chair and table do not fit than where there is a good fit.
5. Different amounts of time devoted to eating and to verbal socializing will be found between fit and misfit seating arrangements.
6. The chair-table relationship is particularly important in dining

rooms, recreation rooms, and therapy spaces where a table is essential to the performance of the activity.

ESSENTIAL WHEELCHAIR MODIFICATIONS

This chapter section will describe the essential features and modifications necessary on existing wheelchairs in order to provide the wheelchair sitter optimal stability with minimal restraint (Branton, 1966). Proper postural stability will also assure the optimal spinal posture and pressure distribution for the wheelchair dependent individual, along with minimizing his discomfort, fatigue, and energy expenditure. More details on these essential wheelchair modifications were given recently by Zacharkow (1984a).

Also in this chapter section, the author will give his suggestions for some much needed and overdue improvements in wheelchair design.

Back Height

A wheelchair is too often thought of as a chair solely needed for mobility and for functional activities. Just as important, but often overlooked, is the need for a wheelchair to be a chair of comfort and relaxation. According to Brattgård (1969), "The wheelchair must be used as a work-chair, a resting-chair and a means of transport indoors as well as outdoors." Andersson and Örtengren (1974b) mentioned that a wheelchair should be designed for both work and rest. In the wheelchair surveys reported by Platts (1974) and Fenwick (1977), comfort was a special concern of the wheelchair users.

A relaxing wheelchair should support the thoracic region of the back in addition to the lumbar area (Diffrient et al., 1974). A chair for work, however, should not interfere with arm motion and with propelling the wheelchair. With current wheelchair design, the best compromise with backrest height should depend on the level of functioning of the wheelchair dependent individual.

The minority of wheelchair users with good trunk musculature and propelling a manual wheelchair for great distances during the day should have the backrest extend as high as possible in the thoracic region without covering the scapulae. Therefore, the backrest height should extend to approximately ½ to 1 inch (1.3 to 2.5 cm) below the inferior angle of the scapulae (Figure 130). With higher backrest heights, the wheelchair push handles may interfere with an extreme backward arm motion.

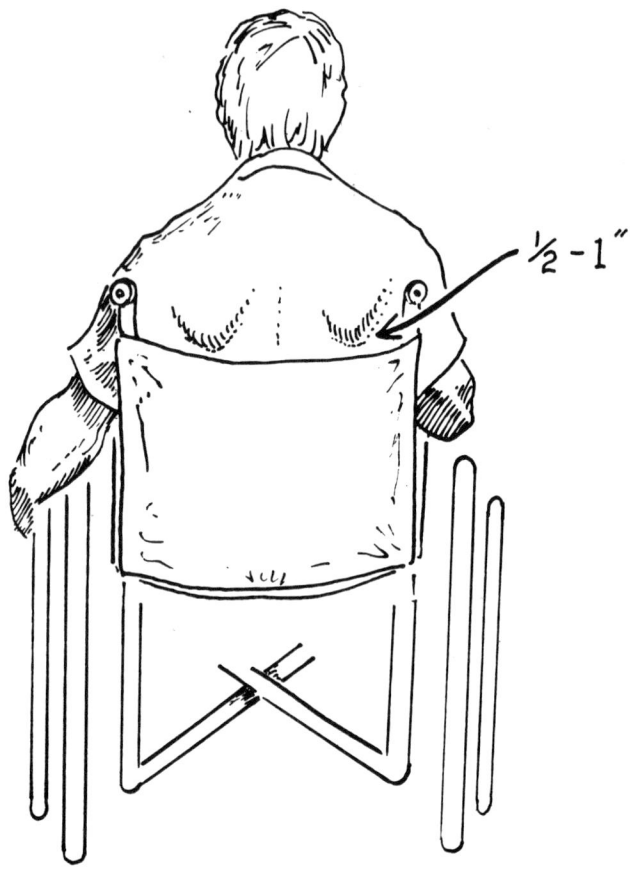

Figure 130. Recommended backrest height for individuals with good trunk musculature, propelling a manual wheelchair for great distances. From Zacharkow, D.: *Wheelchair Posture and Pressure Sores.* Springfield, Thomas, 1984. Reproduced with permission of Charles C Thomas, Publisher.

The vast majority of wheelchair dependent individuals not propelling their manual wheelchair for great distances, along with all individuals in electric wheelchairs, should have the backrest even higher, supporting at least the lower part of the scapulae. This extra backrest height is very critical in order to obtain proper upper trunk support and stabilization from the backrest. This is also the backrest height preferred by able-bodied individuals when in a reclined sitting posture with a backrest inclination of 15 to 18 degrees from upright (Ridder, 1959; Wotzka et al., 1969; Grandjean et al., 1973).

A very important design change needed in wheelchair backrests is the incorporation of a high backrest that is contoured around the shoulder blades to allow free arm mobility (Kroemer, 1982) (Figure 131). Such a design change would provide the critical trunk stabilization and relaxation for many wheelchair users, without interfering with arm motion and wheelchair propulsion. Pressure could then also be avoided over the outer scapulae and shoulders (Taylor, 1917).

Figure 131. A high backrest, contoured around the shoulder blades to allow free arm mobility.

Seat Height

The vast majority of current wheelchairs have seats that are much too high. These current seat heights usually range from 19½ to 21 inches, including standard, electric, and reclining models. There is a great need, however, for all wheelchair models to have seat heights of 17 inches and lower. These lower seat heights will allow proper modifications for a stable sitting base, proper seat inclination, and proper seat cushioning without limiting the wheelchair sitter's accessibility to the environment (tables, desk, vans, etc.).

Seat Boards

According to Rockey and Nelham (1984), "Postural support cannot be effectively provided until pelvic stability has been achieved, since the trunk cannot be held in the correct position if the pelvis is moving about." Pelvic stability can first be provided by establishing a stable sitting base in the wheelchair. Therefore, a seat board should be placed over the hammock seat.

The seat board will correct the postural hip adduction and internal rotation resulting from the hammock seat, thereby providing a wide base of support for increased sitting stability. (Zacharkow, 1984a; DHSS, 1980). In addition, a position of moderate hip abduction will facilitate achieving a lumbar lordosis when sitting, due to the increased hip flexion possible in this position (Mandal, 1982; Rockey and Nelham, 1984).

A dropped seat with suspension hooks is one technique used for lowering the seat height on wheelchairs. However, two points of caution are worth mentioning in regards to the use of a dropped seat:

1. With a dropped seat, the normal flexibility of the wheelchair frame will be lost (Jay, 1983). This can have an adverse effect regarding the increased trauma to the buttocks and lumbar spine from road shock and vibration.
2. The exposure of the crossbraces of the wheelchair frame above the dropped seat can be a cause of trochanteric pressure sores by either direct pressure or shearing, even after the addition of a wheelchair cushion (Figure 132).

Armrests

Adjustable, well-padded armrests should be a standard feature on all wheelchairs. Swearingen et al. (1962) found that the addition of chair arms resulted in a 12.4 percent reduction of body weight on the sitting surface. Brattgård and Severinsson (1978) mentioned that proper adjustment of the armrests will reduce pressures under the ischial tuberosities by about 25 to 30 percent.

Due to the function of the armrests in supporting the upper trunk and contributing to overall sitting stability, it is critical that they be the correct height. For current armrest design, with the upper arm vertical and the forearm horizontal, the armrest should be positioned approximately 1 to 1.2 inches higher than the tip of the elbow (Bunch and Keagy,

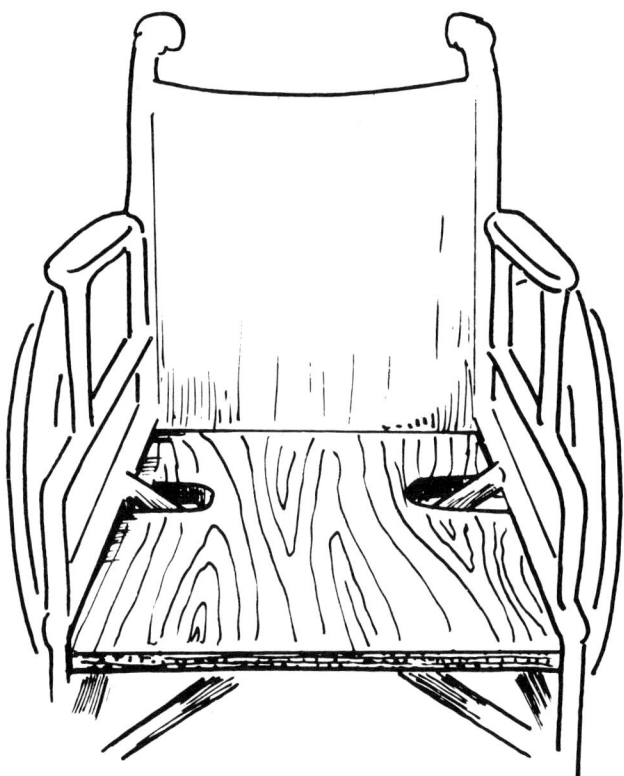

Figure 132. Exposure of the crossbraces of the wheelchair frame above the dropped seat. From Zacharkow, D.: *Wheelchair Posture and Pressure Sores.* Springfield, Thomas, 1984. Reproduced with permission of Charles C Thomas, Publisher.

1976; Torrance, 1983; Andersson and Örtengren, 1974b; Ridder, 1959). This will provide the proper armrest height for the "anthropometric position," sitting erect with the upper arms vertical and the forearms horizontal. However, as discussed in Chapter Four, this is a posture assumed by very few people when sitting. The individual's arm posture will also be influenced by the specific task being performed while sitting in the wheelchair (Drury and Coury, 1982).

This reflects the need for another critical change in wheelchair design, that of a four-way adjustable wheelchair armrest. The four different armrest adjustments needed are in height, angle of inclination, fore-aft adjustment, and adjustability in distance between the armrests (Aarås, 1983; Diffrient, 1984). These adjustments would allow proper arm support in the commonly observed work postures with the shoulders flexed and the forearms elevated.

The adjustability in distance between the armrests would prevent the individual's arm from hitting the armrest when propelling the wheelchair. It would also help reduce the excessive shoulder abduction needed by a female with wide hips and narrow shoulders to obtain arm support.

Seat Inclination

An inclined seat in the wheelchair is necessary to help prevent the individual's buttocks from sliding forward on the seat, resulting in a "deleterious shearing force, backward rotation of the pelvis, poor pressure distribution, and a kyphotic posture of the lumbar spine" (Zacharkow, 1984a). The proper seat inclination will help keep the individual's back against the backrest and pelvic-sacral support, resulting in greater trunk stabilization.

However, for any given backrest inclination, a greater seat inclination will be necessary for a wheelchair dependent individual compared to an able-bodied individual. The reasons are as follows:

1. Due to the lower extremity weakness or paralysis of most wheelchair dependent individuals, the pressure of the feet on the wheelchair footrests will not be as effective in counteracting the forward slide on the seat.
2. The use of rough, textured fabrics for cushion covers to help counteract the sliding force on the seat cannot be used with many wheelchair dependent individuals (Åkerblom, 1948; Branton, 1969, 1971; Asatekin, 1975; Schaedel, 1977). These fabrics are contraindicated for pressure sore prevention, as they will result in excessive friction or interface shear on the buttocks (Chow, 1974). As a material with a lower surface friction is needed for pressure sore prevention, a greater seat inclination is therefore warranted for the wheelchair.
3. With sudden acceleration or deceleration of the wheelchair, a greater seat inclination will aid in retention of the individual in his proper sitting posture on the wheelchair seat (Jacobs et al., 1980).
4. With movement of the wheelchair, a greater seat inclination will help prevent sliding forward on the seat due to road shock and vibration (Diffrient et al., 1974).

Another important change needed in current wheelchair design would be to have an adjustable seat inclination. In order to obtain the proper seat inclination with current wheelchair design, a tapered wedge of very firm foam can be placed under the wheelchair cushion, between the cushion and seat board (Zacharkow, 1984a) (Figure 133).

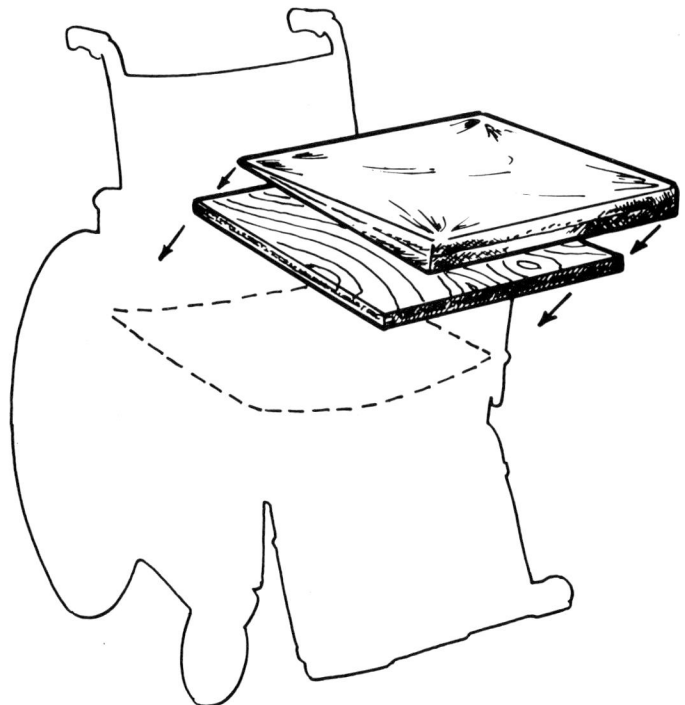

Figure 133. Tapered wedge placed on top of seat board. From Zacharkow, D.: *Wheelchair Posture and Pressure Sores.* Springfield, Thomas, 1984. Reproduced with permission of Charles C Thomas, Publisher.

Backrest Inclination

In order to obtain proper trunk stabilization and support from the backrest, the wheelchair dependent individual often needs to sit with a trunk inclination approaching 10 to 15 degrees from vertical (Zacharkow, 1984a). The greater the trunk muscle paralysis and sitting instability, the greater the trunk inclination required. With the trunk properly stabilized, bimanual hand use is facilitated (Andersson and Örtengren, 1974b; Bunch and Keagy, 1976; Do et al., 1985). With able-bodied seating, a backrest inclination up to 20 degrees is still considered an alert, working posture (Croney, 1981; Diffrient et al., 1974).

Swearingen et al. (1962) found that the addition of a backrest inclined 15 degrees from vertical will reduce body weight on the sitting area by 4.4 percent. However, in this part of Swearingen's study, the seat was not inclined and a pelvic-sacral support and foot support were not used, factors which will accentuate the use of the backrest. So, the reduction in body weight on the sitting surface will be much greater than 4.4 percent with these additional features.

Although many wheelchair dependent individuals will benefit from the improved upper trunk support of a high, inclined backrest with proper pelvic-sacral and thoracic support, one group worth discussing further are those individuals with weakness or paralysis of the serratus anterior muscles. With proper support of the scapulae against the backrest, these individuals will have a marked improvement in arm functioning when sitting.

An adjustable backrest inclination is a needed improvement in current wheelchair design. This point was emphasized in both British wheelchair users surveys (Platts, 1974; Fenwick, 1977).

Back Cushion

An overlooked concern of many wheelchair users is the need for a back cushion in the wheelchair (Platts, 1974; Fenwick, 1977; Snyder and Ostrander, 1972). A back cushion is essential to improve the comfort and pressure distribution of the backrest (Zacharkow, 1984a).

In able-bodied studies, a softer foam cushion is preferred for the backrest of a chair, as opposed to a firmer seat cushion (Ohuchi and Hayashi, 1969). Therefore, except for a firm pelvic-sacral support and firm lower thoracic support, a softer foam cushion should also be used for the wheelchair backrest (Zacharkow, 1984a).

A common mistake often made with wheelchair modifications is to add a solid wooden back insert with little, if any, foam padding, in order to provide better back support. These inserts will often have the adverse effect of pushing the individual's shoulders and upper trunk forward, thereby preventing proper upper trunk support against the backrest. These inserts also fail to provide any lateral trunk stability.

Thigh to Trunk Angle

The critical seat and backrest angles of the wheelchair are those angles resulting when the wheelchair is occupied by the sitter, with the seat and backrest cushions being compressed by the weight of the sitter (Branton, 1966; Schaedel, 1977).

For a stable wheelchair sitting posture, Zacharkow (1984a) recommended a thigh to trunk angle of approximately 95 degrees. If the thigh to trunk angle is any less, such as 90 degrees or 85 degrees, there will be much difficulty in obtaining the proper lumbar lordosis due to the excessive tension in the hip extensors that will tend to rotate the pelvis backwards and flex the lumbar spine. With greater thigh to trunk angles, especially beyond 100 degrees, sliding forward on the seat may become a problem for many wheelchair dependent individuals.

Pelvic-Sacral Support

A pelvic-sacral support, made of a very firm foam and placed behind a softer back cushion, is essential for wheelchair sitting. For wheelchair dependent individuals, the pelvic-sacral support will reduce lumbar disc pressure, improve cervical spinal posture and diaphragmatic breathing, and help preserve soft tissue tightness in the lower back. In addition, forward weight shifts will require minimal effort.

As detailed in Chapter Five, the following changes in pressure distribution will occur with a lumbar lordosis:

1. The coccyx will not bear weight on the seat (Bennett, 1928; Howorth, 1978).
2. There will be better support for the upper trunk against the backrest, with a greater percentage of the body weight being taken by the backrest (Strasser, 1913; Diebschlag and Müller-Limmroth, 1980; Majeske and Buchanan, 1983, 1984; Zacharkow, 1984a).
3. With a lumbar lordosis and the proper thigh to trunk angle, the resting position for the ischial tuberosities will be further posterior on the seat, and the weight line of the trunk will be shifted anterior to the ischial tuberosities (Strasser, 1913; Bennett, 1928; Kamijo et al., 1982; Watkin, 1983; Zacharkow, 1984a; Shields, 1986).

Placement of Pelvic-Sacral Support

Proper placement of the pelvic-sacral support is critical. A line between the highest parts of the iliac crests will usually cross the lumbar spine at about the lower half of the body of the L4 vertebra, or at the level of the disc between L4 and L5 (Macgibbon and Farfan, 1979; Snell, 1978; Basmajian, 1977; Hoppenfeld, 1976) (Figure 134). Therefore, the greatest pelvic-sacral support should be located below the highest part of the posterior iliac crests, in order to provide proper support to the upper sacrum, pelvis, and lower lumbar spine. If the support is placed too high, it will not support the sacrum and pelvis. If the support is placed too low, it will cause the individual to slide forward on the wheelchair seat, resulting in a loss of contact with the lower backrest, kyphosis of the lumbar spine, and poor pressure distribution on the seat.

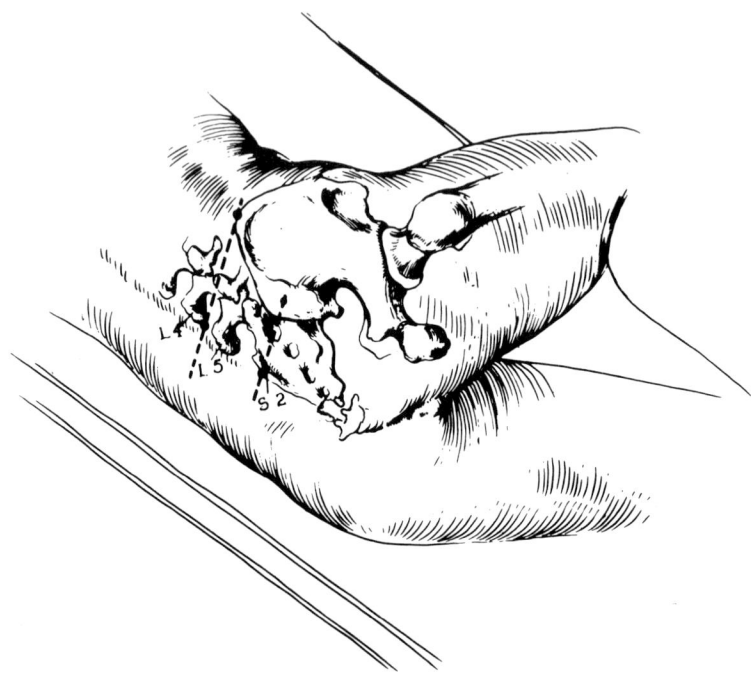

Figure 134. A line connecting the tops of the posterior iliac crests will cross the spine between the L4 and L5 vertebrae. A line between the posterior superior iliac spines will cross S2 and the center of the sacroiliac joint. From Hoppenfeld, S.: *Physical Examination of the Spine and Extremities.* New York, Appleton-Century-Crofts, 1976. Reproduced with permission of Appleton and Lange.

A common mistake made by clinicians is to add a lumbar (pelvic-sacral) support to a wheelchair with a low backrest height. As this support will result in both a backward leaning trunk posture and an increase in sitting height (Majeske and Buchanan, 1983, 1984), a low backrest height will then destabilize the sitter. Therefore, in order for proper trunk stabilization and support of the upper trunk against the backrest, the backrest must be high enough to provide lower scapular support for the vast majority of wheelchair dependent individuals.

A group of wheelchair dependent individuals where a high backrest providing proper scapular support is extremely important are those individuals with one or more of the following musculoskeletal characteristics:

a. Abdominal muscle paralysis, particularly a flaccid paralysis.
b. Paralysis of the serratus anterior.
c. A very long trunk (Lucas and Bresler, 1961; Zacharkow, 1984a; Pope, 1985a, 1985b).

Without proper scapular support and abdominal support (binder or corset), these individuals will have great difficulty in bringing the upper trunk forward over the hips in a lordotic, reclined sitting posture.

Lumbar Kyphosis and Back Pain

Earlier in this chapter it was mentioned that a basic misconception regarding wheelchair sitting posture is that comfort can be ignored, as many wheelchair dependent individuals lack normal sensation over their lower backs, buttocks, and legs. Zacharkow (1984a), however, observed that many wheelchair dependent individuals "complain of pain in their lower backs from prolonged sitting, even though they have complete spinal cord injuries and lack normal sensation to the lower back. However, just by changing the individual's sitting posture with a lumbar pad and inclining the seat and backrest, the back pain often resolves."

A more recent study by Schmoyer (1984) on low back pain following spinal cord injury lends support to these observations. According to Schmoyer (1984):

"While a substantial number of cord injured patients complain of low back pain, these complaints are often assumed to result from central nerve damage. The findings of this study indicate that back pain may often result from mechanical lesions possibly related to or aggravated by prolonged periods of lumbar kyphosis either in bed or in the wheelchair."

Out of twenty spinal cord injured individuals with probable mechanical back pain, including three individuals with neurologically complete cervical injuries, seventeen individuals reported substantial pain relief following positioning in lumbar extension or treatment with lumbar extension exercises. The three individuals that continued to have intermittent problems were all ambulatory and with neurologically incomplete injuries (Schmoyer, 1984).

Lower Thoracic Support

Besides firm support for the upper sacrum, posterior iliac crests, and lower lumbar vertebrae, a second critical region for firm back support is just below the inferior angles of the scapulae, at the lower thoracic spine and thoracolumbar junction. Support in this region is critical to promote spinal extension and proper stabilization of the thorax, along with maintaining the normal axial relationship of the pelvis and thorax that is characteristic of erect standing posture.

In a lordotic wheelchair sitting posture without proper thoracic support, the shoulders will end up projecting behind the hip joint line. The result will be a backward carriage of the shoulders and upper trunk. What is desirable, is a forward carriage of the chest and upper trunk, achieved through lower thoracic extension (Mosher, 1899; Checkley, 1909).

Leg Position

A basic misconception regarding leg position in the wheelchair is that in order to reduce pressure under the ischial tuberosities, the feet of a person in a wheelchair should bear little or no weight (Lindan et al., 1965; Bush, 1969). This philosophy on leg position avoids the fact that sitting is not a static activity, but a dynamic activity. Sitting with the legs hanging freely or with minimal foot support can have the following detrimental effects for the wheelchair dependent individual:

1. It will be a very unstable sitting posture, as the weight of the unsupported legs will destabilize the trunk (Darcus and Weddell, 1947; Branton, 1966).

2. It will be a fatiguing sitting posture, resulting in an increase in back muscle activity in an attempt to stabilize the trunk (Lundervold, 1951a).

3. The weight of the unsupported legs will result in a force causing the buttocks to slide forward on the seat. The individual will end up in a slumped, kyphotic sitting posture, with an increase in pressure and shearing forces over and posterior to the ischial tuberosities (Zacharkow, 1984a; Murrell, 1965; Miller and Sachs, 1974).

4. Without proper foot support, the seated individual will obtain less support from the wheelchair backrest. The study by Swearingen et al. (1962) first studied individuals sitting on a horizontal platform without back support or foot support. Swearingen et al. (1962) then found that

proper foot support would reduce 18.4 percent of the body weight from the seat. They further found that the addition of a backrest inclined 15 degrees from upright, but with the legs dangling and without foot support, reduced 4.4 percent of the body weight from the seat.

The combination of foot support and a backrest with a 15 degree inclination resulted in a 31.3 percent reduction in body weight from the seat. As the arithmetic addition of these two chair features separately would only involve a 22.8 percent reduction in body weight from the seat, the increased reduction in body weight on the seat when these two chair features are combined can be attributed to the greater pressure exerted on the backrest with proper foot support.

5. With the feet bearing little if any weight, there will be a very high cut-off pressure at the distal posterior thigh. The compression of the posterior thighs in this posture will obstruct the venous blood flow from the lower legs (Guyton et al., 1959; Morimoto, 1973; Pottier et al., 1969; Åkerblom, 1948, 1954).

In regards to the cardiovascular demands of sitting for able-bodied individuals, Ward et al. (1966) investigated the cardiovascular parameters of twenty healthy individuals 5 minutes after changing from a supine to a sitting posture. The peripheral pooling of blood from sitting resulted in a 20 percent decrease in stroke volume, an 18 percent increase in heart rate, and a 10 percent decrease in cardiac output.

Shvartz et al. (1982) found that after 5 hours of quiet sitting for six able-bodied young men, there was a 19.4 percent increase in venous pooling in the calf, and a 15.6 percent decrease in calf blood flow.

The peripheral pooling of blood and cardiovascular stress of sitting will be even greater for many wheelchair dependent individuals with various degrees of paralysis. Therefore, improper footrest position resulting in a high cut-off pressure at the distal posterior thigh can have detrimental effects on the cardiac output, fatigue, discomfort, and energy expenditure of the sitter (Glassford, 1977; Winkel and Jørgensen, 1986). Obstruction of the venous blood flow may also be a causative factor in venous thrombosis (Homans, 1954).

The wheelchair footrests should therefore be adjusted to provide firm foot support, with no pressure being exerted over at least the distal one-fourth of the posterior thighs (Bennett, 1928; Carlsöö, 1972).

ADDITIONAL CONSIDERATIONS
IN WHEELCHAIR PRESCRIPTION

Seat Width

According to Brattgård (1969), "The width of the seat should be the minimum acceptable for the individual disabled person." Zacharkow (1984a) recommended a clearance of no more than approximately ½ inch (1.3 cm) to the side of each hip (Figure 135). With additional clearance, not only will the overall width of the wheelchair increase, but the individual's shoulders will have to be more abducted for propelling the wheelchair (Brattgård, 1969; Zacharkow, 1984a). Wider seat widths will also facilitate a lateral migration of the pelvis on the seat.

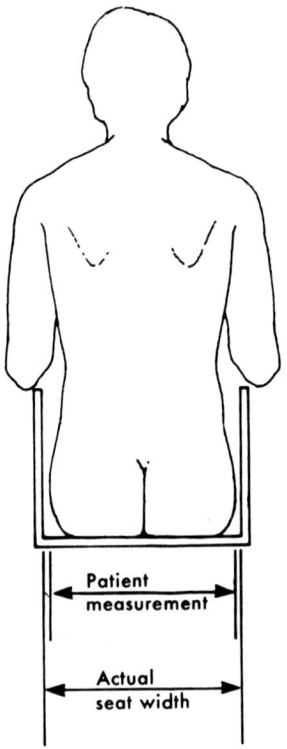

Figure 135. There should be a clearance of no more than approximately ½ inch to the side of each hip. From Bunch, W.H., and Keagy, R.D.: *Principles of Orthotic Treatment.* St. Louis, Mosby, 1976. Reproduced with permission of W. H. Bunch, M.D.

Seat Depth

The pressure against the posterior calf with an excessive seat depth will force the sitter away from the pelvic-sacral support of the backrest, and into a slumped, kyphotic posture (Figure 136).

An insufficient seat depth will result in an unstable sitting posture, as the individual will lose the additional stability provided by the upper posterior thigh support (Brattgård, 1969; Brattgård et al., 1983). Wheelchair transfers will also be more difficult for the wheelchair dependent individual with an insufficient seat depth.

With proper seat depth, there should be approximately 2 to 3 inches of clearance from the front edge of the seat to the back of the knee (Figure 137).

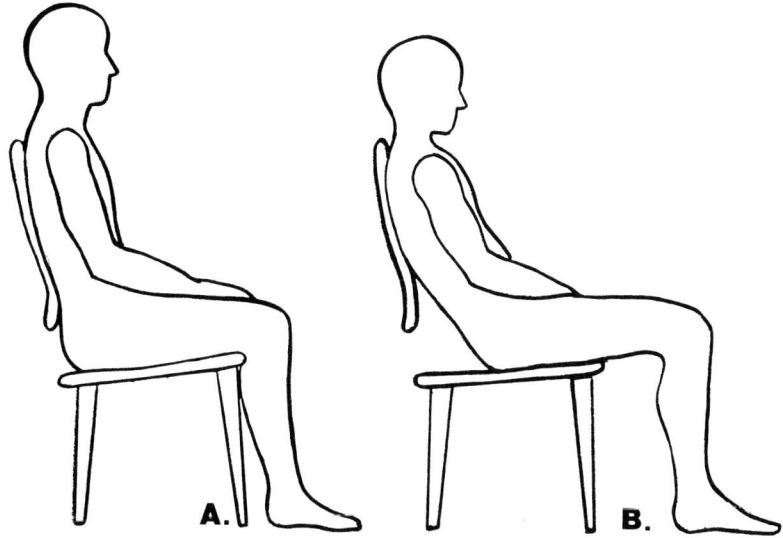

Figure 136.
    A. Excessive seat depth, resulting in pressure behind the knees.
    B. Slumped posture adopted in order to remove pressure from behind the knees.

Tires

Unless a wheelchair is being used exclusively indoors, it should be equipped with pneumatic tires (Lauridsen and Lund, 1964). Road shock and vibrations will be more easily transmitted to the seated individual in a wheelchair with solid tires. As will be discussed later in this chapter,

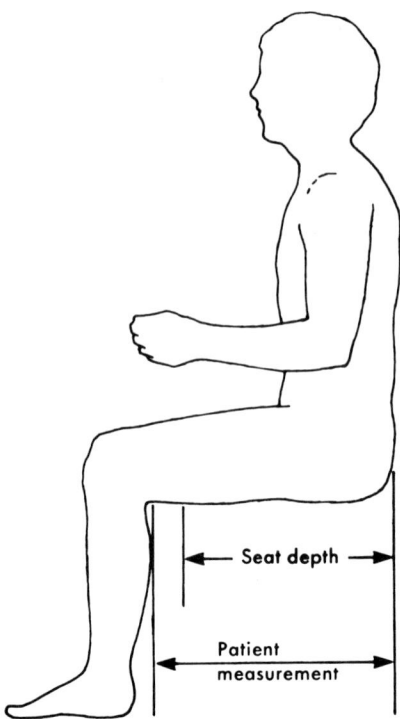

Figure 137. Proper seat depth should allow approximately 2 to 3 inches of clearance from the front edge of the seat to the back of the knee. From Bunch, W.H., and Keagy, R.D.: *Principles of Orthotic Treatment.* St. Louis, Mosby, 1976. Reproduced with permission of W.H. Bunch, M.D.

road shock and vibration are major factors overlooked in the etiology of pressure sores.

ADDITIONAL CONSIDERATIONS IN WHEELCHAIR DESIGN

Lightweight wheelchairs ("sports" wheelchairs) have been shown to have substantially less posterior stability than conventional wheelchairs (Loane and Kirby, 1985). Unfortunately, these wheelchair designs are very popular not only for athletic activities, but also for everyday use.

Besides the danger of tipping over backwards, these lightweight wheelchairs do not promote a healthy sitting posture. One of the beneficial effects of sitting in a wheelchair with an inclined seat, inclined backrest, and pelvic-sacral support is that the individual's upper trunk will receive greater support from the backrest. In a lightweight chair, however, this posture will result in an even greater posterior instability to the chair. In order to improve the rear stability in lightweight chairs, individuals in these chairs spontaneously adopt a kyphotic sitting posture, with the upper trunk and shoulders hunched forward, away from the backrest.

Changing the rear axle position of a lightweight chair in order to increase the seat and backrest inclinations will not have the same beneficial effect on sitting posture as incorporating a seat wedge and back wedge to a conventional wheelchair. This change in axle position will make the lightweight chair even more unstable posteriorly, and will actually work against the sitter receiving increased support and stabilization from the backrest.

Lever Drive Wheelchairs

In regards to future wheelchair designs, a lever drive system has been found to be a more efficient method of wheelchair propulsion compared to the use of handrims (Engel and Hildebrandt, 1974; Brubaker et al., 1984; van der Woude et al., 1986).

Compared to the use of handrims, a lever drive system would probably promote a more stable sitting posture with wheelchair propulsion, with the individual obtaining increased support from the backrest and a reduction in shearing forces over the buttocks (Pope, 1985b; Brubaker et al., 1984).

POSTURAL RISK FACTORS AND PRESSURE SORES

General Postural Risk Factors

Due to differences in body build, certain wheelchair dependent individuals will have a greater risk of pressure sore formation.

Thin Individuals

A thin individual with minimal subcutaneous fat in the buttocks region and thin gluteal musculature will produce higher peak pressures under the ischial tuberosities than a heavier individual, who will transmit his body weight over a larger area of the buttocks (Reswick et al., 1964; Lindan et al., 1965; Garber and Krouskop, 1982; Minns et al., 1984). Among able-bodied subjects, Dillon (1981) and Kadaba et al. (1984) both found that the deepest indentation profile at the buttocks-cushion interface was produced by thin individuals of low body weight.

Among twenty variables studied, Williams (1972) found body weight to be the most effective predictor of pressure sores, with thin individuals being more likely to develop pressure sores.

Tissue Atrophy in Buttocks Region

The preceding studies would also indicate an increased risk for pressure sores among those wheelchair dependent individuals with a reduced sitting area due to gluteal atrophy. With an ischiobarograph, Minns and Sutton (1982) found that spinal cord injured individuals with pressure sores presented a much reduced contact area on the flat, rigid sitting surface of the ischiobarograph, compared with able-bodied individuals of similar body weight. The contact area of the spinal cord injured individuals was usually less than one-third the contact area of the able-bodied individuals of similar body weight.

In a study in Bombay, India by Dhami et al. (1985), plaster of Paris impression casts were taken over the buttocks of paraplegic and able-bodied individuals while in the seated position. Among the paraplegic individuals, Dhami et al. (1985) reported "a significant alteration in the contour of the gluteal impressions. The contour in normal subjects is a gentle convexity while in paraplegics it is more conical and has a sharper projection." In paraplegic individuals with ischial pressure sores, the average depth of the ischial tuberosity impression was greater than in both the able-bodied individuals and the paraplegic individuals without pressure sores.

The decreased sitting area in the geriatric population is also an important factor in the susceptibility of this wheelchair dependent population to pressure sores. Comparing different age groups among males, Swearingen et al. (1962) found that there was a steady decline in the average sitting area per age group beyond age forty.

When sitting on a hard seat, Bennett and associates (Bennett et al., 1981, 1984; Bennett and Lee, 1984, 1985) reported that both paraplegic individuals and ill, geriatric individuals developed approximately three times the median shear load at a point 2 to 3 cm lateral to the ischial tuberosity compared to a group of young, able-bodied individuals. The buttocks tissue atrophy may account for the increased shear loads developed just lateral to the ischial tuberosities in the paraplegic and geriatric groups studied.

Flaccid Paralysis of Buttocks

The gluteal muscle atrophy resulting from a flaccid paralysis of the buttocks will result in even higher peak pressures and sharper pressure gradients on the sitting surface. Watson (1983) commented that pressure sores are most prevalent in spinal cord injured individuals with "flaccid paralysis and wasted buttocks." Bedbrook (1981) mentioned the increased risk of pressure sores among the spinal cord injured population as a result of surgical procedures or intrathecal alcohol injections performed to eliminate lower extremity spasticity.

Noble (1978a, 1979) and Scull and Noble (1977) have provided the most documentation in regards to the increased risk of pressure sores in individuals with a flaccid paralysis of the buttocks. Based on pressure measurements taken under the ischial tuberosities and greater trochanters of 151 spinal cord injured individuals while seated on a standard foam cushion, Noble (1979) found significantly higher ischial and trochanteric pressures in individuals with a flaccid paralysis versus a spastic paralysis of the buttocks.

In regards to sex differences, the male spinal cord injured individuals had significantly higher ischial and trochanteric pressures than the female spinal cord injured individuals. In addition, although the flaccid male individuals tended to develop high pressures over both the ischial tuberosities and trochanters, flaccid female individuals tended to develop high pressures over only the ischial tuberosities and not over the greater trochanters (Noble, 1979).

Based on the preceding pressure data regarding sex differences and spasticity-flaccidity, Noble (1981) considered the inherent risk for pressure sores over the buttocks among the spinal cord injured population to be in the following order:

1. Flaccid male individual = greatest risk.
2. Flaccid female individual.
3. Spastic male individual.
4. Spastic female individual = least risk.

Using a pneumatic cell pressure transducer, Yang et al. (1984) measured the pressures under the ischial tuberosities of 39 adults while sitting on an unpadded wooden chair. The mean ischial tuberosity pressure was 79.9 mm Hg for females and 109.0 mm Hg for males.

The tendency for females to show lower peak pressures under the buttocks, as documented by Noble (1978a, 1979), Scull and Noble (1977), and Yang et al. (1984) can be attributed to the following:

1. The female's greater subcutaneous tissue in the buttocks and hips regions.
2. Females are usually less heavily built above the pelvis compared to males.
3. The ischial tuberosities and acetabula are wider apart in the female pelvis, with the ischials also being more everted (Ellis, 1977; Warwick and Williams, 1973). Therefore, the female pelvic structure will favor a wider indentation profile and a wider pressure distribution (Figure 138).

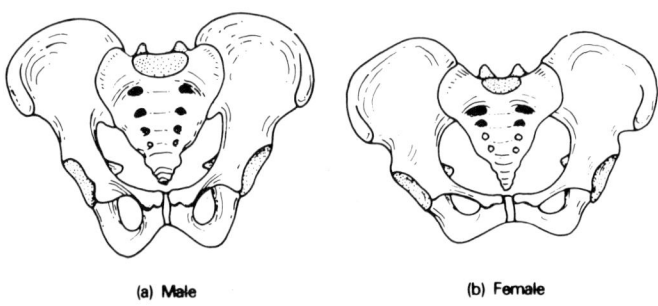

Figure 138. Comparison of the male and female pelvis. From Ellis, H.: *Clinical Anatomy*, 6th ed. Oxford, Blackwell Scientific, 1977. Reproduced with permission of Blackwell Scientific Publications Ltd.

However, there is an increased backward tilt of the sacrum in the female, with the coccyx being in a more upright position and less curved forward compared to the male (Johnson, 1981; Warwick and Williams, 1973). Therefore, in a slumped sitting posture, the female sacrococcygeal region will be more exposed to greater localized pressures and shearing forces. This may be one reason for the greater incidence of sacrococcygeal pressure sores observed in the female geriatric population, where slumped, kyphotic sitting postures are commonly observed (Barton and Barton, 1981).

Specific Postural Risk Factors

Two specific postural risk factors of major importance in the development of pressure sores over the buttocks are (1) sitting with a pelvic obliquity and/or (2) the loss of lumbar lordosis when sitting (Drummond et al., 1982a, 1982b, 1983, 1985; Zacharkow, 1984a, 1984c, 1985).

Comparing the seated pressure distribution of fifteen paraplegic individuals with ischial or coccygeal pressure sores to fifteen able-bodied individuals, Drummond et al. (1982b) reported a significant shift in pressure distribution posteriorly in the paraplegic individuals with pressure sores, due to the inability to sit with a lumbar lordosis. These individuals with pressure sores sat with 60.3 percent of the sitting pressure distributed over the ischial tuberosities and sacrococcygeal region, compared to an average of 39 percent in able-bodied individuals able to sit with a lumbar lordosis. In addition, twelve out of the fifteen paraplegic individuals with pressure sores (80 percent) sat with a pelvic obliquity and a marked asymmetrical loading of the ischial tuberosities.

Earlier, Burke et al. (1980) analyzed the pressure distribution in relation to pelvic obliquity in over 100 individuals:

> "The normal pressure distribution in patients with no scoliosis and no pelvic obliquity was found to be equally distributed between the ischial tuberosities. With increase in pelvic obliquity the pressure was found to be unequally divided until, at 10 degrees, the pressure was taken entirely under one ischial tuberosity and the patient's sitting became unstable. With further increase in pelvic obliquity the pressure became distributed between the ischial tuberosity and the greater trochanter with improvement in stability. With still further increase in pelvic obliquity the pressure was found to be taken entirely under the greater trochanter with deterioration in stability."

In addition, when a person sits with a pelvic obliquity, the pelvis may drift toward the side of the wheelchair seat opposite the lower side of the pelvis. This can add a deleterious shearing force to the buttocks (Zacharkow, 1984a). In his experiments on shear stress, Chow (1974) used a 6 inch (15.2 cm) diameter gel hemisphere to simulate the human buttock, with a wooden core to simulate the ischial tuberosity. Chow found the shear stress due to sideways movement to be the most critical at the inside of the gel hemisphere near the simulated ischial tuberosity.

As critical as these two risk factors are in pressure sore formation, it is important to discuss the various intrinsic and extrinsic causes of a loss of lordosis and a pelvic obliquity when sitting.

Reasons for Loss of Lumbar Lordosis

Limited Hip Mobility in Flexion

Among the reasons for limited hip mobility in flexion when sitting are the following:

1. Tightness and/or spasticity of the hip extensors, primarily the hamstrings, but also at times the gluteus maximus (Stokes and Abery, 1980; Rang et al., 1981; Bowen et al., 1981; Szalay et al., 1986).

2. Degenerative changes at the hip joint such as due to osteoarthritis or rheumatoid arthritis (Rosemeyer, 1973).

3. Limited hip flexion due to heterotopic ossification around the hips. This complication is frequently found in individuals with spinal cord injuries and traumatic head injuries (Tibone et al., 1978; Garland et al., 1985).

4. Certain clothing, such as very tight jeans, can also limit hip flexion to the point where one cannot sit with a lumbar lordosis (Fisk, 1986).

5. Sitting in a wheelchair with the knees in an extended position will limit hip mobility due to the pull of the hamstring muscles. This will result in a marked backward rotation of the pelvis and kyphosis of the lumbar spine (Floyd and Roberts, 1958; Stokes and Abery, 1980; Brunswic, 1984a, 1984b).

Decreased Spinal Mobility in Extension

Among the reasons for a loss of lordosis when sitting due to decreased spinal mobility in extension are the following:

1. Age-related degenerative changes in the spine. Milne and Lauder (1974) found the loss of lumbar lordosis to be increasingly common in individuals age sixty-five and older.

2. Tightness and/or spasticity of the rectus abdominis muscle (Zacharkow, 1984a).

3. Lack of proper thoracolumbar support with bed positioning, particularly with supine and sitting bed postures (Zacharkow, 1984a).

4. Spinal stabilization procedures that result in a loss of lumbar lordosis by not contouring the spinal instrumentation (Luque et al., 1982; Luque, 1986; Moe et al., 1978; Cummine et al., 1979; Kostuik and Hall, 1983; Osebold et al., 1982).

5. Improper physical therapy, resulting in overstretching of the posterior ligaments and musculature of the thoracolumbar spine (Nickel, 1957; Adkins et al., 1960). This also includes improper wheelchair

positioning and physical therapy activities for an individual with a cervical spine injury and trunk muscle paralysis, who is stabilized in a halo vest. As the halo vest does not extend low enough to support the lumbar spine, upright sitting and mat activities in a halo vest will quickly overstretch the lumbar spine, resulting in a marked kyphotic sitting posture (Zacharkow, 1984a).

Wheelchair-Related Factors

As previously discussed in this chapter, many wheelchair-related factors can also result in a lumbar kyphosis when sitting. Among these factors are the following:

1. Armrests too low.
2. Hammock backrest upholstery.
3. Backrest height too low.
4. Backrest angle too upright.
5. Seat depth too long.
6. An inclined backrest with a horizontal seat.

Reasons for Pelvic Obliquity

Intrinsic Factors

1. Long C curve scoliosis. When the individual with a non-structural paralytic scoliosis is supine, the spine is relatively straight and the pelvis can be easily levelled. However, in a sitting position, due to the paralysis of the trunk musculature and the effect of gravity, the individual shows a thoracolumbar C curved scoliosis with pelvic obliquity, the lower side of the pelvis being on the convex side of the curve (Bennett, 1955; James, 1956, 1976; Garrett et al., 1961; Lancourt et al., 1981; Brown ct al., 1984) (Figure 139).

2. Scoliosis due to trunk muscle imbalance and/or spasticity imbalance. The trunk muscles usually found to have the imbalance are the main hip hikers: the quadratus lumborum, the internal oblique abdominals, and the external oblique abdominals (Lowman, 1932; Mayer, 1936; James, 1956) (Figure 140). An imbalance of the latissimus dorsi may also be involved in causing a pelvic obliquity. The stronger muscles will be located on the high side of the pelvis, the concave side of the spinal curve.

3. An atrophied unilateral buttock (James, 1956, 1976; Bennett, 1955, 1961).

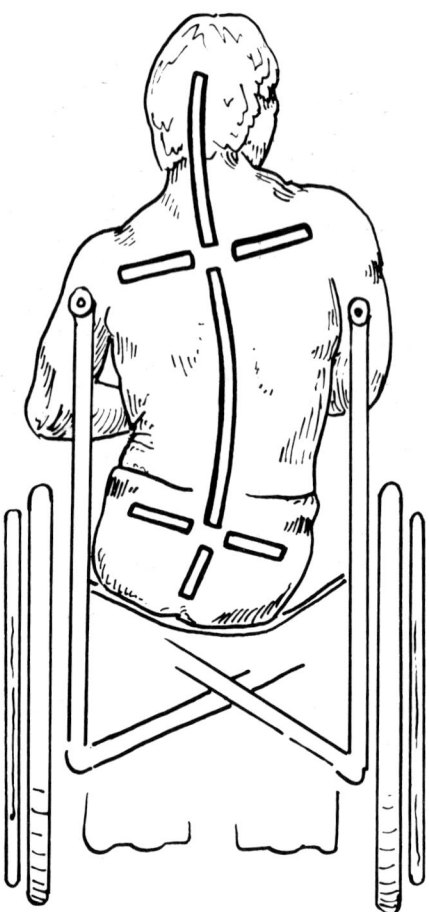

Figure 139. Long C curve scoliosis. From Zacharkow, D.: *Wheelchair Posture and Pressure Sores.* Springfield, Thomas, 1984. Reproduced with permission of Charles C Thomas, Publisher.

4. A small hemipelvis (Ingelmark and Lindström, 1963; Simons and Travell, 1983; Schafer, 1983).

5. Heterotopic bone formation limiting hip flexion unilaterally (Hassard, 1975). The pelvis will be lower on the side with the nonrestricted hip flexion. Of course, hip flexion may also be limited unilaterally due to hamstring tightness or degenerative hip disease.

6. A unilateral ischiectomy (Figure 141). If a total or subtotal ischiectomy is performed, the individual will then sit with a pelvic obliquity, the lower side of the pelvis being on the side of the ischiectomy (Stark, 1962). This will be true even if a muscle flap has been used to fill the defect left

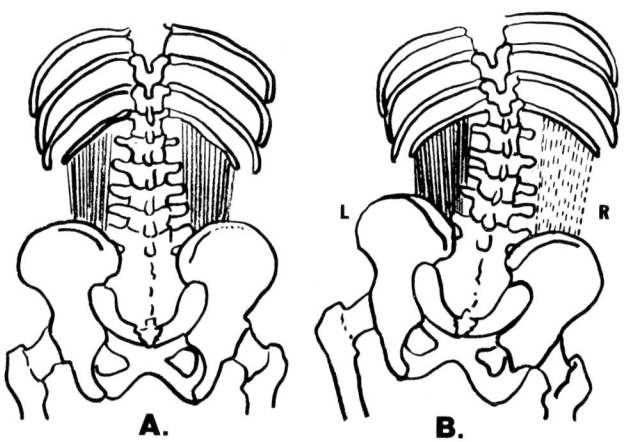

Figure 140.
A. Equal strength of quadratus lumborum muscles.
B. Paralysis of the right quadratus lumborum, resulting in a pelvic obliquity.
Adapted from Mayer, L.: Further studies of fixed paralytic pelvic obliquity. *The Journal of Bone and Joint Surgery, 18:*87–100, 1936. With permission of The Journal of Bone and Joint Surgery.

by the ischiectomy, due to the ease with which muscle will deform under pressure, and the approximately 50 percent reduction of bulk when a muscle is released from its insertion (McCraw et al., 1977; Lopez, 1983). In a clinical series of myocutaneous flaps for pressure sore coverage, Daniel and Faibisoff (1982) found nearly total muscle atrophy one to two years after surgery.

7. A unilateral hip disarticulation or a high AK amputation. A major problem with sitting balance will occur after a hip disarticulation or a high AK amputation, since the proximal femur is extremely critical for sitting stability (Spira and Hardy, 1963; Grundy and Silver, 1984). The individual will have a tendency to fall to the side of the amputation. If he has the voluntary musculature to do so, the individual will shift his body weight over the non-amputated leg in order to balance, resulting in increased pressure on the non-amputated side (Ohry et al., 1983; Zacharkow, 1984a).

An individual's sitting balance and pressure distribution can also be affected by a subluxated or dislocated hip. Among a chronic spinal cord injured population with non-septic hip instability (dislocation), Baird et al. (1986) noted a marked preponderance of pressure sores on the buttock opposite the dislocated hip.

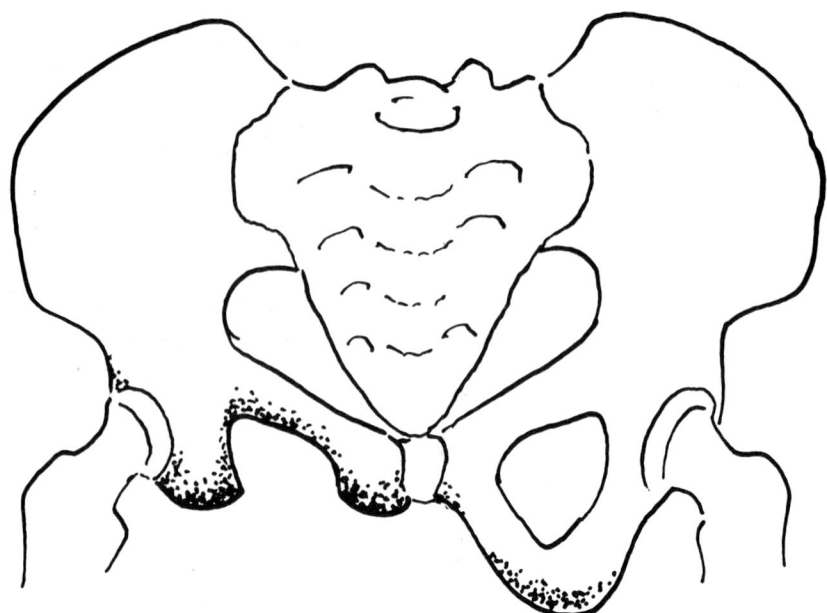

Figure 141. Unilateral ischiectomy. From Zacharkow, D.: *Wheelchair Posture and Pressure Sores.* Springfield, Thomas, 1984. Reproduced with permission of Charles C Thomas, Publisher.

Extrinsic Factors

Asymmetrical sitting postures are very common among able-bodied individuals, especially if a chair does not have the proper stabilizing features (Schoberth, 1962). Similarly, a wheelchair without the proper stabilizing features will often cause the wheelchair dependent individual to assume an asymmetrical sitting posture. Among these extrinsic factors are:

1. Using a hammock wheelchair seat without a seat board. This, by itself, can result in a pelvic obliquity when sitting. It can also make a pre-existing pelvic obliquity worse.

2. A wheelchair seat that is too wide will force the individual to lean to one side for arm support (Gibson and Wilkins, 1975).

3. Armrests that are too low will result in excessive leaning to one side in order to obtain some arm support (Diffrient et al., 1974).

4. The lateral instability resulting from certain air-filled and water-filled wheelchair cushions with thick membranes will often result in a postural pelvic obliquity and asymmetrical sitting posture (Zacharkow, 1984a; Perkash et al., 1984).

5. Having the drive control of the electric wheelchair positioned at the side of the armrest will often result in an asymmetrical sitting posture. This can be corrected by moving the drive control to a more central position (Gibson et al., 1975).

This can be a major problem for a C5 spared quadriplegic individual using an electric wheelchair. Individuals at this lesion level will often have good functioning of the rhomboidei and levator scapulae, but paralysis of the serratus anterior muscle. With movement of the drive control, the unopposed action of the rhomboidei and levator scapulae will raise the medial border of the scapula, resulting in downward rotation of the glenoid cavity. Without proper lateral trunk stabilization, the individual's upper trunk will fall to the side of the drive control, resulting in a marked asymmetrical sitting posture.

Other extrinsic factors causing a pelvic obliquity and asymmetrical sitting posture include the following:

6. Prolonged side lying bed positioning on one side can contribute to the development of a scoliosis and pelvic drop on the side the person constantly lies on (Fitz, 1898; Browne, 1916).

7. Poor workstation design can easily result in an asymmetrical sitting posture as discussed in Chapter Eight. In addition, the typical writing position will often end up being an asymmetrical posture (Alexander, 1966; Schoberth, 1962). The increased pressure will be on the buttock opposite the writing arm.

8. Sitting with the legs crossed will result in both a pelvic obliquity and a lumbar kyphosis, with a marked increase in pressure on the buttock of the uncrossed leg (Roth, 1899; Strasser, 1913; Schoberth, 1962; Dempsey, 1962; Bromley, 1976).

Wheelchair Modifications for Additional Pelvic and Spinal Stability

Gluteal Pad

A gluteal pad can be placed beneath the wheelchair cushion, under the buttock on the lower side of the pelvis. It can be effective in levelling the pelvis in individuals with an atrophied unilateral buttock, small hemipelvis, and a trunk muscle imbalance involving the quadratus lumborum and lateral abdominals (James, 1956, 1976; Bennett, 1955; Simons and Travell, 1983; Schafer, 1983) (Figure 142).

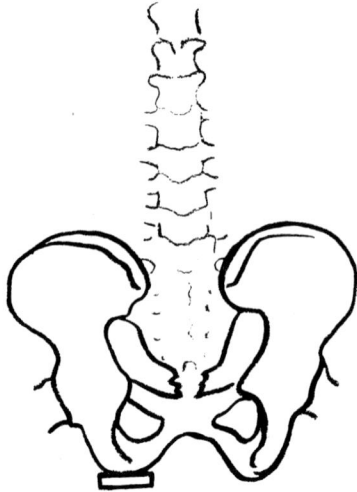

Figure 142. Gluteal pad.

However, if there is a contracture present in the trunk musculature on the high side of the pelvis (the concave side of the lumbar curve), the gluteal pad will shift the entire trunk laterally to the concave side while levelling the pelvis (Glancy, 1978). This is obviously unacceptable since it will throw the individual off balance, and the line of gravity will not fall between the buttocks. The contracture must be reduced before the gluteal pad can be used.

Pelvic Side Supports

Additional lateral pelvic stability can be achieved with side supports on the wheelchair seat (Figure 143). Side supports are critical when correcting a scoliosis, as they will prevent lateral migration of the pelvis.

Contoured Base

Pelvic stability can also be enhanced by placing a slightly contoured base under the wheelchair cushion (Grandjean et al., 1973; Ridder, 1959).

Low Pelvic Lap Belt

In individuals with marked hip extensor spasticity, a low pelvic lap belt will be necessary to provide additional pelvic stabilization and prevent anterior migration of the pelvis on the seat (Carlson et al., 1986; Petersen et al., 1987). Belt malposition, particularly upwards displacement,

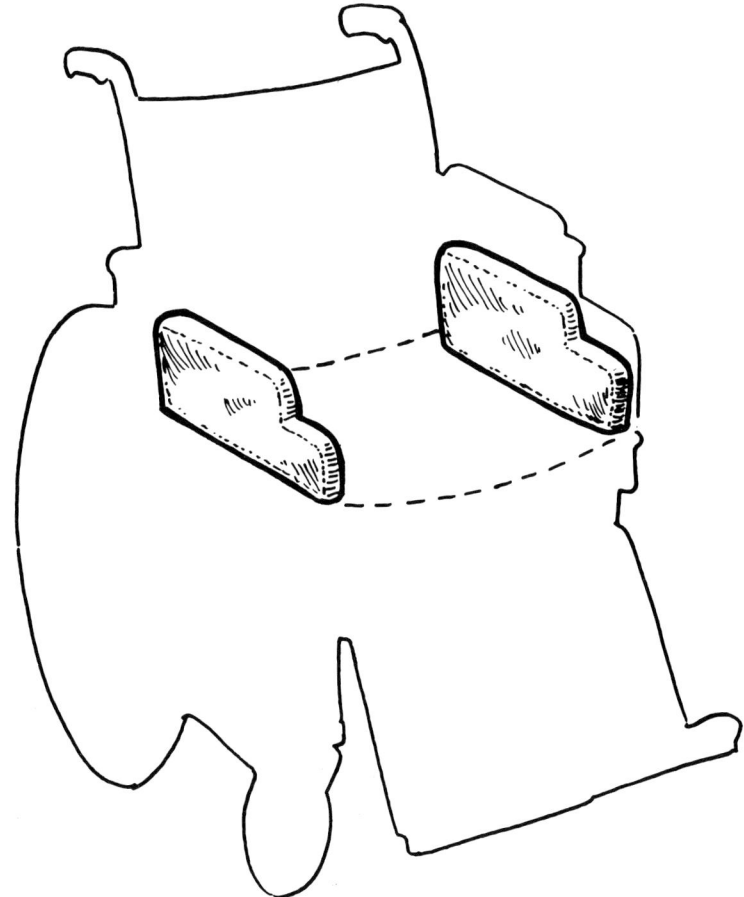

Figure 143. Pelvic side supports. From Zacharkow, D.: *Wheelchair Posture and Pressure Sores.* Springfield, Thomas, 1984. Reproduced with permission of Charles C Thomas, Publisher.

can easily occur from a low belt angle with the horizontal (Wells et al., 1986). Angles of 45 to 60 degrees from the horizontal have been suggested for the low pelvic lap belt (Cooper, 1987; Trefler et al., 1978). The lap belt should be positioned below the anterior superior iliac spines, with the line of pull in a posterior-inferior direction (Margolis et al., 1985; Wells et al., 1986).

More elaborate pelvic stabilization techniques were recently reviewed by Cooper (1987).

Trunk Supports

Mid-line trunk stabilization can be provided with lateral trunk supports (McKenzie and Rogers, 1973). In addition, with a paralytic scoliosis, corrective or supportive forces to the thoracic spine can be applied with trunk supports. However, the corrective forces must be along the line of the ribs in order to transmit the supportive forces to the spine (Nelham, 1981, 1984) (Figure 144).

With the three point pressure system, one of the trunk supports is positioned under the apex of the curve on the convex side. The other trunk support is placed high under the axilla on the concave side. The third point in the three point pressure system involves the pelvic side support on the concave side of the curve (Trefler and Taylor, 1984) (See Figure 144).

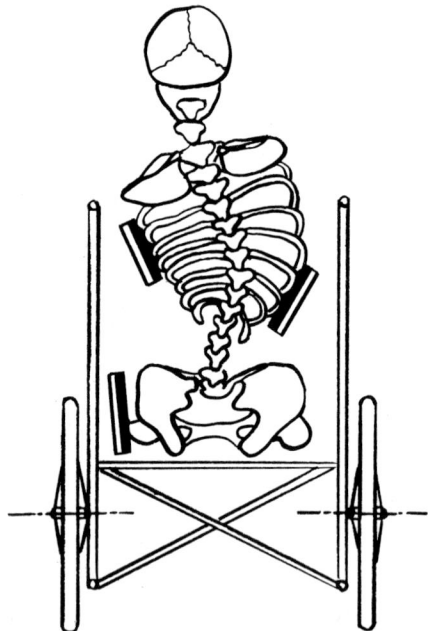

Figure 144. Correct application of supportive or corrective forces. From Nelham, R.L.: Seating for the chairbound disabled person—a survey of seating equipment in the United Kingdom. *Journal of Biomedical Engineering, 3:*267–274, October 1981. Reproduced with permission of Butterworth Scientific Ltd.

Abdominal Support

The increase in the resting intra-abdominal pressure resulting from an abdominal binder or corset can have several beneficial effects on wheelchair dependent individuals with paralysis or weakness of the transversus and oblique abdominal muscles (Lucas and Bresler, 1961; Bartelink, 1957; Morris et al., 1961; Morris, 1973; Grillner et al., 1978; Grew and Deane, 1982):

1. Increased trunk stability. Compression of the abdominal cavity will result in a distending force acting upward on the diaphragm and downwards on the pelvic floor, tending to extend the spine and support the upper trunk.

2. The proper resting intra-abdominal pressure will help obtain the normal axial relationship of the thorax and pelvis.

3. The decrease in intradiscal pressure resulting from the reduced axial loading of the lower lumbar spine may help relieve low back pain in some wheelchair dependent individuals.

4. The increase in resting intra-abdominal pressure may also help reduce the spinal stress from road shock and vibration.

5. Proper abdominal support can facilitate diaphragmatic breathing by restricting the descent of the diaphragm when sitting, thereby resulting in a better diaphragmatic excursion, improved rib cage expansion, and an increased vital capacity (Dail and Affeldt, 1957; Kirby et al., 1966; Goldman et al., 1986).

6. In individuals prone to postural hypotension, such as high level quadriplegic individuals, proper abdominal support can prevent the pooling of blood in the splanchnic region (Emerson, 1911; Sieker et al., 1956; Rowell et al., 1972; Krebs et al., 1983).

7. The beneficial effects of proper abdominal support on trunk stability, cardiac output, and diaphragmatic breathing will all contribute to reducing the fatigue and energy expenditure of many wheelchair dependent individuals.

If the abdominal support is placed too high, the beneficial effects will be lost, as the lower abdomen will protrude beneath the lower end of the support. Therefore, the inferior border of the abdominal support should be approximately 1/2 inch above the symphysis pubis, and the superior border should be approximately 1/2 inch below the xiphoid process (Berger and Lusskin, 1975). The abdominal support should fit the snuggest at the bottom, diminishing in pressure from below upwards. The upper part of the abdomen and the lower rib cage should be free from constriction (Lovett, 1916; Goldman et al., 1986).

Systemic Risk Factors

Many systemic factors will greatly increase the risk of developing pressure sores for an individual already displaying postural risk factors. Some of these factors are discussed in the following paragraphs.

Poor Nutrition

Specific nutritional factors considered critical in reducing the body's resistance to external pressure are protein malnutrition, ascorbic acid deficiency, and zinc deficiency (Mulholland et al., 1943; Rudd, 1962; Moolten, 1972, 1977; Agarwal et al., 1985; Husain, 1953; Ringsdorf and Cheraskin, 1982; Burr, 1973; Natow, 1983; Cerrato, 1986; Pinchcofsky-Devin and Kaminski, 1986).

Infection

In animal studies on the etiology of pressure sores, Groth (1942) and Husain (1953) both found that bacteria tended to be localized at sites of tissue compression.

According to Robson and Krizek (1978):

> "Skin and soft tissue that has been injured by denervation or pressure-ischemia can become infected from endogenous sources by serving as a locus minoris resistentiae. Known sites of infection distant in the body may seed these locally resistance-depressed areas. The urinary tract is the prime source in the paraplegic patient. Therefore prevention and eradication of these infections is paramount if skin breakdown is to be avoided."

It is the impression of Daniel et al. (1985) that "deep cavitary pressure sores occur because of a systemic bacteremia with localization to a pressure-damaged area, followed by infection and eventual rupture through the skin."

Low Blood Pressure

Individuals with low blood pressure will tolerate less external pressure on the sitting surface compared to individuals with normal blood pressure (Holstein et al., 1977; Larsen et al., 1979; Matsen et al., 1981). Due to its effect on the systemic blood pressure, dehydration is considered an important risk factor (Andersen and Kvorning, 1982; Holstein et al., 1979).

Occlusive Arterial Disease

Due to the reduced arterial blood supply, occlusive arterial disease will decrease the tissue tolerance for external pressure (Husain, 1953; Matsen et al., 1981).

Fever

Fever will increase the risk of pressure sores by raising the body's metabolic rate and its demand for oxygen (Torrance, 1981; Schell and Wolcott, 1966).

Edema

Edema will impair oxygen delivery and metabolic exchange due to the increased distance from the capillaries to the cells (Tepperman et al., 1977; Schell and Wolcott, 1966). In regards to nutritional factors, edema can result from hypoproteinemia (Ruschhaupt, 1983).

Smoking

Experimental studies have shown a reduced blood flow from smoking, attributed to the vasoconstrictive effects of nicotine (Mosely et al., 1978; Kaufman et al., 1984; Reus et al., 1984).

In a study among spinal cord injured individuals, cigarette smoking was found to be positively correlated with both a higher incidence of pressure sores and more extensive pressure sores (Lamid and El Ghatit, 1983). A survey by Barton and Barton (1981) found male smokers four times more likely to develop pressure sores on the heels than male non-smokers.

Massage

A common misconception is to massage the reddened skin areas over pressure points on the buttocks (Shannon, 1984). According to Torrance (1983), "Non-blanching hyperaemia is suggestive of damage to the microcirculation and massage will only increase the mechanical trauma to the disrupted vessels."

Among a geriatric population, Dyson (1978) reported a 38 percent reduction in pressure sores when the practice of rubbing over pressure areas on the buttocks was discontinued.

Moisture

Due to incontinence or perspiration, moisture can increase the frictional damage to the skin as well as prolonging any shearing force on the buttocks (Sulzberger et al., 1966; Lowthian, 1976).

Maceration of the skin can also result from incontinence, making the skin more susceptible to tearing and to infection (Lowthian, 1975, 1982). Besides the maceration associated with fecal incontinence, the exposure of skin to the bacteria and toxins in stool may be an important factor in pressure sore development (Allman et al., 1986).

THE ROLE OF PRESSURE IN PRESSURE SORE ETIOLOGY

Point Pressure Versus Uniform Pressure Distribution

A common finding found from a review of the literature on pressure sores is that the primary causative factor in pressure sore formation is external pressure exceeding capillary pressure on the weight-bearing surface. By microinjection technique, Landis (1930) found the mean capillary blood pressure to be 32 mm Hg in the arteriolar limb of the capillary loop in human skin (at the base of the fingernail), 20 mm Hg at the summit of the loop, and 12 mm Hg in the venous limb of the capillary loop. These figures are cited often in the literature.

For example, Houle (1969) in his evaluation of wheelchair seats and cushions referred to a prolonged sitting pressure of 32 mm Hg or more as the main factor in producing pressure sores. Souther et al. (1974) also based their evaluation of wheelchair cushions on mean capillary pressure. Pressures in their study were measured under the right ischial tuberosity. Agris and Spira (1979) mentioned that a prolonged external pressure applied to the skin greater than 32 mm Hg would result in tissue anoxia and cell death. A book on pressure sores mentions that a pressure as low as 11 mm Hg will impair cutaneous blood flow (Constantian and Jones, 1980) and that a prolonged pressure above 11 mm Hg could result in irreversible microcirculatory damage (Eriksson, 1980).

Most articles in the literature, when referring to the relationship between pressure sore formation and capillary blood pressure, cite Kosiak's research. In his paper written in 1959, Kosiak mentioned that it would appear reasonable that complete tissue ischemia might occur when external pressures approximating capillary blood pressure (13 to 32 mm Hg) are applied to the body. He concluded that temporary relief of external pressure is necessary every few minutes.

Using dogs, Kosiak reported microscopic pathological changes in tissues to which a pressure of 60 mm Hg had been applied for only one hour. However, it must be remembered that very localized point pressures were applied in Kosiak's study. The pressure was applied over a dog's femoral trochanter and ischial tuberosity by the inverted plunger of a 20 cc hypodermic syringe.

There is, however, a great difference between a point pressure application and the potential area for pressure application on the human buttocks and upper posterior thighs. Husain (1953) stressed that "It is obvious, therefore, that in dealing with pressure effects one must be careful to distinguish between evenly distributed pressure, which is tolerated to a remarkable extent, and localised or point pressure, whereby a pressure gradient is induced in tissues and vascular compression follows."

According to Bennett and Lee (1985):

> "The issue of localized versus a uniform pressure is a significant one in terms of shear importance: a localized pressure usually entails the generation of large shear values, whereas a uniform pressure distribution does not. Given a simple loading, the resultant shear stress is proportional to the pressure gradient.
>
> It follows that as the pressure gradient becomes steep, the corresponding shear stress becomes large. When a piston is pushed directly into tissue, we would anticipate a high shear stress in proximity to the piston perimeter because here the applied pressure changes from full value to zero."

Chow and Odell (1978) concluded that:

> "Localized pressure causes deformation, mechanical damage and blockage of blood vessels. Hydrostatic pressures (with equal components in all directions) cause little or no deformation.
>
> The state of stress observed in the buttocks can be decomposed into a combination of shear stress and hydrostatic stress. If one accepts the observation that hydrostatic pressure is relatively harmless to biological tissues, then the harmful stress must be some form of shear. Shear stress is involved in uniaxial pressure, localized pressure, any nonuniform pressure distribution or any pressure that causes distortion."

Scales (1980) actually referred to pressure sores as tissue distortion sores.

In the sitting position, any tissue deformation and shearing forces will be particularly harmful to the adipose tissue for the following reasons:

1. Compared to muscle with its thick fascial covering, and skin with its extremely thick capillary walls, adipose tissue lacks significant tensile strength. Adipose tissue is therefore the most vulnerable to shearing forces that may compromise its blood supply by the angular stretching of the vessels (Shea, 1975; Reichel, 1958).

The arterial and venous capillaries in the subcutaneous fat have walls that are only 0.1 to 0.3 um thick, whereas the capillary walls in the dermis are 2 to 3 um thick (Braverman and Keh-Yen, 1981).

According to Braverman and Keh-Yen (1981):

> "The cutaneous microcirculatory blood vessels are unique. They have extremely thick walls, a highly developed system of endothelial cell-related external filaments that most likely have an anchoring function, and a distinctive internal and external pattern of elastic fiber deployment. All three features would be protective against the constant external shearing forces to which these vessels are subjected."

2. The most critical areas for shear stress will be over and just adjacent to the weight bearing bony prominences (Chow, 1974; Le et al., 1984). Using a 6 inch (15.2 cm) diameter gel hemisphere to simulate the human buttock, Chow (1974) found the critical area for shear stress to be 1/2 to 1 inch (1.3 to 2.5 cm) from the tip of the hemisphere. This would correspond to the region just adjacent to the tip of the ischial tuberosity. However, the deep soft tissue coverage of the ischial tuberosity (along with the sacrum and greater trochanter of the femur) in the sitting position does not include muscle, but only adipose tissue (Daniel and Faibisoff, 1982; Minami et al., 1977).

Studies Involving a More Uniform Pressure Application

Other studies may have more application to the potential tolerance for external pressure when sitting, especially in association with a cushion providing a good uniform pressure distribution. These human studies have avoided a severe point pressure application and thereby minimized the tissue deformation and shear forces.

Holstein et al. (1977) used an isotope washout technique to measure the external pressure required to stop skin blood flow. The external pressure was applied by an 11 cm by 11 cm plastic cushion placed beneath a blood pressure cuff. The external pressure on the calf required to stop the skin blood flow in normal subjects was found to lie between the diastolic and mean arterial pressure as measured in the femoral artery. (It was on the average 10.8 mm Hg lower than the systemic mean arterial blood pressure.) During induced variations in the individual's systemic blood pressure, the external pressure required to stop skin blood flow was found to vary parallel to the systemic mean arterial blood pressure.

An isotope washout study by Larsen et al. (1979) compared the external pressure required to stop skin blood flow on the lower back in normotensive individuals, hypertensive individuals, and individuals with paraplegia and quadriplegia. An 11 cm by 11 cm plastic cushion was also used for the pressure application. In all groups, the external pressure required to stop skin blood flow was slightly below the auscultatory mean arm blood pressure.

Dahn et al. (1967) applied external pressure to a segment of the calf with an air-filled, 17 cm wide plethysmograph. By an isotope clearance method, they found that muscle blood flow stopped in the five normal subjects when the external pressure was near the diastolic blood pressure, measured in the femoral artery of the same leg.

Nielsen (1983a) reported the cessation of blood flow in both subcutaneous tissue and muscle at an external pressure close to the local diastolic blood pressure. An isotope washout technique was also used, and tissue pressure was increased by a 50 cm wide cuff around the calf.

It would appear from the preceding studies that with a more evenly distributed external pressure, the pressure required to stop blood flow is much greater than the capillary blood pressure, being somewhere between the diastolic and mean blood pressure.

External Pressure and Dependency

According to Holstein (1982), "the potential of tissue perfussion is determined by the driving pressure, i.e., the difference between the actual local arterial blood pressure and the local venous pressure."

Based on the arteriovenous gradient theory, with a uniformly applied external pressure the increased tissue pressure will result in an increase in local venous pressure. As a result, the local arteriovenous gradient will be diminished and, therefore, local blood flow will be reduced.

As Matsen (1980) explained:

> "Some reduction in local arteriovenous gradient can be compensated for by changes in local vascular resistance. This process, known as autoregulation, maintains local blood flow over a range of arteriovenous gradients. However, when the arteriovenous gradient is significantly

reduced, autoregulation becomes relatively ineffective. At this point local blood flow is determined primarily by the local arteriovenous gradient."

However, if a limb is placed below heart level in a dependent position, there will be an increase in both the local arterial and venous pressures. The increase in blood pressure will be .77 mm Hg for every centimeter below heart level (Ganong, 1979). Matsen et al. (1980) have demonstrated that a given external pressure will have less of an effect on the local arteriovenous gradient if the limb is in a dependent position, as opposed to being at heart level (Figure 145). This is an important point to consider in regards to the tolerance for external pressure in the sitting position, where the buttocks are not at heart level but in a dependent position.

Nielsen (1982, 1983b) determined the change in muscle and subcutaneous tissue blood flow when external pressure was applied by a whole leg cuff to the mid-calf region in both the horizontal position and a position 45 cm below heart level. With the mid-calf region at heart level, an external pressure of 60 mm Hg reduced muscle blood flow by 59 percent and subcutaneous blood flow by 79 percent (Nielsen, 1982). However, with the mid-calf region in a dependent position of 45 cm below heart level, an external pressure of 60 mm Hg reduced muscle blood flow by only 19 percent and subcutaneous blood flow by only 45 percent (Nielsen, 1983b). The greater reduction in fat blood flow compared to muscle blood flow in both the horizontal and dependent positions can most likely be attributed to the greater deformation of the subcutaneous blood vessels from the external cuff compression.

Beneficial Effect of External Pressure

As early as 1930, Trumble discussed the beneficial effect of moderate pressure. By assisting lymph flow and preventing edema, Trumble (1930) felt that moderate pressure would be an important factor in the prevention and treatment of pressure sores. Krouskop et al. (1978) considered lymph vessel occlusion to be a major factor in pressure sore formation. The resulting accumulation of metabolic waste products will result in tissue necrosis.

In experiments with dogs, Miller and Seale (1981, 1985) found the terminal lymphatic clearance of sulfur colloid from the subcutaneous tissue to increase with increased external pressures up to 60 mm Hg.

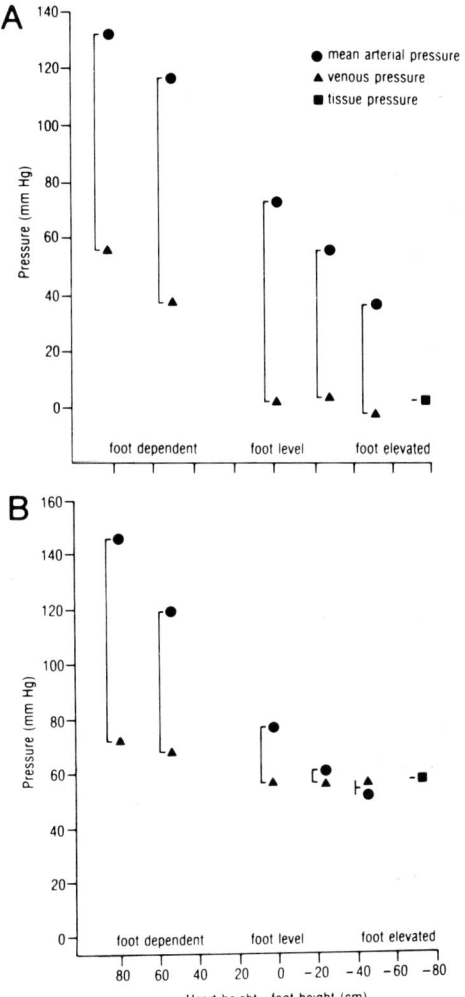

Figure 145.
A. The effect of foot position on the local arteriovenous gradient in a limb with normal tissue pressure. The pressures within the dorsalis pedis artery and a dorsal foot vein were measured by direct vessel cannulation. The position of the limb relative to the heart was manipulated by tilting the subject on a circle bed. The local arteriovenous gradient is represented by the vertical lines.
B. The effect of foot position on the local arteriovenous gradient in a limb with increased tissue pressure. Tissue pressure was increased to 60 mm Hg by the external application of pressure. Notice that with the limb in a dependent position, there will be less of an effect on the local arteriovenous gradient.
From Matsen, F.A.: *Compartmental Syndromes.* New York, Grune and Stratton, 1980. Reproduced with permission of Grune and Stratton, Inc.

According to the authors (1985):

> "The increase in the lymphatic clearance rate can be attributed to greater pressure gradients between the interstitial fluid and the open lymphatic vessels. External pressure applied to the surface of the skin causes an increase in the interstitial fluid pressure of the subcutaneous tissue, which drives the lymph through the terminal vessels into the larger lymph vessels."

The lymphatic clearance of sulfur colloid from the subcutaneous tissue stopped at external pressures of approximately 75 mm Hg. This was attributed to the collapse and occlusion of the terminal lymph vessels at this external pressure. However, the bottom of the cylinder used to apply pressure to the dog's thigh was a plastic disc 4.5 cm in diameter. Therefore, greater subcutaneous tissue deformation would be involved compared to a more uniform application of external pressure.

WHEELCHAIR CUSHION SELECTION

Regardless of the cushion eventually selected for the wheelchair dependent individual, it is obviously advantageous to first remove as much superincumbent body weight as possible from the sitting surface. Proper fitting chair arms, the proper backrest height and inclination with pelvic-sacral and thoracic support, and proper foot support are therefore all critical (Swearingen et al., 1962; Brattgård et al., 1978; Zacharkow, 1984a).

Next, it is advantageous to remove as many bony prominences as possible from the sitting surface. The use of a pelvic-sacral support with the proper thigh to trunk angle will remove the coccyx and sacrum from the seat. In addition, more weight will be shifted anterior to the ischial tuberosities and over the upper posterior thighs.

The wheelchair cushion should always be considered as one component of the total seating system. This system includes the wheelchair, the wheelchair cushion, the individual sitting, and his posture. Posture must be analyzed regarding the influence of intrinsic factors as well as extrinsic factors (the wheelchair and wheelchair cushion).

Two essential functions of a good wheelchair seating system are (1) to provide a uniform pressure distribution and (2) to provide a stable sitting surface (Graebe, 1979; Cousins et al., 1983).

Uniform Pressure Distribution

The objective of seat cushioning for able-bodied individuals is to obtain a relatively wider distribution of pressure at pressure points (Asatekin, 1975). However, in wheelchair seating for pressure sore prevention, the objective is an overall uniform pressure distribution.

Pressure Transducers

Wheelchair cushion selection is often based on single pressure readings using a pressure transducer under certain bony prominences. Usually the closer the pressure reading under each ischial tuberosity is to 20 to 32 mm Hg (representing mean capillary blood pressure), the better the cushion is considered to be.

However, a single pressure reading taken under each ischial tuberosity reveals little regarding the overall pressure distribution of the cushion (Chow and Odell, 1978). These pressure measurements are affected by many variables, and their relationship to actual tissue pressures has not been determined (Cochran and Palmieri, 1980; Palmieri et al., 1980). Patterson and Fisher (1979) reported a large degree of error in pressure measurements from small transducers when they were used over an uneven bony surface, similar to the ischial tuberosity.

A major problem with any pressure transducer is the compliance problem at the cushion-buttocks interface (DeLateur et al., 1976). As Grant (1985) stated, "Any pressure transducer will, to some extent, distort the interface which is to be measured."

Instead of a small sensor for a single pressure reading, a large pressure pad may be used, consisting of many transducers that covers the entire sitting surface. However, these pads will produce major alterations of the cushion-buttocks interface (Barbenel, 1983). The added surface tension from these pressure evaluation pads will interfere the most with cushions that have the best uniform pressure distributing qualities (Graebe, 1977).

When wheelchair cushion selection is based solely on pressure readings, two critical factors in pressure sore etiology are ignored:

1. Shear stress is ignored.
2. Wheelchair sitting as a dynamic activity is ignored.

Shear Stress

According to Noble (1977):

"There are basically two shear stress components acting at the skin/cushion interface. The shear stress component acting perpendicular to the skin is generated essentially by the non-uniformity of the pressure distribution and so where contact pressures change rapidly the shear stress is highest.

The other shear stress component acts parallel to the shear/support interface and is directly influenced by frictional conditions. This stress is increased by any movement of the body parallel to the support surface and reaches quite high levels during such activities as the propulsion of wheelchairs, repositioning and transferring."

Sitting as a Dynamic Activity

The reliance on pressure measurements taken at one point in time assumes sitting to be a static activity. However, the sitting body is in a "dynamic state of continuous activity" (Branton, 1966). Even slight body movements will introduce "large variations in pressure and shear gradients" (Black and Reed, 1983).

Patterson and Fisher (1980) made a continuous recording of the ischial tuberosity pressures of twelve paraplegic individuals over an entire day. The oscillatory changes in pressure that were observed throughout the day were attributed to shearing forces (Figure 146). According to Patterson (1984), "these oscillations in pressure may be caused by either motion of the individual or by roughness in the road surface that is transmitted to the body."

Besides the shearing forces due to propelling the wheelchair, the oscillations in pressure and shearing forces from road shock and vibration would affect all individuals both in their wheelchairs and in automobiles (See Figure 146). These forces would be the most detrimental to individuals sitting in a slumped, kyphotic posture without proper pelvic stabilization and upper sacral support from the backrest. Such a posture in itself results in high pressures and shearing forces over and posterior to the ischial tuberosities. Branton (1969) referred to the possibility of "relatively fast oscillatory movements of the pelvis rocking over the tuberosities. This possibility arises in all sitting postures in which the top of the sacrum is not resting against a back rest."

Patterson (1984) stressed the need to focus on the dynamic and shear relieving properties of wheelchair cushions rather than the static pressure relieving property.

According to Noble (1978b), an important factor in cushion design is "the response of materials to transient dynamic or inertial loadings. Whilst a cushion may result in acceptable contact pressures during static

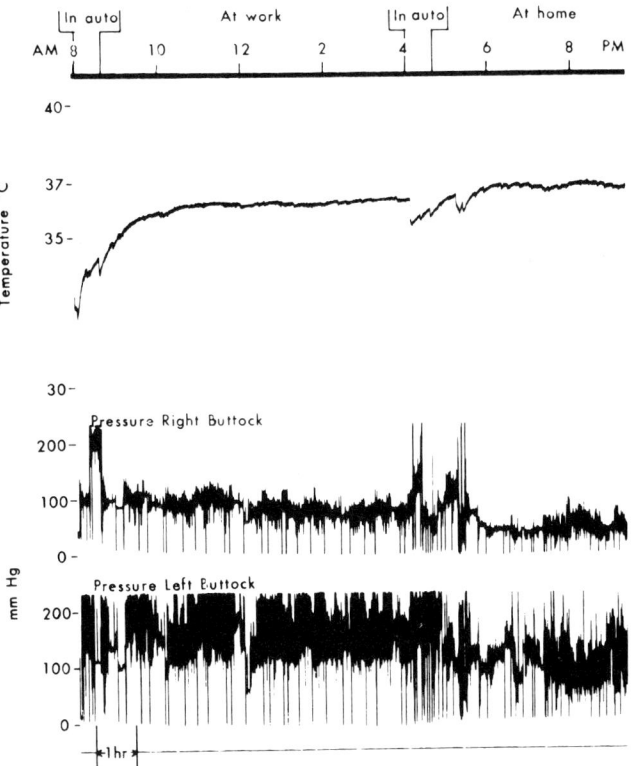

Figure 146. A typical pressure record from one subject over a twelve hour period. Note in particular the oscillations in pressure when in an automobile. From Patterson, R.P., and Fisher, S.V.: Pressure and temperature patterns under the ischial tuberosities. *Bulletin of Prosthetics Research, 10-34(Vol. 17, No. 2)*:5–11, Fall 1980.

sitting, it may cause irreversible tissue trauma during movement through a lack of resilience or shock absorbing capacity. A good example of this is a moulded plaster seat which is completely incapable of adopting to the changes of weight distribution and anatomical shape accompanying the postural movements associated with daily activities."

Regarding the shock absorbing capacity of wheelchair cushions, Kauzlarich (1981) studied the effect of various cushions in reducing the force transmitted to the wheelchair user when going over a bump. The Roho® cushion, an air-filled cushion with interconnected air cells, performed the best in reducing the user acceleration due to the wheelchair going over a bump.

The worst cushion in regards to shock absorption was a viscoelastic foam cushion. When loaded, viscoelastic foam will slowly mold to the contours of the sitter's buttocks. However, it has a high stiffness to rapid movement.

According to Kauzlarich (1981), the viscoelastic foam cushion "distributed the load of the seated patient much slower than all of the other cushions, and for sudden impacts there was almost no cushion deflection." The cushion "acts like a board placed across the seat so that the force transmitted to the patient when going over a bump is about the same as when no cushion was used" (Kauzlarich, 1981).

Viscoelastic foam cushions were found by Krouskop et al. (1983) to have "an unusually high association with skin irritation and tissue breakdown." This observation is possibly related to the poor dynamic and shear relieving properties of viscoelastic foam.

Cushion-Covering Membranes

Uniform pressure distribution is directly related to the envelope property of the cushion. The envelope property is "the geometrical wrapping of the cushion around the buttocks" (Chow, 1974). A good enveloping cushion will provide a large contact area for the buttocks (Chow, 1974).

Many fluid flotation cushions (water, air) and gel cushions lose their ability to provide a uniform pressure distribution due to their enclosing membranes. The increased surface tension of the membrane will result in a "hammock effect" (Chow, 1974; Chow et al., 1976), and adversely affect the envelope property and pressure distribution of the cushion. The stiffer the membrane, the greater the "hammock effect" (Figure 147).

As taut, nonstretch covers will increase the surface tension at the cushion-buttocks interface, a loose-fitting two way stretch cover should be used over wheelchair cushions. In addition, a "hammock effect" can also be caused by the individual wearing tight, nonstretch pants (Manley et al., 1977).

In addition to the "hammock effect," friction or interface shear on the skin is also produced by the membrane covering the cushion (Chow, 1974). The lesser the coefficient of friction of the membrane, the lesser the interface shear on the skin (Chow et al., 1976). However, some interface shear is necessary for sitting stability.

Cushion Stability

In regards to able-bodied seating, Branton (1966) commented that if the seat features fail to stabilize the sitter, then the individual must

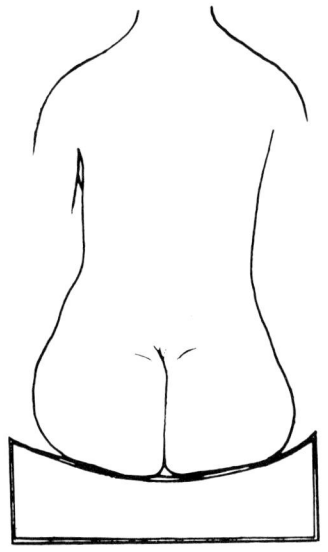 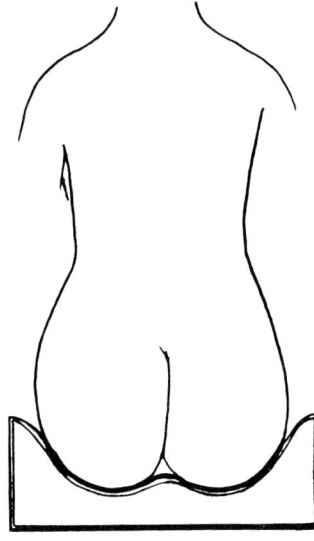

Figure 147.
    Left: The hammock effect. The body is supported by tension in the cushion-covering membrane.
    Right: A cushion-covering membrane that conforms sufficiently to the buttocks.
    From Torrance, C.: *Pressure Sores. Aetiology, Treatment and Prevention.* London, Croom Helm, 1983. Reproduced with permission of Croom Helm Ltd.

stabilize himself. "The greater the instability, the less likely is muscular relaxation and hence the greater is discomfort" (Branton, 1966).

Kohara and Sugi (1972) also considered stability to be the most important factor influencing seat comfort.

Branton (1966) commented in regards to cushioning for able-bodied individuals that "a state can easily be reached when cushioning, while relieving pressure, deprives the body structure of support altogether and greatly increases instability."

These comments have even greater implications for wheelchair dependent individuals, with disabilities involving various degrees of muscle weakness or paralysis, along with balancing problems. Instability in regards to cushioning can markedly increase the fatigue and energy expenditure of the wheelchair sitter. Due to cushion instability, slumped and asymmetrical postures will often be adopted. These are postures associated with a greatly increased risk of pressure sores.

The cushions considered to result in the most unstable sitting platform are water-filled cushions, air-filled cushions, and gel cushions (Mooney et al., 1971; Bolton, 1972; DHSS, 1980; Cochrane and Wilshere,

1982; Jay, 1983). The resulting sitting instability with these cushions will present additional problems with propelling the wheelchair and transferring.

Most foam cushions will result in a very stable sitting platform. Although an air-filled cushion, the Roho® Dry Floatation Cushion also provides a fairly stable sitting surface (Nelham, 1981; Jay, 1983). This is due to the restricted air flow between the interconnected air cells, along with the avoidance of the surface tension problems associated with other fluid-filled cushions.

Energy Management

Energy management is often neglected by clinicians when prescribing wheelchair cushions and evaluating the total seating system. Sitting stability is a critical component of energy management (Graebe, 1979). The more stably an individual sits, the less fatigued he will be after a full day of sitting.

The weight of the wheelchair cushion is another important consideration in energy management. According to Graebe (1979), "a lightweight cushion is important, not only in the ease of handling the cushion but it also reduces the work expended in the acceleration and deceleration of a wheelchair. The less mass the person must move the less energy they expend."

Other critical components of energy management involve the frequency and technique of weight shifting, and the temperature effects of the wheelchair cushion.

Weight Shifts

Agris and Spira (1979) mentioned that the wheelchair dependent individual should "shift his weight or elevate himself for approximately 1 minute out of every 15 minutes." Schell and Wolcott (1966) advocated a wheelchair push-up every ten to fifteen minutes for individuals with major disabilities.

However, for many wheelchair dependent individuals, it can be extremely fatiguing to shift one's weight every 15 minutes. This is especially true if the wheelchair push-up or extreme side-to-side weight shift is advocated by the clinician for pressure relief.

Attention-getting weight shifts such as push-ups and extreme side-to-side pressure reliefs will probably not be done when the individual finds himself in social situations outside of the hospital (Zacharkow, 1984a; Jones, 1979). Interestingly, no correlation has been found in studies between the frequency of wheelchair push-up pressure reliefs and a history of pressure sores in spinal cord injured individuals (Patterson and Fisher, 1980; Fisher and Patterson, 1983; Merbitz et al., 1985).

Occasional weight shifts, however, are still beneficial for the following reasons:

1. To prevent the build-up of temperature and humidity under the buttocks (Clark, 1974; Patterson and Fisher, 1980).
2. To promote proper disc nutrition (Krämer, 1977, 1981; Krämer et al., 1985).
3. To enhance lymph flow through variations in total tissue pressure (Casley-Smith, 1983; Olszewski and Engeset, 1980; Szabó et al., 1980).

These benefits of weight shifting can be accomplished most easily with a forward weight shift (Figure 148). This type of weight shift is not as conspicuous and much less fatiguing than wheelchair push-ups. With the forearms resting on the thighs or wheelchair armrests, the resulting anterior sitting posture with a forward weight shift will be much more stable compared to the postures assumed with other weight shifting techniques.

In addition, if the individual is sitting in a lordotic posture with a pelvic-sacral support in place, very minimal hip flexion is needed with a forward weight shift to completely relieve the ischial tuberosities of pressure. The forward weight shift will also be facilitated with a firm cushion (Brand, 1980; Shah, 1981; Black et al., 1981; Ayoub, 1972). When performing the forward weight shift, it is extremely critical that the motion occurs at the hips and not by spinal flexion (Goldthwait, 1909; Mosher, 1919; Schurmeier, 1927; Garner, 1936; Pope, 1985b).

In regards to the beneficial effect of a forward weight shift on fatigue and energy expenditure, Hanson and Jones (1970) reported a significant decrease in heart rate with a forward leaning sitting posture with fore-arms resting on the thighs, compared to a more upright sitting posture.

In an interesting study by Bardsley et al. (1983), movement recordings using a force platform were made of twenty-four spina bifida children while they watched television. A significant association was found between a history of pressure sores and the direction of pressure relief movements. A greater proportion of children who aligned their pressure relief movement in an anteroposterior direction had a history of no pressure sores

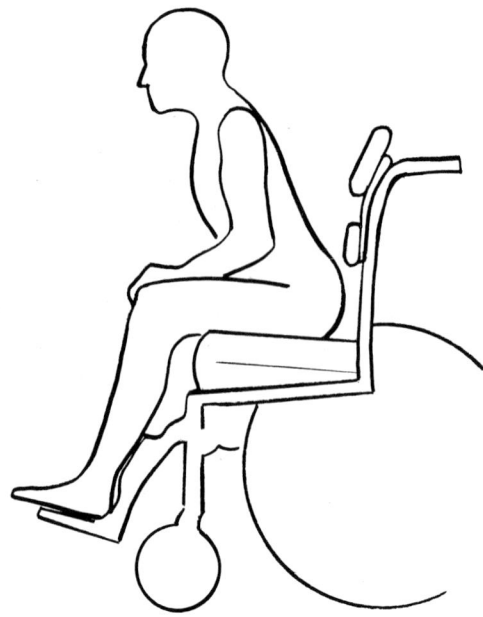

Figure 148. Forward weight shift. Adapted from Zacharkow, D.: *Wheelchair Posture and Pressure Sores.* Springfield, Thomas, 1984. With permission of Charles C Thomas, Publisher.

compared to children who aligned their pressure relief movement in a lateral direction.

According to the authors, anteroposterior movements "appear to offer a more effective quality of pressure relief than do lateral movements. The results can be understood if the relief is obtained by the patient rocking rather than lifting. Rocking sideways to reduce the pressure over high-risk sites, such as the ischial and femoral tuberosities of one side, will increase the pressures on the contralateral side. Rocking forward or backward allows of relief of the high-risk areas on both sides, probably by transferring the load to the thighs."

Temperature

The temperature effects of wheelchair cushions can also have important implications regarding the energy expenditure of the wheelchair dependent individual. For example, one advantage given for the use of water-filled seat cushions is that they will cause a beneficial drop in skin temperature under the ischial tuberosities (Fisher et al., 1978).

However, there can be major problems associated with the use of a cold cushion:

1. A cold cushion will conduct heat away from the body (Grandjean, 1980a). This loss of heat not only can increase the energy expenditure of the individual, but it can also be very uncomfortable (Grandjean, 1980a; Cousins et al., 1983).
2. A drop in skin temperature can result in vasoconstriction of the skin vessels, and thereby decrease the blood flow (Walther et al., 1971; Mooney et al., 1971; Crewe, 1983; Ek et al., 1985).
3. In individuals with lower extremity spasticity, sitting on a cold cushion may cause an increase in leg spasms (Guttmann, 1976; DHSS, 1980).

The vibration associated with movement of the wheelchair can also increase the leg spasms when sitting on a cold cushion. The tonic vibration reflex has been shown to be much stronger when a muscle is cold, whereas vibration reflexes are depressed when a muscle is warm (Eklund and Hagbarth, 1966; Carlsöö, 1982).

According to the philosophy for using a cold wheelchair cushion, the skin temperature gives an indication of the tissue temperature, and therefore of the metabolic need of the tissue beneath the skin surface. An increase in tissue temperature in conjunction with tissue ischemia may increase the susceptibility to pressure sores as tissue metabolism and oxygen demand increase at a rate of 10 percent per centigrade degree rise in temperature.

However, this philosophy for using a cold cushion assumes that tissue ischemia will automatically result from sitting on a wheelchair cushion. Preliminary studies by Holloway et al. (1985) involved the use of laser Doppler flowmetry to continuously monitor the skin blood flow over the ischial tuberosities of paraplegic individuals while seated on various wheelchair cushions. The results often revealed blood flow levels indicative of a nonischemic state. Therefore, an increase in skin temperature may also be indicating an increase in skin blood flow and tissue perfussion (Kroese, 1977; Hutchison et al., 1983; Barton and Barton, 1981).

Humidity

Another important consideration is the humidity at the cushion-buttocks interface, as there will be a marked decrease in tensile strength of the skin with an increase in relative humidity (Wildnauer et al., 1971). A large increase in relative humidity will occur with the use of non-porous covers such as plastic and vinyl (Stewart et al., 1980; Brattgård et al., 1976). The individual's clothing, such as non-porous underwear, can also adversely affect the humidity of the sitting area (Brattgård et al., 1976; Crewe, 1983). In the wheelchair users survey reported by Fenwick (1977), the most complaints of discomfort involved plastic covered cushions, which were described as being hot and clammy.

In regards to the use of sheepskin covers, a polyester fabric simulation of sheepskin was found to be very poor for absorbing water vapor. However, resilient natural sheepskin was found to absorb nearly fifteen times as much water as a polyester fabric simulation of sheepskin (Denne, 1979a, 1979b).

Specific Cushion Recommendations

With the various claims being made by cushion manufacturers as to the superiority of their product based on pressure measurements or clinical trials, it is easy for the clinician to become overwhelmed when trying to decide on the best cushion for a wheelchair dependent individual.

However, based on the concepts of uniform pressure distribution, stability, energy management, and the deleterious effects of road shock and vibration as presented in this chapter, the cushion needs of the vast majority of wheelchair dependent individuals can be met with two types of wheelchair cushions:

1. The Roho® Dry Floatation cushion (high profile model).
2. Latex foam cushions.

The Roho® cushion, with its ability to reduce high pressure gradients and surface tension problems would be the cushion of choice for the high-risk individuals as described in this chapter (Zacharkow, 1984a; Noble, 1978b; Shaw and Snowdon, 1979).

In regards to foam cushions, latex foams are preferred. A major difference between latex foam and polyurethane foam is in their ability to undergo local deformation, with polyurethane foams being more prone to shape distortions.

As Marchant (1972) observed:

"If an indenting force is applied to a latex foam, the foam is evenly compressed throughout its thickness. With polyether foam the compression is confined to a localised area near the indentor."

Cochran and Slater (1973), in their evaluation of wheelchair cushions, also found latex foams to have excellent pressure distributing qualities.

However, test results with polyurethane foams would imply that "relatively small changes in applied load or indentation, generated by motion, could produce large changes in sitting pressure or possible 'bottoming.'"

Regarding cushion covers, a two-way stretch cover that is both loose fitting and porous will help minimize both surface tension and humidity problems at the cushion-buttocks interface.

Problems With Foam Cutout Cushions

The danger with foam cutout cushions is that they can result in high shear forces at the edges of the cutout (Brand, 1978; Aronovitz et al., 1963). Bennett (1975) mentioned that shear stress would be at a maximum with a sharp increase or decrease in compressive stress. The cutout edge is often only 1/2 to 1 inch (1.3 to 2.5 cm) lateral to the ischial tuberosity (Figures 149 and 150).

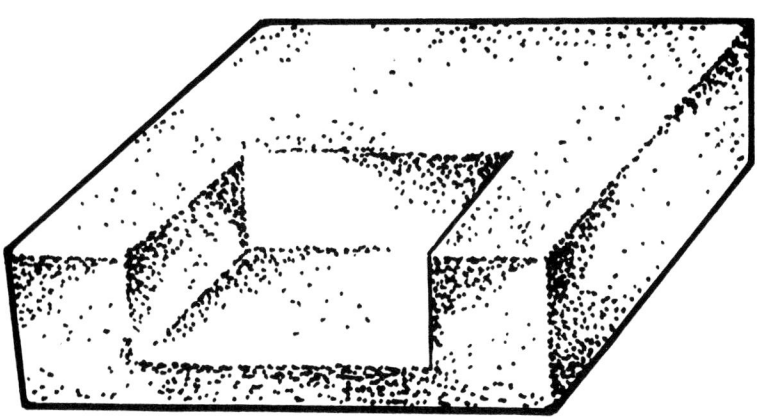

Figure 149. Foam cutout cushion. From Zacharkow, D.: *Wheelchair Posture and Pressure Sores.* Springfield, Thomas, 1984. Reproduced with permission of Charles C Thomas, Publisher.

Too often cutout cushions are constructed with ischial pressures at 0 mm Hg. This can result in edema formation in the ischial region and increase the potential for pressure sores due to the impaired nutrition of the cells (Trumble, 1930).

In addition to the high shear forces at the edges of the cutout, other problems associated with use of the cutout cushion include the following:

1. The wheelchair dependent individual must consistently sit with his ischial tuberosities in exact alignment within the cutout area of the cushion. This will obviously be difficult if not impossible for many

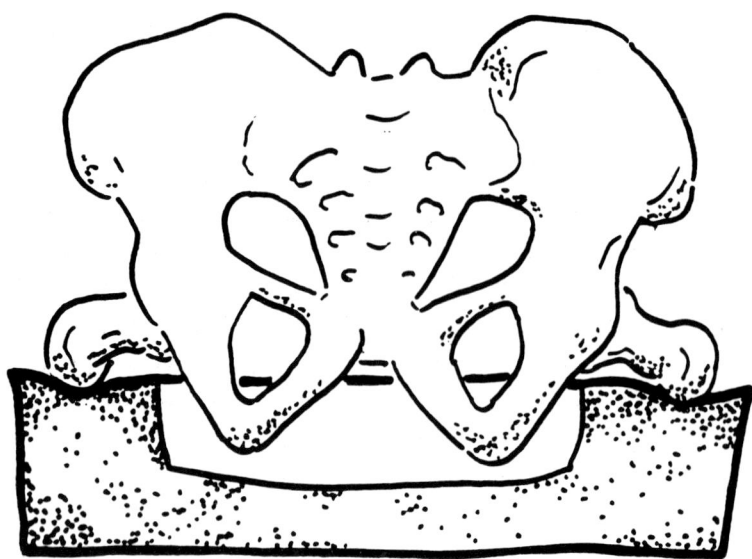

Figure 150. A foam cutout cushion can result in high shear forces at the edges of the cutout. There will also be a marked shift in weight bearing to the greater trochanters. From Zacharkow, D.: *Wheelchair Posture and Pressure Sores.* Springfield, Thomas, 1984. Reproduced with permission of Charles C Thomas, Publisher.

  wheelchair dependent individuals with impaired sensation to the buttocks.

2. The cutout dimensions of the cushion will change if the cushion is used on a hammock wheelchair seat (Manley et al., 1977; Ferguson-Pell et al., 1980).

3. With the use of cutout cushions there will be a marked shift in weight bearing to the greater trochanters. However, based on their structure and function, the trochanters are completely unsuited for bearing this much weight in the sitting position (Helbig, 1978). With able-bodied individuals, trochanteric weight bearing will quickly lead to discomfort (Hertzberg, 1972; Hockenberry, 1977).

Cutout foam cushions are often used with spinal cord injured individuals. However, this population is the least suited for a marked increase in trochanteric weight bearing, with the commonly noted cartilage atrophy in the hip joint, and the loss of bone mineral resulting in osteoporosis (Abramson, 1948; Chantraine, 1978; Griffiths et al., 1972; Anderson and Breidahl, 1981).

POSTURE PROBLEMS OF SPECIFIC POPULATIONS

Lower Limb Amputee

As mentioned earlier in this chapter, the major problem with sitting balance will occur after a hip disarticulation or a high AK amputation, since the proximal femur is extremely critical for sitting stability (Spira and Hardy, 1963; Grundy and Silver, 1984). The individual will have a tendency to fall to the side of the amputation. If he has the voluntary musculature to do so, the individual will shift his body weight over the nonamputated leg in order to balance, resulting in increased pressure on the nonamputated side (Ohry et al., 1983; Zacharkow, 1984a). With the resulting asymmetrical sitting posture, back pain is a common complaint (Ohry et al., 1983).

Regarding essential wheelchair modifications, a gluteal pad will be necessary if an ischiectomy has been performed on the amputated side. The individual will also require trunk supports or a molded backrest shell to balance and sit symmetrically (Zacharkow, 1984a). Otherwise, the individual will continue to balance over his narrowed, unilateral base of support, thereby increasing the risk of pressure sores over the buttock of the nonamputated side.

Prostheses should also be considered for individuals with high unilateral or bilateral amputations, as they can markedly improve the individual's sitting balance and stabilization in the wheelchair (Chari and Kirby, 1985, 1986; Grundy and Silver, 1984; Darcus and Weddell, 1947).

Multiple Sclerosis

Individuals with multiple sclerosis have a very high incidence of pressure sores. In the Greater Glasgow Health Board area survey, the incidence of pressure sores among individuals with multiple sclerosis was 17.8 percent (Jordan and Clark, 1977). In a ten year retrospective Scottish survey of patients with pressure sores in a regional plastic surgery unit, 46 percent of all individuals presenting with ischial sores had multiple sclerosis (McGregor, 1984). Petersen and Bittmann (1971) estimated that in Denmark approximately 20 percent of nonambulatory individuals with multiple sclerosis have pressure sores.

With multiple sclerosis, factors contributing to pelvic obliquity such as trunk muscle paralysis, trunk muscle imbalances, and spasticity imbalances are very similar to the spinal cord injured population, along with the loss or impairment of sensation. In addition to muscle weakness, the individual's state of general fatigue becomes important when considering the possible need for a high, inclined backrest providing proper scapular and thoracic support, and a headrest. With weakness of the diaphragm, the use of an abdominal binder or corset will not only facilitate diaphragmatic breathing, but it can also improve phonation.

Hemiplegia

Among individuals with hemiplegia, a 15.1 percent incidence of pressure sores was found in the Greater Glasgow Health Board area survey (Jordan and Clark, 1977).

In an investigation of the sitting pressure distributions of various groups of disabled individuals, Nicol and Koerner (1985) found extremely asymmetrical pressure distributions among the hemiplegic individuals studied. The unilateral trunk muscle paralysis in many hemiplegic individuals will contribute to a pelvic obliquity, with these people usually bearing more weight on the paralyzed side when sitting. Additional factors such as the buttock and leg atrophy on the paralyzed side, and the poor sitting balance due to spatial disorientation or neglect, will contribute to an increased risk of pressure sores (Batchelor and Farmelo, 1975; Landin et al., 1977).

The most posturally unstable hemiplegic individuals need all the stabilizing features of a good geriatric reclining wheelchair, with proper scapular and thoracic support, and headrest; along with the proper seat and backrest inclinations. The use of a gluteal pad to level the pelvis may be necessary for some of these individuals. For very disoriented individuals, trunk supports may also be required.

The need for a low seat height without an excessive seat depth is critical for hemiplegic individuals who propel the wheelchair with one foot (Staas and Lamantia, 1983). Otherwise, the individual will scoot forward in the chair in order for the foot to reach the floor, and end up in a slumped posture. The resulting posture can produce very high shearing forces over the buttocks and sacrococcygeal region when propelling the wheelchair.

Cancer Patients

Among individuals with malignant neoplasms, the incidence of pressure sores reported in the Greater Glasgow Health Board area survey was 12.3 percent (Jordan and Clark, 1977). The emaciated condition of many cancer patients with advanced disease, along with nutritional deficiencies, will make them very prone to pressure sores, with the development of very high pressure gradients and shearing forces on the sitting surface (Andersen and Kvorning, 1982; Berjian et al., 1983). In addition, radiation therapy can also result in an impairment of the vascular system (Moustafa and Hopewell, 1979). Therefore, even with normal sensation, a wheelchair cushion with good uniform pressure distributing qualities is critical.

The need for the wheelchair to be a chair of comfort and relaxation is critical for this patient population. A reclining wheelchair with a high backrest to provide good support to the thoracic spine and scapulae, along with a headrest, should be considered (Zacharkow, 1984a).

Geriatric Population

With the percentage of the United States population age sixty-five and over projected to increase from 11.4 percent in 1981 to 13.1 percent in the year 2000 and 21.7 percent in the year 2050 (Bureau of the Census, 1982), special attention must be given to the wheelchair dependent geriatric population.

The vast majority of pressure sores occur in individuals over age seventy (Andersen and Kvorning, 1982). In the Greater Glasgow Health Board area survey, 70 percent of those individuals with pressure sores were age seventy and over (Barbenel et al., 1977). In an English survey involving 14,448 hospital patients, David et al. (1985) reported that 85 percent of those individuals with pressure sores were over age sixty-five. A Swedish survey by Ek and Boman (1982) found that 83 percent of those individuals with pressure sores were over age sixty-five, most frequently in the 80 to 84 year old age group.

In the Greater Glasgow Health Board area survey, the incidence of pressure sores was 11.6 percent in individuals age seventy and over, compared to an incidence of 6 percent in individuals age sixty-nine and under (Clark et al., 1978). A survey by Norton et al. (1975) of 250 patients admitted to the geriatric unit of a hospital found that 24 percent of these individuals developed pressure sores after admission. Almost twice as many individuals over age eight-five developed pressure sores compared to those individuals under age seventy-five. In a one-day survey of pressure sores in an orthopaedic hospital in England, Lowthian (1979) reported a 6.5 percent overall incidence of pressure sores, but a 13.8 percent incidence in individuals age seventy and over.

Regarding nursing facilities for the elderly, one study from a skilled nursing facility reported a 24 percent incidence of pressure sores (Michocki and Lamy, 1976). The pressure sores were either present on admission or developed after admission. Approximately 20 to 35 percent of individuals admitted to extended care facilities from acute care hospitals may have pressure sores on admission (Reed, 1981; Shepard et al., 1987). In a one-day survey involving seven skilled care nursing homes, Garibaldi et al. (1981) found the most common site of infection to be pressure sores.

Among studies where the majority of individuals were age sixty-five and over, the sacral pressure sore was the most common (Jordan and Clark, 1977; Jordan, Nicol, and Melrose, 1977). For example, a 30.7 percent incidence of sacral sores was reported by David et al. (1985), a 35 percent incidence by Ek and Boman (1982), a 43 percent incidence by Petersen and Bittmann (1971), and a 51 percent incidence of sacral sores was reported by Andersen and Kvorning (1982).

Many sacral sores are the result of wheelchair sitting and not bedrest. Petersen and Bittmann (1971) reported that 36 percent of individuals with sacral pressure sores were not confined to bed rest but were up in wheelchairs. Hartigan (1982) feels that a large percentage of lower sacral pressure sores in nursing homes are due to the wheelchair. Bailey (1967) referred to the "characteristic geriatric low sacral sore" from sitting in a reclining chair. Barton and Barton (1981) attribute over 50 percent of sacral pressure sores to the sitting position.

Lowthian (1975) reported on a survey involving ninety-five patients who had recently been admitted to hospital geriatric units. Forty-five of these individuals were nursed in sitting positions for prolonged time periods, while either in bed or in a geriatric chair. Subsequently, nineteen of these forty-five individuals (42 percent) developed sacral/buttocks pressure sores.

Data from Barton and Barton (1981) revealed that 57 percent of sacral sores occurred in women as compared to 43 percent in men. The greater thoracic kyphosis present in the female geriatric population may possibly predispose them to sit in a more slumped, kyphotic posture (Cowan,

1965; Fon et al., 1980; Milne and Williamson, 1983). There is also an increased backward tilt of the sacrum in the female, with the coccyx being in a more upright position and less curved forward compared to the male (Johnson, 1981; Warwick and Williams, 1973; Derry, 1912). As a result, in a slumped, kyphotic posture the female sacrococcygeal region will be exposed to greater localized pressures and shearing forces.

Major factors that will make the geriatric population more prone to pressure sores are occlusive arterial disease, muscle atrophy and loss of subcutaneous tissue, and the decrease in skin elasticity and resiliency with age (Dick, 1951; Kirk and Chieffi, 1962; Howell, 1969; Daly and Odland, 1979; Montagna and Carlisle, 1979; Torrance, 1981; Bouissou et al., 1984). In a light and electron microscopic study, Braverman and Fonferko (1982a) found a marked degradation of elastic fibers in the buttock skin after age seventy, with abnormalities being present in the majority of fibers. Biopsies from the buttock skin of four individuals ages eighty to ninety-three years old also indicated abnormally thin-walled microcirculatory blood vessels (Braverman and Fonferko, 1982b). Bennett et al. (1981) found the average shear values produced by a geriatric hospitalized group when sitting to be three times greater than the shear values in a young healthy group studied. Therefore, even with normal sensation, a wheelchair cushion with good uniform pressure distributing qualities is critical for these wheelchair dependent individuals.

In the geriatric units of hospitals, the majority of individuals have been found to be chairfast, as opposed to being confined to bed rest (Wells, 1980). Sitting comfort must therefore not be overlooked regarding the wheelchair dependent geriatric population. Reclining wheelchairs are very often necessary, with high backrests and headrests. An inclined seat is critical to help prevent the individual from sliding forward on the wheelchair seat and developing a dangerous shearing force on the buttocks and sacrum. (Similarly, sitting in a semireclined position in bed necessitates elevating the thigh section of the bed.)

In addition, some of the poorest fitting wheelchairs are often found in nursing homes. These are wheelchairs with overstretched and hammock seat and backrest upholstery, and nonadjustable armrests. Upper back support and head support, both critical for many individuals in nursing homes, are often overlooked (Hartigan, 1982). Wheelchair cushions are nonexistent, or in some cases a pillow or blanket from the bed is used on the wheelchair seat.

Chapter Twelve

EXERCISE AND DESIGN IMPLICATIONS

"In regard to the sitting posture, I believe the time will come when we will have to conform our chairs to the individual rather than the individual to the chair" (Meisenbach, 1917).

Proper exercise is critical to counteract the potentially harmful effects of prolonged, improper sitting. Exercises should be designed to bring into use muscles that are inactive or overstrained at the chair and desk (McKenzie, 1915). When working in a sitting position, it is characteristic to find a kyphotic spinal posture, with "the chest more or less contracted, the upper part of the body leaning forward against the desk, the thorax bent forward and downward, pressing downward upon the abdominal organs" (Enebuske, 1892).

Sargent (1906) considered most occupations, including that of the student, to overwork the flexor muscles. Taylor (1901) stressed that "nine-tenths of the ordinary movements of life are flexions. Little more than one-tenth involve definite extensions except it may be in the case of the legs." Taylor (1901) considered many postural faults to result from a lack of extensor tone. He observed that "extensor power may gradually lessen or become almost lost from disuse and yet flexor power remains."

IMPORTANCE OF TRUNK MUSCULATURE

The spinal column sustains the weight of the head, shoulders, arms, and thorax (Skarstrom, 1909). The trunk musculature is of primary importance in maintaining stability of the spine (Lucas and Bresler, 1961; Smeathers and Biggs, 1980). Besides the back extensor musculature, the "lower abdominals" are also important in erect trunk posture by providing the proper resting intra-abdominal pressure (Roaf, 1977; Armstrong, 1965).

Stockton (1913) stressed that:

"A certain amount of intra-abdominal pressure is necessary if the viscera of that region are to be held in their proper places and proper relations. This is possible only when the body is erect in sitting and standing, when the chest is kept habitually raised in its normal position,

311

and when the abdominal muscles are strong, and are not allowed to relax, pouch out, and thus favor the descent of the organs."

Kellogg (1927) made some interesting observations in regards to weak trunk musculature:

"In consequence of the weakness of the muscles that support the trunk, and especially weakness of the waist muscles, an ungraceful and unnatural carriage of the body appears, not only in walking and standing but in sitting. The weak-waisted woman is comfortable only when sitting in a rocking or easy chair. She cannot be comfortable unless the back is supported. Consequently, in sitting, the muscles of the trunk are completely relaxed, thus causing collapse of the waist and protrusion of the lower abdomen by the compression at the waist occasioned by the depression of the ribs."

The trunk is considered to be the center, and the starting point, for all of man's movements (Mercier, 1888; Knudsen, 1947). Proper trunk stabilization will allow limb movements to occur with maximum efficiency (Armstrong, 1965).

Two important aspects of training the trunk musculature involve endurance and proper muscular coordination of the trunk.

Trunk Muscle Endurance

Endurance training is particularly important for the trunk muscles, as they are the major restraints of the spinal column, which sustains the weight of the head, shoulders, arms, and thorax. Skarstrom (1909) considered the efficiency of the trunk muscles, as regards their supporting function, to depend on their tone and endurance. He advocated prolonged, static contractions to improve the muscles' supporting function. Rathbone (1934) also stressed the importance of static positions held frequently in posture training.

Jokl (1984) considered the paraspinal and abdominal musculature to be primarily involved in "static antigravity functions and activities that require prolonged muscular contraction." The endurance of the trunk musculature required for their postural function is best trained by prolonged, submaximal isometric contractions (Young et al., 1985).

Poor isometric endurance of the back musculature has been implicated as a major cause of low back pain (Biering-Sørensen, 1984a, 1984b; Nicolaisen and Jørgensen, 1985; deVries, 1968).

Proper Muscular Coordination of the Trunk

According to Roaf (1977), "Many people have very little conscious awareness of their trunk. An important part of the development of good control of posture is developing the body image of the trunk." Equally important would be developing the conscious awareness of proper head posture.

Thomas and Goldthwait (1922) expressed the following view on muscular coordination of the trunk:

"The most fundamental and at present a most unappreciated fact is the value of the proper mechanical working of the body. Until a person understands good body mechanics and has gained the correct muscular coordination, chiefly of the abdomen, chest, and back, exercise is of little benefit and often is harmful. If a person is taught how to correct his own bad posture and aims to make a good posture habitual, he has gained more than the benefits from many strenuous exercises taken with the body poorly poised."

Checkley (1890) considered a major problem with most individuals to be a lack of joint awareness, particularly in differentiating between flexion of the spine and flexion at the hips. As a result, the individual bends his spine "as if there were no hip joints in his anatomy."

Proper posture training must start in childhood (Taylor, 1901; Goldthwait, 1916). Handley (1986) emphasized that:

"If bad habits become established, they are difficult to extinguish, especially during adulthood. For example, if a child in primary school is allowed to sit and write with head bowed, neck twisted, shoulders hunched and back twisted and rounded, this will become an established pattern of movement. This bad habit will then feel right, and as a result, any postural adjustments made to correct this position will feel strange or awkward."

There is still obviously great difficulty in obtaining properly designed chairs and desks. Therefore, individuals of all ages need to be shown how to sit correctly on all furniture, rather than letting their spines passively conform to improperly contoured chairs.

HEAD, NECK, AND UPPER THORACIC SPINE

The greatest burden for the neck and upper thoracic spine involves supporting the shoulders and arms. The shoulders are suspended from those trapezius muscle fibers arising from the lower three to four cervical vertebrae and inserting on the acromion process. Therefore, the heavy pull of the shoulders and arms on the lower half of the cervical spine must be counteracted by the back extensors in the region of the lower cervical and upper thoracic spine (Knudsen, 1947).

According to Knudsen (1947):

> "It is on the carriage of the upper 5–6 dorsal vertebrae that the shape of the neck and the position of the head depend first and foremost.
>
> The upper half of the dorsal spine is the stiffest part of the whole spinal column, and it is consequently in that part that round back generally begins to develop. By the increased curving of the dorsal spine caused by round back, the upper 5–6 dorsal vertebrae will be inclined forward, the cervical column will move with it into the same inclined position, but in order to be able to look forward one will raise one's head by bending the upper part of the cervical spine backwards; its slight physiological curve will therefore be increased, i.e., lordosis of the neck will set in as seen in all people with round backs."

With prolonged sitting at a horizontal desk, a slumped head posture will easily become habitual. As a result, when the eyes are raised to look ahead, it has to be accomplished by an increased lordosis in the upper cervical spine.

Head posture is considered critical in providing "the reference for the parts of the body in relation to one another and in relation to the surrounding space" (Laville, 1985). By altering the proper head-neck relationship, faulty posture at the desk may contribute to the development of a defective kinaesthetic system (Alexander, 1918). Cohen's (1961) work showed that the neck proprioceptors are critical for orienting the head in relation to the body, and that neck proprioception "plays a very important role in maintaining proper orientation, balance and therefore motor coordination of the body."

Excessive flexion of the upper thoracic spine is also considered to be a major cause of round shoulders (Appleton, 1946; Knudsen, 1947).

> "When the back is rounded, the upper 4 or 5 dorsal vertebrae and the lower 3 or 4 cervical vertebrae are moved forward. From the cervical vertebrae mentioned, the shoulders and arms are suspended as these vertebrae form the origin of portator scapulae, the strongest part of trapezius. The shoulders will tend to place themselves vertically under their points of suspension. If these are moved forward, as in round back, the shoulders will move forward, too. This will affect the pectoral muscles. Their origin and insertion are now brought nearer together. They will adjust their lengths accordingly; they become shorter and

they will fix the shoulders in the forward position. In this case round back is the cause of round shoulders, as commonly found in young people with sedentary work during which they sit with rounded backs" (Knudsen, 1947).

Design Implications

The proper desk inclination and chair-desk relationship are critical to assure an erect posture of the head, neck, and trun' (Bendix and Hagberg, 1984; Weber et al., 1986). The neck muscle tension forces and cervical compression forces will be reduced with an inclined desk (Less and Eickelberg, 1976).

Backrest designs that push the shoulders forward will increase the upper thoracic kyphosis and result in a forward position of the head (Hawley, 1937).

Arm Support

Proper arm support is critical to promote extension of the upper thoracic spine, along with a more erect head and neck posture. In addition, proper arm support can significantly reduce the trapezius muscle load (Mahlamäki et al., 1986; Granström et al., 1985; Kvarnström, 1983; Avon and Schmitt, 1975). This is critical for supporting elevated arms in various work situations, thereby reducing the static load on the shoulder and neck muscles (Westgaard and Aarås, 1984).

There are several ways of achieving proper arm support, such as:

a. On the desk, if there is a proper chair-desk relationship.
b. From the forearm support of a keyboard.
c. From armrests attached to the work surface (Rohmert and Luczak, 1978).
d. From armrests on the chair.
e. With the hands in the lap.

It is critical that armrests are adjustable in order to provide proper support for the specific task requirements. This is a major fault in current armrest design. Armrest adjustments are needed in height, angle of inclination, fore-aft adjustment, and adjustability in distance between the armrests (Aarås, 1983).

Exercise Implications

A forward carriage of the head can result in increased stress and fatigue of the posterior neck and upper thoracic spinal musculature, with eventual weakening of these same muscles. According to Hawley (1937), this "tends to become a vicious circle whereby unnecessarily great demands are made upon the extensors of the upper thoracic spine which, for that reason, become gradually weaker and less able to function adequately." However, if the upper thoracic spine is held in correct posture, the muscular effort required for erect carriage of the head is slight.

Correction of both a forward head malposture and a round-shouldered posture lies in *improving the tone of the upper thoracic back extensors* (Skarstrom, 1909; Hawley, 1937; Knudsen, 1947).

In the prone extension exercises discussed in this section, it is extremely critical that the head is extended *with the chin drawn in*. With this technique, the backward movement of the head and neck will result in extension of the upper thoracic spine (Skarstrom, 1908, 1909; Hawley, 1937; Knudsen, 1947).

Prone Head Extension (Figure 151)

From a prone lying position on the floor, with the arms along the sides of the body, the chin is drawn in to prevent a forward and upward tilt of the chin. The head is then raised vertically upward, and held at the top position for five to ten seconds. When correctly performed, only a few inches of elevation of the head will be possible (Hawley, 1937).

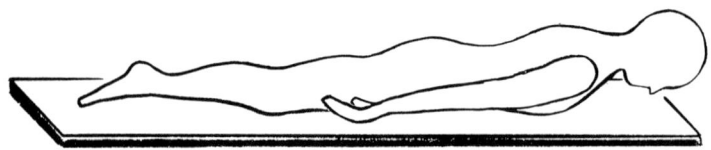

Figure 151. Prone head extension. Adapted from Hawley, G.: *The Kinesiology of Corrective Exercise*. Philadelphia, Lea and Febiger, 1937. With permission of Lea and Febiger.

Neck Range of Motion at Work

Postural immobilization of the head is characteristic of the high speed execution of simple tasks, such as certain VDT work (Laville, 1983, 1985).

Therefore, it is important that gentle neck range of motion exercises be carried out at intervals throughout the working day while sitting.

LOWER THORACIC SPINE (THORACOLUMBAR JUNCTION)

The approximation of the thorax to the pelvis in slumped, kyphotic sitting postures will localize stress to the lower thoracic spine and thoracolumbar junction.

In erect standing and sitting, the line of gravity for the upper body is already anterior to the thoracic spine, resulting in a force tending to increase the thoracic kyphosis. In upright sitting, Vulcan et al. (1970) calculated the center of gravity of the upper torso to be approximately 0.75 inches anterior to the ninth thoracic vertebra (Figure 152). Therefore, "the weight of the chest wall, the thoracic contents, and part of the arm and shoulder complex exert a bending moment on the spine via the rib cage" (Vulcan et al., 1970). The lower thoracic spine has been referred to as the "hinge area" for spinal flexion (Latham, 1957).

In a slumped, kyphotic anterior sitting posture, the gravity line will be even further anterior to the thoracic spine. The site of maximum spinal stress will be the lower thoracic spine, as this is the region maximally offset from the gravity line (Chandler, 1933; Alexander, 1977). The thoracic kyphosis has also been found to increase with age, which is hardly a healthy trend (Fon et al., 1980; Asmussen and Klausen, 1962).

Habitually slouched standing and sitting postures will also have a tendency to result in weakness and overstretching of the lower abdominal muscles, along with a decrease in the normal resting intra-abdominal pressure (Stockton, 1913).

Design Implications

Thomas and Goldthwait (1922) considered three factors to be critical in maintaining an erect trunk posture when sitting:

1. Maintain the normal axial relationship of the thorax and pelvis.
2. The ribs and chest must be raised to their normal position.
3. The head must be held in its normal position.

As head posture has already been mentioned, the first two points will now be discussed. The critical chair design feature needed to fulfill these first two factors is proper backrest support over the T9 thru L1 region of

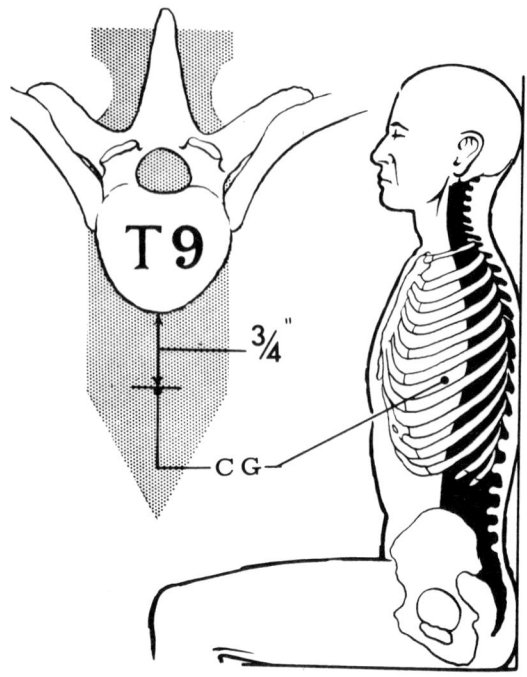

Figure 152. Center of gravity of the upper torso in upright sitting. From Vulcan, A.P., King, A.I., and Nakamura, G.S.: Effects of bending on the vertebral column during +Gz acceleration. *Aerospace Medicine, 41:*294–300, 1970. Reproduced with permission of the Aerospace Medical Association.

the spine. Proper back support will promote spinal extension along with stabilization of the thorax.

The inferior angle of the scapula is located opposite the eighth thoracic vertebra (Lovett, 1916; Basmajian, 1977). Therefore, critical back support for the T9 thru L1 region should be located just below the shoulder blades. (Some chair backrests may be modified to provide proper lower thoracic support. See Figure 153.)

Rathbone (1934) felt that proper chair design with back support to this spinal region could also improve standing posture. "While the chair is holding the trunk in extended position, the neuromuscular system is being patterned in a desirable posture which can carry over into standing and into movements."

However, without proper extension of the lower thoracic spine, along with activation of the lower abdominals, there will be a decreased resting intra-abdominal pressure and a lowering of the abdominal viscera. The resulting relaxed and protruding lower abdomen will then cause the

Figure 153. This chair has been modified with a firm support for the lower thoracic spine. The open space for the buttocks will also allow proper pelvic-sacral support for many individuals.

pelvis to gradually migrate forward on the seat. The proper axial relationship of the thorax and pelvis will be lost, along with the loss of pelvic and trunk stabilization. The sitter will then spontaneously search for other less healthy means of postural stabilization (Mosher, 1899). A similar process will occur when standing, with the loss of lower thoracic extension and intra-abdominal pressure resulting in a forward migration of the pelvis and a backward leaning trunk posture.

Exercise Implications

Rathbone (1934) considered the back extensors to need the most strengthening in the region of the tenth thoracic vertebra to the second lumbar vertebra. According to Wiles (1937), "The weakest part of the extensor mechanism of the spine is in the upper lumbar and lower dorsal regions. The mass of the erector spinae is getting smaller, and the change of curve from concave to convex is taking place, making it mechanically a vulnerable spot."

Knudsen (1947) observed that:

"The joints of the dorsal spine are those that first lose mobility in civilised life because of lack of use. Bendings backward are unconsciously performed in the more mobile loin. Even little children may have dorsal curves that cannot be stretched to the normal. On the other hand, it is seldom the dorsal spine has lost its ability to bend forward, to increase the curve. When the dorsal spine has lost its mobility, one is not able to control its extensors as one controls the extensors of the neck, or of the loin, or all other muscles of the skeleton."

Lovett (1916) stressed the potential harm from prolonged stretching of the back muscles, due to the continued maintenance of flexion of the spine, along with asymmetrical attitudes. He advocated symmetrical back extension exercises to counteract the deleterious effects of flexed and rotated sitting postures, especially in regards to the thoracic spine (Lovett, 1900).

The critical postural abdominal muscles can be grouped together as the "lower abdominal" muscles. Included here are the lower fibers of the transversus abdominis, the lower anterior fibers of the internal obliques, and the lower (lateral) fibers of the external obliques. These lower abdominal muscles should be considered as part of man's normal postural antigravity muscles.

The role of these lower abdominal muscles includes:

1. Compression and support of the abdominal viscera (Ono, 1958; Hatami, 1961; Kendall et al., 1971).
2. Maintaining the proper axial relationship of the thorax and pelvis. The lower (lateral) fibers of the external obliques are considered the most important regarding this function (Kendall et al., 1971).

Rathbone (1934) stressed that the lower abdominal muscles should be grouped with the extensors of the body, as they "take part in all trunk movements, due to a reflex connection with the other muscles of the torso. They are intimately connected in the reflex of extension and are contracted whenever the body assumes the completely extended position."

Rathbone (1936) also commented upon the relationship between weak abdominal muscles and weak erector spinae: "It has been our observation for several years that the lower abdominal wall and the erector spinae muscles are associated reflexly in all vigorous extensions of the human body. Therefore, with sagging abdominal walls one would expect to find weak backs."

This relationship between the erector spinae and lower abdominals has also been noted in more recent studies. Woodhull-McNeal (1986) found a positive correlation between activity in the erector spinae and external obliques when standing. Deusinger and Rose (1985) reported a high degree of correlation between back extensor force and external oblique EMG activity with prone back extensions. Carlsöö (1980) also reported strong activity of the lower (lateral) external obliques during strength testing of the back extensors.

Prone Back Extension With Arms at Sides (Figure 154)

In a prone position, the arms should be placed along the sides of the body, with the palms facing the ceiling. Keeping the chin drawn in, this exercise starts as a head extension. The movement continues by further raising the upper trunk until the upper sternum just clears the floor. This top position is then held for five to ten seconds.

It is important to maintain proper head posture throughout the exercise. Avoid bringing the chin forward and upward, and hyperextending the neck. By limiting the back extension to a slight raising of the upper trunk, this exercise will localize the work to the lower thoracic and upper lumbar back extensors (Stafford, 1928; Knudsen, 1947), as opposed to the lower lumbar spine.

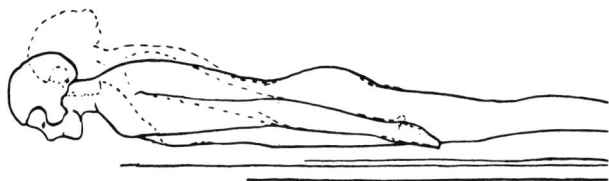

Figure 154. Prone back extension with arms at sides. From Drew, L.C.: *Individual Gymnastics,* 3rd ed. Philadelphia, Lea and Febiger, 1926. Reproduced with permission of Lea and Febiger.

Prone Back Extension With Arms Extended Forward (Figure 155)

The starting position for this exercise involves lying prone with the arms extended forward. This exercise then starts as a head extension with the chin drawn in, and continues until the shoulders and upper chest are raised off the floor. The arms are then kept in an extended position and raised off the floor, in line with the upper trunk. This position is then held for five to ten seconds.

One should concentrate on reaching forward with the arms, and think of lengthening the entire body during the exercise.

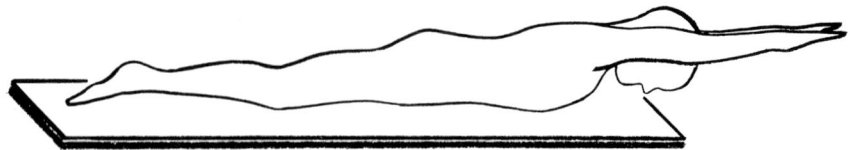

Figure 155. Prone back extension with arms extended forward. Adapted from Hawley, G.: *The Kinesiology of Corrective Exercise.* Philadelphia, Lea and Febiger, 1937. With permission of Lea and Febiger.

Sitting Trunk Extension (Figure 156)

This is an excellent exercise to counteract the "postural depression" and related musculoskeletal stresses from prolonged sitting. It should be done at regular intervals throughout the sitting period at work.

The chair used for this exercise should have a backrest that extends to approximately just below the shoulder blades. The chair should also be stable enough to support the individual when leaning heavily against, and over the backrest.

The exercise starts by taking a deep inspiration and reaching with extended arms toward the ceiling. At this stage the exercise is similar to the spontaneous "stretching and yawning" one often does when waking up, which will activate the back extensors and facilitate extension of the thoracic spine (Mercier, 1888; Lee, 1922; Knudsen, 1947). From this vertical stretch position, one should then lean back slightly over the chair backrest, avoiding any neck hyperextension in the process. Then, slowly return to the upright sitting position and exhale. (This exercise may also be performed with the hands placed behind the head.)

According to Knudsen (1920):

> "The higher and the farther back the arms are to be brought the more strongly must the shoulder-blades be rotated, and, therefore, the more must pectoralis minor and major be stretched; as these muscles are attached to the ribs they will, by a strong pressing backward of the arms in the stretch position, pull the ribs with them — that is, lift the thorax. On account of the small amount of movement possible in the joints between the ribs and the vertebrae the result of this pull is that the thoracic curve is straightened, which straightening is helped by the head being held high with the chin drawn in."

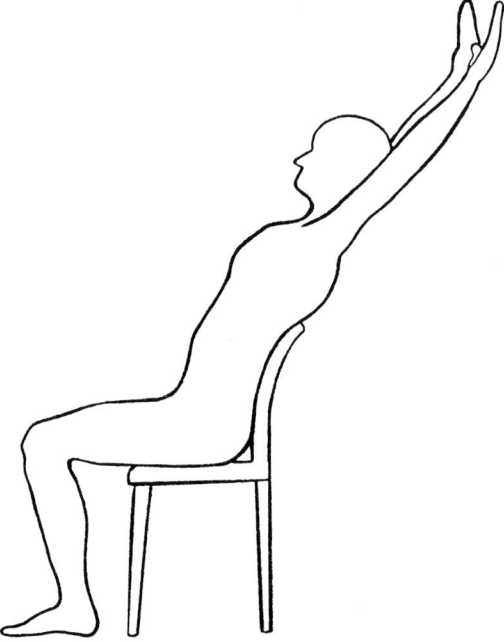

Figure 156. Sitting trunk extension.

Posse (1890) listed the benefits of such a trunk extension exercise as follows:

1. To straighten the thoracic spine.
2. To cultivate the extensibility of the superior region of the abdomen (stretch the upper rectus muscle).
3. To expand the inferior portion of the chest.

The backrest of the chair will help in this exercise to localize the trunk extension movement to the thoracic spine, as opposed to the lower lumbar spine.

Knudsen (1947) explained that:

". . . . the slightly mobile dorsal spine is situated just above the very mobile lumbar spine. It is seldom that people in daily life have to perform movements which necessitate a stretching of the dorsal spine; on the other hand the occasions on which they have to bend forward are numerous. The necessary movements backward are generally performed in the lumbar region alone. As a consequence the control over the dorsal erector spinae is gradually lost. People are able to control and work the cervical and lumbar portions of the extensors of the back as easily as, e.g., the flexors of the arms, for these muscles are used daily and thereby are kept under control."

With standing back extension exercises, it is very difficult to localize the movement to the thoracic spine (Figure 157). Too often, the exercise results solely in a marked hyperextension of the lower lumbar spine, with the hips extended and the pelvis deviated anteriorly and upwardly rotated (Skarstrom, 1909; Knudsen, 1947). This posture results in relaxation of the erector spinae, along with little, if any, extension of the thoracic spine (Clemmesen, 1951).

Figure 157. Standing back extension. From Posse, N.: *Handbook of School-Gymnastics of the Swedish System,* revised ed. Boston, Lothrop, Lee and Shepard, 1902.

Standing back extensions may cause low back pain for some individuals. This will most likely occur in those individuals who habitually stand in a slumped, fatigue posture with a backward leaning of the upper trunk. Some of these individuals may already be standing with the lower lumbar spine in maximum extension (Goldthwait et al., 1934).

Passive press-up exercises as advocated by McKenzie (1981) have actually been shown to be an active exercise (Rose et al., 1981). A large part of this exercise's reported benefits may possibly be due to activation of the lower thoracic and upper lumbar back extensors, along with stretching of the upper rectus muscle (Figure 158).

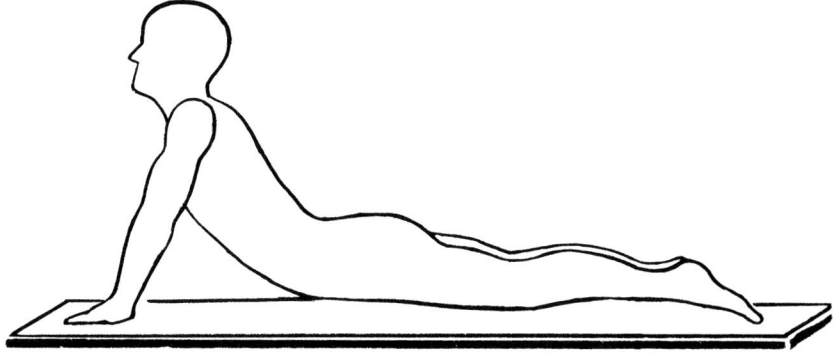

Figure 158. Press-up exercise.

Single Leg Raising (Figure 159)

In regards to abdominal exercises, Bowen (1919) commented that "Movements that bring the lower parts of the abdominal wall into contraction first and most strongly are of course to be preferred, and for this reason leg-raising is better than trunk-raising."

Isometric leg raises will best strengthen the postural "lower abdominal" muscles for their two important functions of supporting the abdominal viscera and maintaining the proper axial relationship of the thorax and pelvis (Zacharkow, 1984b; Strohl et al., 1981; Kendall et al., 1971; Ono, 1958; Floyd and Silver, 1950).

As will be discussed in the forthcoming section on sit-ups, many abdominal exercises have a tendency to contract the chest (Posse, 1890). Therefore, to prevent any depression of the chest by the pull of the abdominal muscles, it is advantageous to take a deep inspiration at the beginning of the leg raise movement, while raising the arms overhead (Skarstrom, 1909). This inspiration and overhead arm raising will also promote extension of the thoracic spine (Dally, 1908; Knudsen, 1947).

Isometric double leg raising is a very difficult exercise for most individuals to perform properly (Zacharkow, 1984b). As proper technique is extremely critical with this exercise and all other exercises, an isometric single leg raise will therefore be described.

For the isometric single leg raise, one should lie on the back with both knees bent and the feet flat on the floor. Taking a deep breath, both arms are then raised overhead. The lower back must then be pressed down against the floor to prevent it from arching during the leg raising (McMillan, 1932; Knudsen, 1947; Zacharkow, 1984b). Then, one leg is raised off the floor with the knee straight, the ankle plantarflexed, and the heel coming approximately four to six inches off the floor. This position is held for five to ten seconds. After lowering the leg and relaxing, the leg raising can then be repeated with the opposite leg.

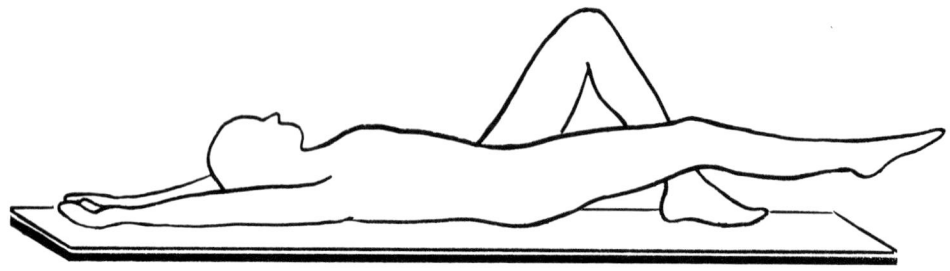

Figure 159. Single leg raise.

PELVIC–SACRAL SUPPORT

The main factor resulting in the backward rotation of the pelvis when sitting is the posterior rocking over the ischial tuberosities that occurs after the buttocks are resting on the seat. With ischial support only, the sitting position will be one of unstable equilibrium, resembling that of "a rocking-chair on a pair of very short rockers" (Hope et al., 1913). In a posterior sitting posture with backward rotation of the pelvis, an additional supporting surface will be provided by the posterior buttocks, coccyx, and sacrum.

A second factor contributing to the backward rotation of the pelvis when sitting is the tension of the hip extensors as the hips are flexed (Keegan, 1953, 1964; Carlsöö, 1972). The backward pelvic rotation will be most marked in individuals with tightness of the hamstrings (Stokes and Abery, 1980).

Design Implications

It is critical for the backrest to provide proper pelvic stabilization (Cohn, 1886; Branton, 1969; Schoberth, 1969). This will reduce or prevent backward rotation of the pelvis, along with having a beneficial effect on lumbar spinal posture. As the shape of the lumbar spine when sitting depends directly on the position of the sacrum and pelvis, support must be given to the upper sacrum and posterior iliac crests (Cohn, 1886; Schoberth, 1969; Oxford, 1973; Wilder et al., 1986).

The commonly placed lumbar support, designed to fit the lumbar concavity, will lose contact with the spine as the individual leans forward, away from the backrest. Proper pelvic-sacral support, however, can still provide pelvic stabilization in a forward leaning posture (Cotton, 1904, 1905; Schoberth, 1969) (Figures 160 and 161).

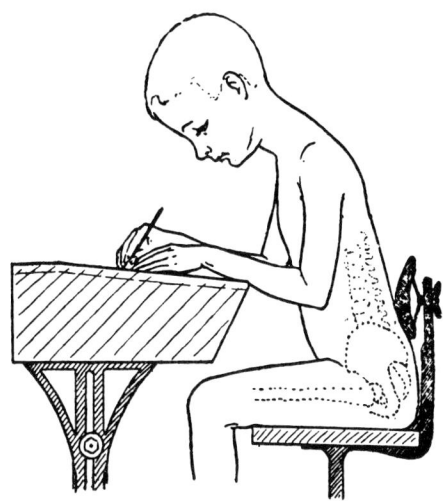

Figure 160. If there is a free space beneath the lower edge of the backrest, proper pelvic-sacral support can still provide pelvic stabilization in a forward leaning posture. From Cotton, F.J.: School-furniture for Boston schools. *American Physical Education Review,* 9:267–284, 1904.

According to Cotton (1904), in leaning forward there is "a slight rocking of the pelvis, and a tendency of the pelvis to slide back (on the yielding flesh of the buttocks) in such a way that the back is still in contact with the support, and may be definitely steadied by this support if it is properly curved. This point seems to have been overlooked. Of course, unless there is a free space beneath the lower edge of the back-rest no such motion occurs—an important reason in favor of leaving such a space free."

Regarding important skeletal landmarks for pelvic-sacral support, a line connecting the highest points of the iliac crests will intersect either the lower part of the L4 vertebral body, or the disc between L4 and L5 (Lovett, 1916; Basmajian, 1977; Hoppenfeld, 1976; Macgibbon and Farfan, 1979). A line connecting the posterior superior iliac spines will cross the

Figure 161. As this individual moves his upper trunk away from the thoracic support, he is still able to obtain pelvic-sacral support. From Cotton, F.J.: School-furniture for Boston schools. *American Physical Education Review, 9:*267–284, 1904.

second sacral vertebra, and the center of the sacroiliac joint (Hoppenfeld, 1976). Therefore, support at this level will also help stabilize the sacroiliac joint.

Additional pelvic stabilization will be provided by the following chair features:

    a. An inclined seat (Åkerblom, 1948; Murrell, 1965; Ayoub, 1972).
    b. A slight concavity to the sitting surface for the buttocks (Kroemer and Robinette, 1968).
    c. Avoidance of seat cushioning that is too soft (Kohara, 1965; Branton, 1966, 1970).
    d. Avoidance of slippery, low surface friction seat covers (Branton, 1969; Schaedel, 1977).
    e. The ability to have both feet firmly supported on the floor (McConnel, 1933).

Exercise Implications

Correct hamstring stretching is important to assure proper pelvic and spinal posture when sitting, in addition to counteracting the tendency for hamstring tightness to develop from prolonged sitting. As will be discussed in the forthcoming section on toe-touching exercises, it is important to minimize the lumbar and thoracic spinal flexion with hamstring stretching techniques. One technique for preventing this spinal stress involves a supine hamstring stretch using a rope or towel (Zacharkow, 1984b) (Figure 162).

The leg not being stretched should be actively extended against the floor to keep the lower back from flexing excessively and the pelvis from rotating upwardly. The rope or towel should be placed around the arch of the other foot. With the knee straight (but not forcefully locked) the leg is raised upwards until a mild stretch is felt behind the knee. The stretch position should be held for approximately two minutes.

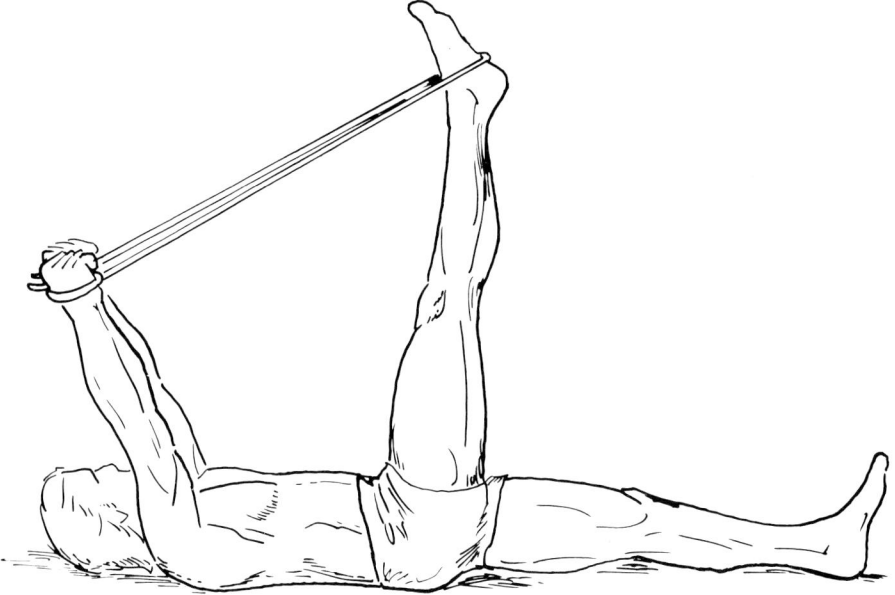

Figure 162. Supine hamstring stretch. From Zacharkow, D.: *The Healthy Lower Back.* Springfield, Thomas, 1984. Reproduced with permission of Charles C Thomas, Publisher.

A greater leg raising will be possible if the ankle is kept in plantarflexion during the stretching (Gajdosik et al., 1985). "Contract-relax" stretching techniques can also be performed with the supine hamstring stretch. This would involve a maximum isometric hamstring contraction against the rope or towel, followed by hamstring relaxation and then passive stretching (Moore and Hutton, 1980; Cornelius and Hinson, 1980; Wallin et al., 1985; Hardy, 1985; Fisk, 1987).

FOOT SUPPORT

Proper foot support when sitting is important for the following reasons:

a. Pelvic stability will be enhanced when the feet are firmly supported on the floor (McConnel, 1933).

b. It will facilitate use of the backrest (Swearingen et al., 1962; Darcus and Weddell, 1947).

c. It is critical for avoiding posterior thigh compression and the obstruction of venous blood flow from the lower legs (Pottier et al., 1967, 1969; Morimoto, 1973).

d. It will facilitate leg position changes, thereby allowing a change in joint angles and muscle tension at the hips, knees, and ankles, along with reducing venous blood stagnation in the lower legs. A change in one's leg position can also temporarily shift pressure from the ischial tuberosities.

Design Implications

Obtaining proper foot support when sitting can best be achieved with an adjustable seat height. When the seat height is properly adjusted, the feet will rest firmly on the floor, with the major weight bearing areas on the seat being the ischial tuberosities and upper half of the posterior thighs (Bennett, 1928; Floyd and Ward, 1967). The distal one-fourth of the posterior thighs should not be loaded when sitting (Carlsöö, 1972).

Problems With Improper Seat Height (Figure 163 B,C,D)

According to Shipley (1980), "Too high a seat leaves feet dangling and unsupported, inducing the sitter to sit forward in order to plant her feet on the ground and so avoiding excessive pressure on the underside of the thighs, but at the expense of back support. Similar problems can arise from seats being too deep."

In regards to schoolchildren, Cohn (1886) observed that when the feet are dangling, "the child soon grows tired. He tries to reach the floor with the tips of his toes at least; and in so doing he bends the thigh downward, slides forward on to the edge of the form and presses his chest on the edge of the table. The necessary result is a further collapse of attitude."

Sitting on the front portion of a high seat will be both an unstable and a fatiguing posture (Kroemer and Robinette, 1968). At the other extreme, the acute angle between the thighs and trunk resulting from a very low seat height will increase the flexion stress to the lumbar and thoracic spine. The approximation of the thorax to the pelvis will also increase the pressure on the abdominal viscera (Aveling, 1879).

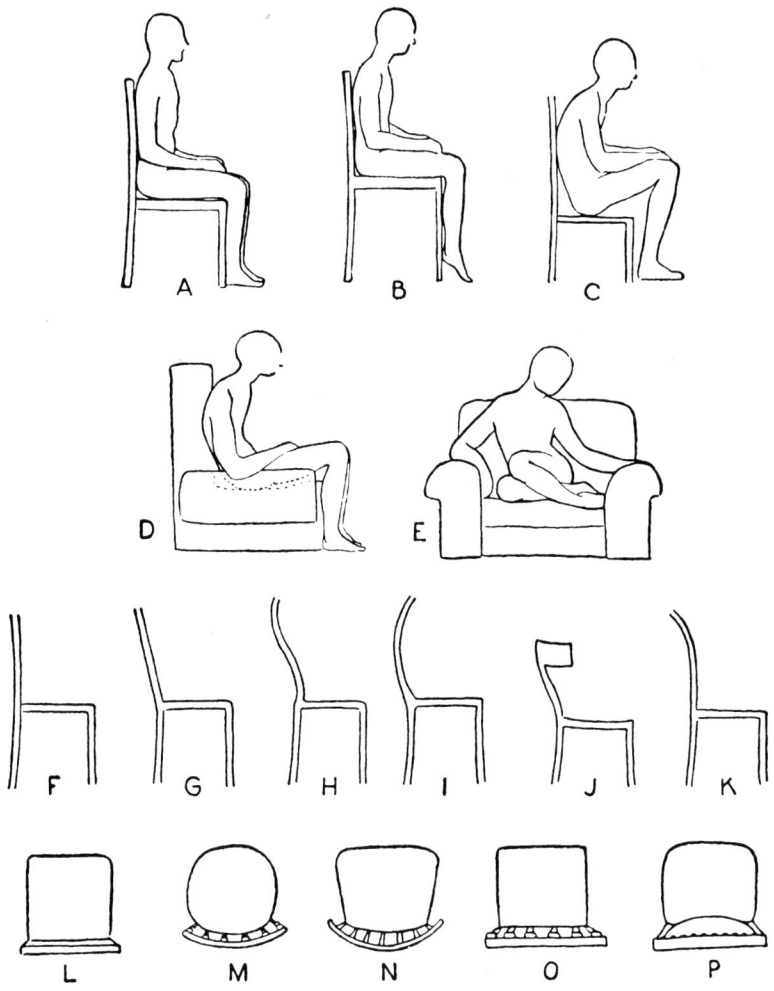

Figure 163.
  B. Seat too high.
  C. & D. Seat too low.
  From Howorth, B.: Dynamic posture. *Journal of the American Medical Association, 131:*1398–1404, August 24, 1946. Copyright 1946, American Medical Association. Reproduced with permission.

Leg Position Changes

Leg position changes will be facilitated by:
  a. The proper seat height (Åkerblom, 1954; Andersson and Örtengren, 1974f).

b. The proper seat depth. Freedom of leg movement will be lost as the seat depth is increased (Ridder, 1959). However, a very short seat depth may feel unstable and also result in a lack of surface for free movement of the legs (Bennett, 1928).

c. A rounded front edge to the seat, which will prevent the front edge from cutting into the distal posterior thighs with leg position changes (Keegan, 1953; Asatekin, 1975; Coe, 1979).

Footrests

Although footrests will be necessary for some individuals in order to obtain proper foot support, they do have some disadvantages:

a. They limit the free movement of the individual's legs. Holding the lower legs continuously in the same position can be very fatiguing (Kotelmann, 1899; Shaw, 1902; Kerr, 1928).

b. The footrests should have an inclination similar to the seat inclination. Otherwise, a very acute angle at the knees may result (Kotelmann, 1899).

Exercise Implications

Intermittent leg exercise is important during prolonged sitting to reduce the swelling and discomfort of the lower legs (Winkel, 1981; Winkel and Jørgensen, 1986). Prolonged passive sitting is also considered a causative factor in venous thrombosis of the lower extremity (Homans, 1954; Makris et al., 1986). Such exercise will have a beneficial effect on the venous blood flow and lymph flow from the lower limb. Of particular importance may be the active contraction of the soleus muscle (Winkel and Jørgensen, 1986).

A beneficial exercise will be one similar to a "dynamic footrest" (Stranden et al., 1983). This exercise involves pressing down alternately with the ball and heel of the foot against the floor (Sauter et al., 1984). After several repetitions, the exercise should be repeated with the other leg. For maximum benefit, this exercise should be repeated intermittently throughout the sitting period.

Casters

The movement of the seated worker at his desk made possible on chairs with casters can help reduce foot swelling (Winkel and Jørgensen, 1986). However, casters can also require increased static muscle work from the legs and back in order to keep the chair in position, especially on hard floors (Lundervold, 1951a; Damodaran et al., 1980; Bell Labs, 1983). Therefore, the type of caster and type of floor covering are both important factors.

SPECIFIC BACKREST ISSUES

"Average" Man Fallacy

According to Hertzberg (1970), in the design of seats and work places:

"The first requirement is to design according to the range of size of the users. Most designers immediately think of using the 'average' man, but this has been shown to be a fallacy. There is really no such thing as an 'average' man or woman. There are men who are average in weight, or in stature, or in sitting height; but the men who are average in two dimensions constitute only about seven percent of the population; those in three, only about three percent; those in four, less than two percent. There are no men average in as few as 10 dimensions. Therefore, the concept of the 'average' man is fundamentally incorrect, because no such creature exists. Work places, to be efficient, should be designed according to the measured range of body size.

But, even when the range is known and expressed in percentiles, the designer must know when to use the high values and when the low. For example, a typist's chair wide enough for a broad-hipped secretary can be quite easily occupied by a small woman with narrow hips; but that same seat, to be acceptable to both women, has to be short from rear to front, so as not to cut into the calves of the small woman. Thus the designer would use the 95th percentile value for seat breadth, and the fifth percentile (or less) for the seat length.

Likewise, the vertical adjustability would have to be enough for at least the central 90 percent of female popliteal height, and, preferably, 95 to 98 percent. If the chair is intended also for export to other countries, where women are usually smaller, the adjustability used should include the range for all populations concerned, because there are large differences between American and foreign body sizes and proportions. For such purposes, of course, good foreign data are needed."

The fallacy of the "average" man concept was well illustrated in a study by Daniels (1952) based on body measurements of over 4,000 Air Force flying personnel. A group of measurements useful in clothing design was used, with an average measurement including approximately the middle 30 percent of the total population.

At the end of ten steps of elimination, there was not one individual remaining who fell within the "average" range for all measurements:

1. Of the original 4063 men, 1055 men were of approximately average *stature.*
2. Of these 1055 men, 302 men were also of approximately average *chest circumference.*
3. Of these 302 men, 143 men were also of approximately average *sleeve length.*
4. Of these 143 men, 73 men were also of approximately average *crotch height.*
5. Of these 73 men, 28 men were also of approximately average *torso circumference.*
6. Of these 28 men, 12 men were also of approximately average *hip circumference.*
7. Of these 12 men, 6 men were also of approximately average *neck circumference.*
8. Of these 6 men, 3 men were also of approximately average *waist circumference.*
9. Of these 3 men, 2 men were also of approximately average *thigh circumference.*
10. Of these 2 men, none were also of approximately average in *crotch length.*

Some anthropometric dimensions for seating design are shown in Figure 164. Of course, these dimensions must be applied with a knowledge of both the dynamics of sitting and the specific task requirements of the sitter. For more detailed anthropometric data, the reader is referred to the following sources: Diffrient et al., 1974; NASA, 1978; Panero and Zelnik, 1979; Pheasant, 1986.

Importance of Backrest Adjustability

Adjustability is especially important in regards to seat height, arm support, and various backrest features.

Branton (1984), for example, in his study on adult back shapes found that:

"From a designer's point of view it is disappointing to find that ranges in the vertical are so great that there is considerable overlap between, say, nape and occiput. Head or neck rests would therefore have to be adjustable over a wide range and even then are not likely to give any satisfaction to more than about 25–30% of sitters. Another disappointing result of these observations is that the horizontal varia-

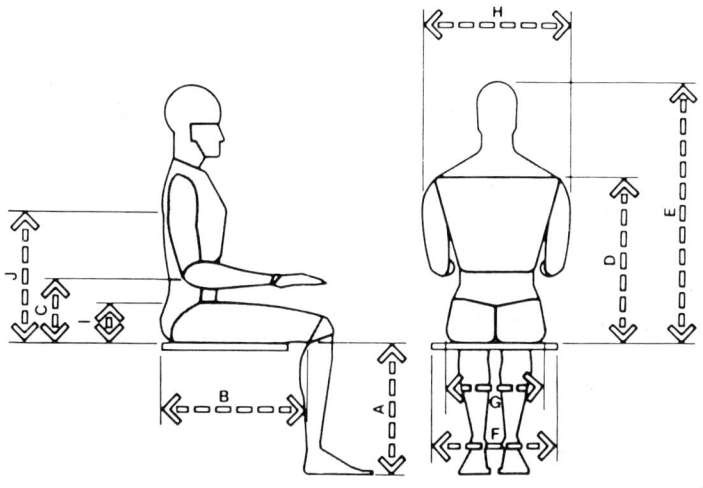

KEY ANTHROPOMETRIC DIMENSIONS FOR CHAIR DESIGN

Measurement	MEN Percentile 5 in	cm	95 in	cm	WOMEN Percentile 5 in	cm	95 in	cm
A Popliteal Height	15.5	39.4	19.3	49.0	14.0	35.6	17.5	44.5
B Buttock-Popliteal Length	17.3	43.9	21.6	54.9	17.0	43.2	21.0	53.3
C Elbow Rest Height	7.4	18.8	11.6	29.5	7.1	18.0	11.0	27.9
D Shoulder Height	21.0	53.3	25.0	63.5	18.0	45.7	25.0	63.5
E Sitting Height Normal	31.6	80.3	36.6	93.0	29.6	75.2	34.7	88.1
F Elbow-to-Elbow Breadth	13.7	34.8	19.9	50.5	12.3	31.2	19.3	49.0
G Hip Breadth	12.2	31.0	15.9	40.4	12.3	31.2	17.1	43.4
H Shoulder Breadth	17.0	43.2	19.0	48.3	13.0	33.0	19.0	48.3
I PSIS Height	4.9	12.5	7.9	20.0	5.1	13.0	7.9	20.0
J Scapular Height	15.9	40.5	18.9	48.0	15.0	38.0	17.7	45.0

Figure 164. Some key anthropometric dimensions for chair design. Note: PSIS Height is from the seat to the posterior superior iliac spine. Scapular Height is from the seat to the inferior angle of the scapula.

Adapted from Panero, J., and Zelnik, M.: *Human Dimension and Interior Space.* New York, Whitney Library of Design, 1979. With permission of Watson-Guptill Publications. (Anthropometric data in I and J is from Pheasant [1986]).

tion of the thoracic curve is from about 4 to 6 cm and thus too large to be compensated for by any softness of cushioning—even if one of acceptable quality could be devised."

Bennett (1925) also considered the problem of a suitable backrest design to be extremely complicated. For example, among 1500 students of all grades, he found a very low correlation between the heights of two important landmarks for back support: the height of the iliac crest from the seat and the height of the inferior angle of the scapula from the iliac crest.

Further complicating the backrest design issue are the following points:

a. The tendency for the thoracic kyphosis to increase with age and the lumbar lordosis to decrease with age (Fon et al., 1980; Milne and Lauder, 1974; Cowan, 1965).
b. The difference in spinal curvature between males and females. There is a tendency for women to have a greater lumbar lordosis than men (Fernand and Fox, 1985; Cunningham, 1886). Beyond age forty, Fon et al. (1980) also found a significantly greater thoracic kyphosis in females compared to males.
c. The difference in the soft tissue contours of the buttocks between the sexes, and also between obese and thin individuals (O'Neill, 1982).

Specific Adjustability Needs

Adjustability of Pelvic-Sacral and Thoracic Support

Of major importance in backrest design is pelvic stabilization and upper sacral support, along with the need for support to the lower thoracic spine (thoracolumbar junction) in order to provide thorax stabilization and an erect trunk posture (Mosher, 1892; Stone, 1900; Taylor, 1917; Rathbone, 1934; Keegan, 1953, 1964; Branton, 1969). In order to meet these requirements for as many individuals as possible, adjustable supports for both of these spinal regions are critical. Although still difficult to find today, adjustable supports were available on some chairs as early as 1889 (Staffel, 1889; Bradford and Stone, 1899; Shaw, 1902; O'Reilly, 1914) (Figure 165).

With motor vehicle seating, easy chairs, and reclining wheelchairs, the use of a headrest would necessitate a third region for adjustable support (Rizzi, 1969) (Figure 166).

Figure 165. Staffel's (1889) chair for schoolchildren with an adjustable back support. Note also the open space for the buttocks and the adjustable forehead restraint. From Staffel, F.: *Die Menschlichen Haltungstypen und ihre Beziehungen zu den Rückgratverkrümmungen.* Wiesbaden, J.F. Bergmann, 1889. Reproduced with permission of J.F. Bergmann Verlag.

Open Space for Buttocks

An open space or recess is necessary beneath the pelvic-sacral support in order to provide space for the posterior protrusion of the buttocks (Hartwell, 1895; Cotton, 1904; Keegan, 1953, 1964; Damon et al., 1966) (Figure 167).

Without the provision of a large space beneath the pelvic-sacral support, the buttocks will be pushed forward on the seat, inducing a slumped sitting posture (Taylor, 1917). Individuals with greater gluteal mass will be pushed away from the backrest, unable to obtain proper support (Dresslar, 1917). Therefore, adjustability of this space is critical, due to the variation in the size of the buttocks among individuals, along with the need to accomodate the bulk of clothes in this region (Cotton, 1904).

An open space beneath the pelvic-sacral support is a much better

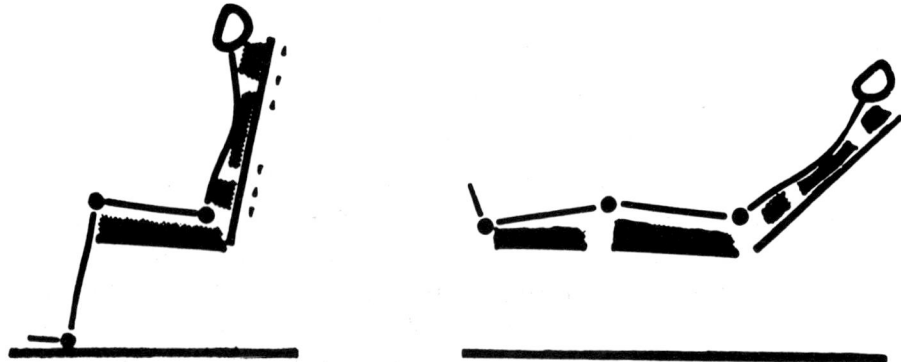

Figure 166. Three adjustable supports will be necessary with motor vehicle seating, easy chairs, and reclining wheelchairs: pelvic-sacral support, thoracic support, and head support. From Rizzi, M.: Entwicklung eines verschiebbaren rückenprofils für auto-und ruhesitze. In Grandjean, E. (Ed.): *Proceedings of the Symposium on Sitting Posture.* London, Taylor and Francis, 1969, pp. 112–119. Reproduced with permission of Taylor and Francis Ltd.

design solution than a continuous backrest having a recess for the buttocks (Hartwell, 1895; Cotton, 1904; Taylor, 1917).

Figure 167. A chair designed by Staffel (1889). Note the open space allowed for the buttocks beneath the adjustable back support. From Staffel, F.: *Die Menschlichen Haltungstypen und ihre Beziehungen zu den Rückgratverkrümmungen.* Wiesbaden, J.F. Bergmann, 1889. Reproduced with permission of J.F. Bergmann Verlag. Note: In this illustration, the back support is located too high for providing proper pelvic-sacral support.

Adjustability of Backrest Inclination

A posture of "alert readiness" with an erect trunk is desirable for all working situations (Bennett, 1928). Important backrest adjustability angles for a slightly reclined, alert sitting posture are considered to be from 5 degrees to 20 degrees (Croney, 1981; Diffrient et al., 1974). The greater backrest inclinations within this range are more important for many disabled individuals, and those able-bodied individuals with very weak trunk musculature. For the majority of individuals, the optimal angle for a slightly reclined, alert sitting posture will be 5 to 10 degrees from vertical.

With a slightly reclined, alert sitting posture, the critical backrest height for able-bodied individuals is just below the inferior angles of the scapulae. Such a backrest height will provide proper support to the lower thoracic spine, without the tendency to push the shoulders forward as with higher backrests (Stone, 1900; Cotton, 1904; Dresslar, 1917; Bennett, 1928).

Swivel Seat

A swivel mechanism on the seat can help prevent flexed, rotated sitting postures. With such a feature, the individual can turn in the chair without rotating the trunk and sacrificing symmetrical back support (Bennett, 1928).

Major Backrest Design Faults

Bancroft (1913) commented that:

"Many chairs literally mold the sitter into most harmful positions. A forward vertical curve in the chair back will thrust the head forward if at the top, or make inevitable a sliding forward and downward in the seat if at the bottom. A lateral curve around the shoulders makes improbable a correct carriage of the shoulder blades for one who sits much in such a chair. The rounded shape thrusts the scapulae forward, and if used habitually becomes a potent cause of round shoulders."

The following backrest design faults all merit discussion:

a. A backrest that pushes the shoulders forward will result in a sitting posture with an increased thoracic kyphosis and a forward head posture (Hawley, 1937). This becomes a potential problem when the backrest begins to extend much above the inferior angles of the scapulae, in order to provide shoulder support. It is a major problem when there is a forward inclination to the upper thoracic support of a high backrest (Figure 168 H, I).

Bennett (1928) considered that "contact against the shoulder blades themselves supports the spine only indirectly through the whole mass of shoulder-girdle structure, and,.... relieves the back muscles only when the spine has sagged below them. Support against the shoulders also interferes with free arm movement."

b. Another major design fault is an excessive horizontal concavity to the upper backrest (Howorth, 1946; Travell, 1955; Oxford, 1973) (Figure 168 M, N).

According to Bennett (1928), "The line across one's back at the shoulders is practically straight, and a curvature in the support here tends to throw the shoulders forward and hinders expansion of the chest and related factors of erect posture." Bennett (1928) felt that "a horizontally straight support below the shoulder blades is not only comfortable but is particularly conducive to expansion of the chest and falling back of the shoulders."

Taylor (1917) stressed that the chair backrest "should be flat from side to side, or only slightly hollowed, so that the shoulders and scapulae may not be pushed forward." Pressure should also be avoided on the outer part of the scapulae and the shoulders (Taylor, 1917).

c. A flexible backrest will rapidly increase one's sitting fatigue as it will require an increase in static muscle activity to stabilize the trunk (Lundervold, 1951a; Kohara, 1965; Kroemer and Robinette, 1968). In addition, leaning against a softly hinged backrest will contribute to a faulty posture of the head and upper thoracic spine (Kendall et al., 1970).

d. The use of attachable convex lumbar supports contoured for the lumbar spine have become very popular for both office seating and motor vehicle seating. However, the use of these supports will not maintain the normal axial relationship of the thorax and pelvis when sitting. The shoulders will end up projecting behind the hip joint line. The result will be a marked posterior displacement of the upper trunk (Figures 169 and 170). There will also be a tendency to overstretching of the postural lower abdominal muscles, along with a decrease in the normal resting intra-abdominal pressure. If the individual attempts to use the upper backrest of the chair, it may be obtained with excessive hyperextension of the lower lumbar spine.

The resulting posture is very different from sitting in a slightly reclined

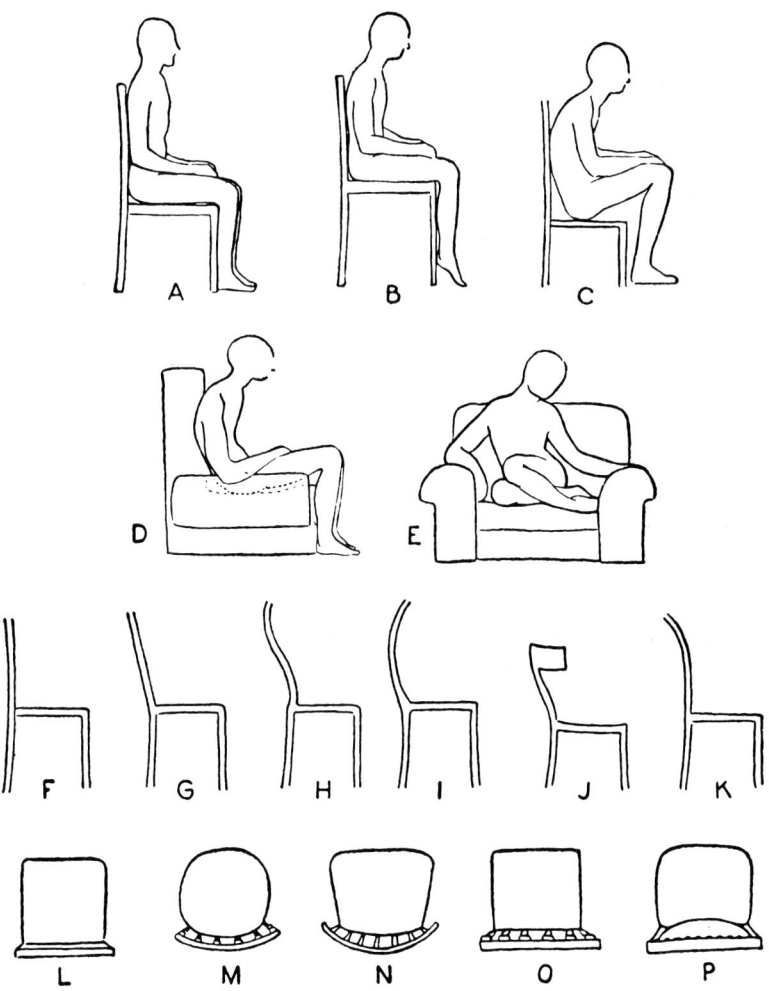

Figure 168.
H & I: Backrest designs that will push the shoulders forward.
M & N: Chairs with an excessive horizontal concavity to the backrest.
From Howorth, B.: Dynamic posture. *Journal of the American Medical Association,* *131:*1398–1404, August 24, 1946. Copyright 1946, American Medical Association. Reproduced with permission.

posture with an erect trunk. *What is desirable is a forward carriage of the chest and upper trunk, achieved through lower thoracic extension in association with proper pelvic-sacral stabilization.*

e. An excessive backrest inclination, along with the lack of proper lower thoracic support, is a common design fault in automobile seats and office chairs. The resulting posture will also distort the normal axial relationship of the thorax and pelvis (Mosher, 1899).

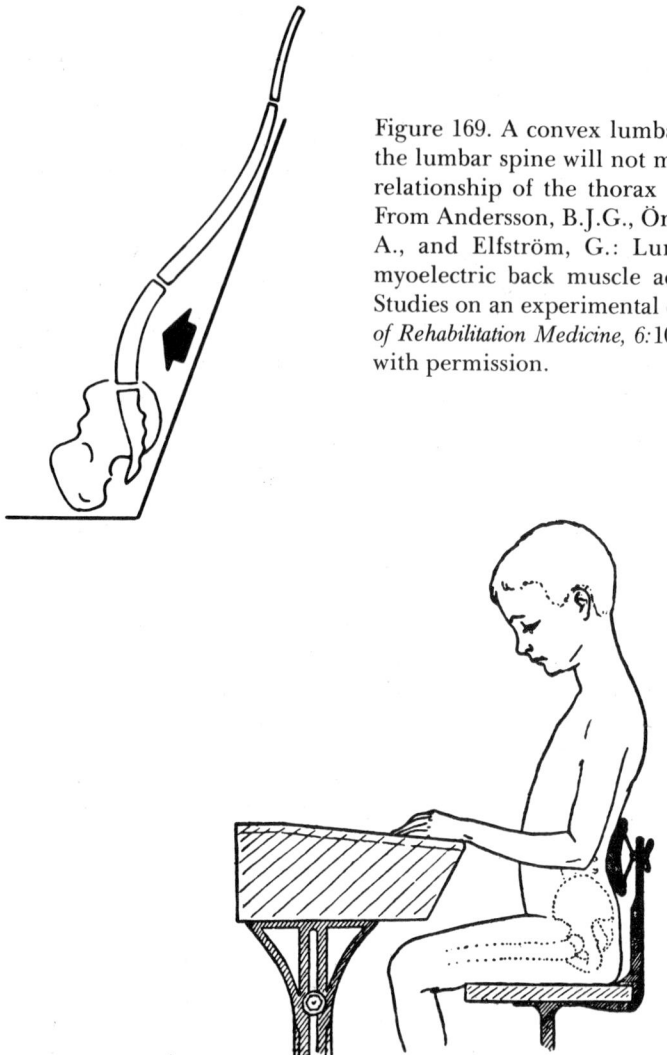

Figure 169. A convex lumbar support contoured for the lumbar spine will not maintain the normal axial relationship of the thorax and pelvis when sitting. From Andersson, B.J.G., Örtengren, R., Nachemson, A., and Elfström, G.: Lumbar disc pressure and myoelectric back muscle activity during sitting. 1. Studies on an experimental chair. *Scandinavian Journal of Rehabilitation Medicine, 6:*104–114, 1974. Reproduced with permission.

Figure 170. The typical convex lumbar support will result in a marked posterior displacement of the upper trunk. There will also be a tendency to overstretching of the postural lower abdominal muscles. From Cotton, F.J.: School-furniture for Boston schools. *American Physical Education Review, 9:*267–284, 1904.

Exercise Implications

Poorly Designed Chairs

The individual sitting should not permit a poorly designed backrest to "mold" his spine into a potentially harmful posture (Bancroft, 1913).

It is far better to keep one's upper trunk slightly forward of such a backrest. With the hips and pelvis pushed back in the chair as far as possible, the thorax can be brought forward through extension of the lower thoracic spine (Drew, 1926). Such an erect trunk posture will also result in a reflex contraction of the lower abdominal antigravity muscles, and an increase in the resting intra-abdominal pressure.

Aveling (1879) considered leaning against the backrest of a chair, with spinal flexion and approximation of the thorax and pelvis, to result in relaxation of the abdominal muscles. As a result, the abdominal muscles' "retaining and supporting influence upon the abdominal contents is consequently lost." Relaxation of the lower abdominal muscles will also occur when leaning forward in a slumped, kyphotic posture.

According to Kellogg (1927):

"A relaxed condition of the abdominal muscles is natural and proper in the horizontal position, but when the muscles of the trunk are relaxed in the erect position, the pressure of the head and shoulders naturally inclines the body either forward or backward, according as the head is before or behind the line of gravity. If the inclination is forward, the result is the shortening of the distance between the sternum and the pubes and the relaxation of the abdominal muscles. This destroys the natural support of the viscera and of the large blood vessels that are found in the lower cavity of the trunk and which have so great a capacity that they are capable of holding all the blood of the body."

Kellogg (1927) added that:

".... so long as the body is in upright position, its several parts — roughly speaking, the head, shoulders, chest and hips — must be maintained in proper relation to one another. This may be accomplished either by energized muscles, as in forcible standing or sitting, or by the aid of a proper support, as when sitting in a chair with a suitably constructed back. The construction of most chairs is such as to make relaxation of the muscles of the trunk unsafe, even dangerous, because of the malpositions into which the trunk is forced by gravity and the resulting effects upon the heart, lungs, diaphragm and other viscera, and the grave disturbance of the functions of respiration and circulation."

In a chair with proper support to both the pelvic-sacral region and the lower thoracic spine, the individual can activate the lower abdominal muscles and increase the resting intra-abdominal pressure by consciously maintaining slight pressure against these two supports.

Unsupported Sitting

In an unsupported sitting posture, such as on a stool or bench, it is important to avoid a slumped, kyphotic spinal posture. It is also important, however, to avoid an overcorrected sitting posture, achieved by a marked lower lumbar hyperextension and anterior pelvic tilt.

Unsupported sitting should involve an upright trunk and "lengthened" spine, with the proper intra-abdominal pressure and proper axial relationship of the head, thorax, and pelvis. The following exercise can be practiced to help achieve this position.

Sitting upright with the chin drawn in and the arms held forward in approximately 135 degrees of shoulder flexion, extend the hips slightly and allow the pelvis to rock posteriorly over the ischial tuberosities. Then, concentrate on holding the pelvis back while the upper trunk is brought slightly forward through extension of the lower thoracic spine (Figure 171). The resulting upright posture will involve activation of the erector spinae and lower abdominals. Sitting stability can then be enhanced by keeping the hands in the lap.

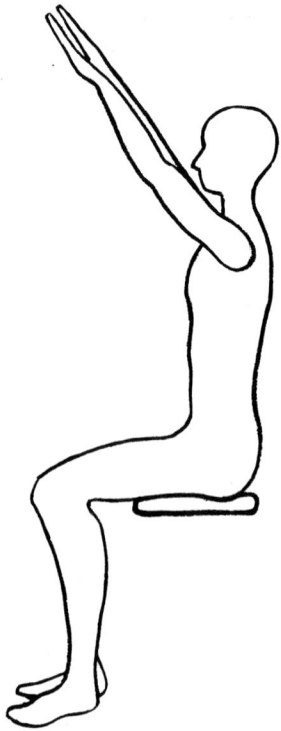

Figure 171. An exercise to achieve proper unsupported sitting posture.

STANDING POSTURE AND EXERCISE IMPLICATIONS

In erect upright standing, gravity will exert a flexion force on the entire spine, with the greatest flexion force occurring to the thoracic spine. The trend to an increased thoracic kyphosis and a lessening of the lumbar lordosis with age was noted by Fon et al. (1980) and Asmussen and Klausen (1962). In her thesis entitled *Trunk Strength and Flexibility as Factors in Posture,* Powell (1930) found the range of thoracic flexibility similar for individuals judged to have good and poor standing postures. However, the poor posture group had more possible thoracic flexion and less possible thoracic extension than the good posture group.

As discussed in Chapter One, the most common postural fault in standing involves a posterior displacement of the upper trunk, with relaxation of the erector spinae. In such a slumped, fatigue posture, the line of gravity will be displaced posteriorly towards the heels (Cureton and Wickens, 1935).

According to Mulliner (1929):

> "In the faulty standing position in which the weight of the body is carried over the heels, the abdominal wall is lax. The inclination of the pelvic girdle is decreased, with a tilting upward in front, so that the plane of the inlet faces in a more nearly horizontal direction than normal.
>
> When not supported by a tonic abdominal wall, the abdominal viscera tend to slip downward, stretching their mesenteries which contain their nerves and blood vessels."

Mulliner (1929) also considered the normal intra-abdominal pressure to be markedly lessened by a flaccid abdominal wall. Of additional consideration is the "frequent jarring of the body from the impact of the heels while walking upon hard pavements. This jar is much more marked when the body weight is carried over the heels instead of further forward as in a normal posture, with result of a persistently downward influence upon the pelvic viscera" (Mulliner, 1929).

Foot Posture

Proper foot posture is critical for both correct pelvic alignment and stabilization, along with activation of the erector spinae. Mosher (1919) observed that in "standing with the body weight upon the heels, the knees relax, the pelvis drops at the back, and the lumbar curve of the spine straightens, carrying the trunk backward beyond its center of gravity. To regain lost equilibrium the shoulders and head are moved forward for ballast and the ribs drop." Relaxation of the erector spinae will also be characteristic of such a standing posture (Floyd and Silver, 1955; Klausen and Rasmussen, 1968).

Swaying forward from the ankles so that less of the body weight is over the heels will help activate the erector spinae of the thoracic and lumbar spine, and bring the pelvis into its proper neutral position (Bancroft, 1913; Mosher, 1913, 1919; Reynolds and Hooten, 1936; Okada, 1972). This shifting of the body weight forward will also facilitate a straighter posture of the upper back (Cureton and Wickens, 1935).

Shoes

Heels on shoes will promote relaxation of the intrinsic postural muscles of the foot, along with the extensor musculature of the trunk (Reynolds and Lovett, 1910; Stewart, 1945). They will also have an adverse effect on spinal posture (Antioch College, 1931; Barker, 1985).

An individual's everyday shoes should ideally be without heels, and they should also be very flexible (Barker, 1985). Walking barefoot should also be encouraged, whenever it is practical (Robbins and Hanna, 1987).

Drew (1926) noted that:

> "In savages and infants the foot is almost fan-shaped and has a wide range of movement, also considerable prehensile power. Savages use the straight foot position, and the manner of walking in the case of races which are unhampered by shoes is by strong flexion of the anterior part of the foot, almost digging the toes into the ground at each step, propelling the weight of the body forward by a strong push with the great toe."

Pronated feet with sagging arches are characteristic of a slumped, fatigue standing posture (Goldthwait et al., 1934). Barefoot walking, flexible shoes without heels, and proper correction of the trunk posture can all be beneficial in helping to activate the intrinsic foot musculature and restoring the medial longitudinal arch of the foot.

Importance of Thoracic Extension

The primary emphasis in standing postural correction is the upper trunk, "which in all relaxed types of poor posture will be found to have sunk backward" (Bancroft, 1913). When the upper trunk is brought forward, "the spinal curves (with the possible exception of the neck), the chest, and the pelvis will all fall into correct position. This whole movement really amounts to bringing the chest forward, but the method of doing it confines the action to the spine, and avoids the artificial distention of the ribs that comes from efforts to lift the chest or protrude it forward" (Bancroft, 1913).

The key area for this spinal correction involves extension of the lower thoracic spine (thoracolumbar junction). Activation of the erector spinae of this region and movement of the thorax forward should result in a spontaneous contraction of the lower abdominals, probably from segmental reflexes at the lower thoracic and upper lumbar levels. With the upper trunk moved forward into its proper position, there will be a lengthening of the rectus abdominis. The lower end of the sternum, as opposed to the lower abdomen, will be the most anterior part of the torso (Checkley, 1890; Kellogg, 1927; Frost, 1938) (Figure 172).

A thesis by King (1932) involved a statistical study of the changes in the position of body segments when "natural" standing position is altered to "best" standing position. The two most important changes in posture improvement involved a straightening of the upper back, and a forward movement of the chest accompanied by a flattening of the abdomen. Her thesis confirmed the importance of giving attention to the relationships between upper and lower back and chest position, rather than focusing on separate body segments.

Arm and Shoulder Posture

The tendency to maintain the arms flexed in front of the trunk will tend to increase the flexion force on the thoracic spine. Thoracic extension will be facilitated, however, if when standing the shoulder "is slightly back of the lateral median line of the body so that the weight is received largely upon the thorax, none of the muscles being in more than slight contraction, and the strain upon the posterior muscles which must occur when the shoulder is held forward is absent" (Goldthwait, 1909).

The typical arm posture when standing involves having the palms facing towards the sides of the body, the elbows pointing posteriorly, and the thumbs facing forwards. A different arm posture, however, will help facilitate the proper position of the chest and thoracic spine: The upper arms should be rotated slightly inwards with the elbows turned slightly outwards and away from the body. The palms should be facing posteriorly,

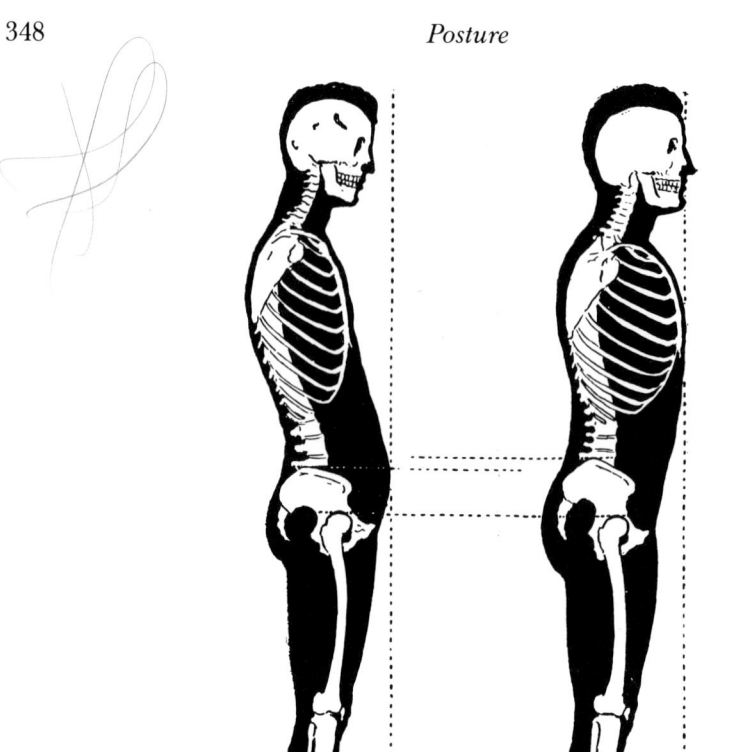

Figure 172.
   Left: Incorrect standing posture with the lower abdomen as the most anterior part of the torso.
   Right: Proper postural correction with the lower end of the sternum as the most anterior part of the torso.
   From Checkley, E.: *A Natural Method of Physical Training.* Brooklyn, William C. Bryant and Co., 1890.

and the thumbs should be facing the sides of the body (Alexander, 1918; Barlow, 1980) (Figure 173). Such an arm posture will also help facilitate the proper position of the scapulae, which should lie "flat and widened across the back of the properly expanded chest" (Barlow, 1980).

Specific Exercises

The following exercises, all previously described in this chapter, will have a beneficial effect on standing posture:

Figure 173. Proper arm posture when standing. From Barlow, W.: *The Alexander Technique.* New York, Warner Books Edition, 1980. Reproduced with permission of Alfred A. Knopf, Inc.

1. Prone back extension exercises will promote extension of the thoracic spine, along with improving the endurance of the erector spinae.

2. Overhead arm raising, as in the sitting trunk extension exercise, will stretch the upper abdominals and the intercostals, besides promoting thoracic extension.

3. The supine isometric single leg raise will improve the strength and endurance of the lower abdominals. Of particular importance for maintaining the proper axial relationship of the thorax and pelvis are the lower (lateral) external obliques (Kendall and McCreary, 1983).

4. As hamstring tightness is often found in a kyphotic standing posture with a decreased pelvic inclination, the supine hamstring stretch will also be a beneficial exercise (Knudsen, 1920, 1947; Salminen, 1984; Vidal, 1984).

The following standing exercises are important for simultaneously obtaining the proper relationship among the body segments:

Forward Weight Shift (Figure 174)

With one foot placed a very short step forward in advance of the other, sway backward from the ankles until the body weight is over the heel of the rear foot. Then, holding the hips and pelvis back, sway gently forward from the ankles until the body weight is shifted to the middle of the rear foot. Proper arm posture should be maintained during this weight shifting exercise, with the palms facing posteriorly and the elbows turned outwards.

Figure 174. Forward weight shift.

With prolonged standing, this forward weight shifting technique will prevent the spinal stress and loss of body symmetry from common asymmetrical standing postures. This technique can be used with many daily activities, such as when standing before a sink, ironing board, or workbench (Sweet, 1947). However, one should always start with the body weight over the rear foot in order to obtain proper pelvic and hip posture, before shifting some body weight to the forward foot.

Modified Bancroft Exercise (Figure 175)

This exercise will help obtain the proper axial relationship of the thorax and pelvis, and facilitate extension of the thoracic spine.

Stand with one foot placed a very short step forward in advance of the other, with the arms held forward in approximately 135 degrees of shoulder flexion. Then, sway backward from the ankles until the body weight is over the heel of the rear foot. In this position, concentrate on holding the hips and pelvis back, but do not allow the trunk to bend forward.

Without altering this position of the hips, pelvis, and thorax, and keeping the chin drawn in slightly, sway forwards from the ankles until the body weight is shifted to the middle of the rear foot (Figure 175). Then, lower the arms to the sides in the correct position with the palms facing posteriorly and the elbows turned outwards.

This exercise, and all standing weight shifting exercises, will be most effective if done barefoot, or wearing shoes without heels.

Figure 175. Modified Bancroft exercise.

Standing Press-Up on Thighs (Figure 176)

This exercise will help reinforce the forward position of the chest, extension of the lower thoracic spine, and the feeling of "lengthening" the spine.

From a standing position with the chin drawn in, one should lean forward from the hips until the palms are resting on the thighs with the elbows slightly bent. The knees should also be only slightly flexed.

Sway backward from the ankles until the body weight is over the heels. Then, while holding the pelvis and hips back, press gently down and back against the thighs, concentrating on the chest coming over the toes and the spine lengthening (Figure 176).

As the upper limb is fixed in this exercise, the resulting upward contraction of the latissimus dorsi will have the effect of extending the lower thoracic spine (Skarstrom, 1909). This is due to its insertion on the lower thoracic spinous processes, and also on the lower three or four ribs.

By bearing the weight of the upper torso on the thighs, spinal compression will be reduced, and lengthening of the spine will be facilitated.

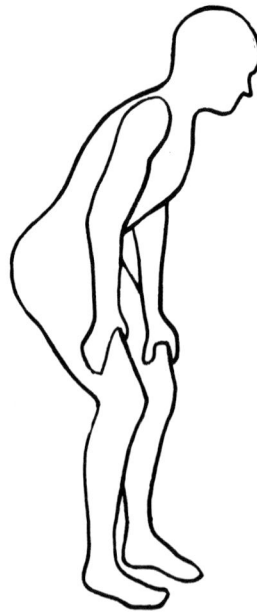

Figure 176. Standing press-up exercise.

Common Errors in Standing Postural Correction

Among the most common mistakes in standing postural correction are the following:

1. Throwing the shoulders back. This will result in a rigid posture with excessive contraction of the scapular retractors. The result will often be a hyperlordotic posture with forward hips and backward shoulders (Checkley, 1890; Taylor, 1901; King, 1932).

2. A forced effort to lift the chest or protrude it forward, resulting in a static distention of the ribs and a hyperlordotic lumbar spine (Checkley, 1890; Bancroft, 1913) (Figure 177).

3. A tendency to bend forward at the waist or hips, instead of swaying forward from the ankles (Bancroft, 1913).

4. The tendency to correct the axial relationship between the pelvis and thorax through an attempted retraction of the abdomen or tightening of the upper rectus. The effect of contracting the upper rectus will be to pull the chest down from above, approximating the sternum and pelvis (Bancroft 1913; Phelps and Kiphuth, 1932; Wiles, 1937). The result will be an increase in the thoracic kyphosis and a flattening of the lumbar lordosis.

Phrases such as "abdomen in" are not necessary for restoring the proper axial relationship of the thorax and pelvis. In postural correction, the thorax is brought forward rather than the abdomen being drawn backward. The increased tension in the lower abdominals will be incidental to getting the upper part of the body in proper position (Bancroft, 1913).

5. An exaggeration of lower lumbar hyperextension and anterior pelvic tilt, as opposed to lower thoracic (thoracolumbar) extension (Skarstrom, 1909) (Figure 178). With proper extension of the lower thoracic spine and movement of the upper trunk forwards, spontaneous contraction of the lower abdominals will result in an increase of the resting intra-abdominal pressure. The result will be a "lengthening" of the lumbar spine in a slight lordosis rather than a hyperlordotic lower lumbar spine.

SCHOOL FITNESS IMPLICATIONS

According to Salminen (1984):

"It seems apparent that, already at school age, habits of posture and physical activities should be more systematically observed. As physical

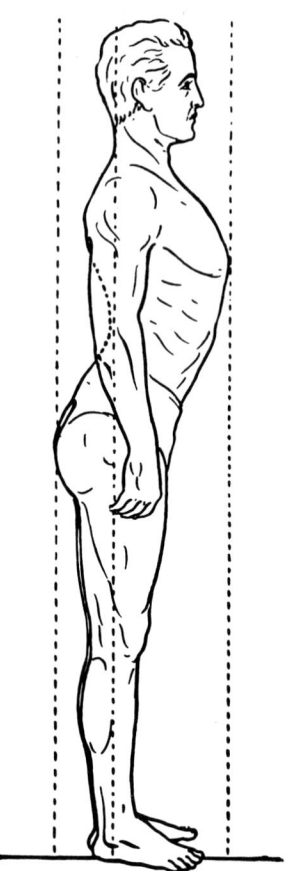

Figure 177. An exaggerated standing posture, with the chest protruded forward and a hyperlordotic lumbar spine. From Checkley, E.: *A Natural Method of Physical Training.* Brooklyn, William C. Bryant and Co., 1890.

education may play a part in forming these habits, a reform of physical education might also play a part in the prophylaxis of back troubles."

Taylor (1901) considered many children's postural problems to arise from a lack of extensor tone: "In many instances there is found normal flexor power with an insufficient or non-compensatory extensor capacity in the muscles."

In 1980, a new fitness test was developed by the American Alliance for Health, Physical Education, Recreation and Dance (AAHPERD) to emphasize health-related physical fitness (Blair et al., 1983). In this test, muscular strength and endurance of the abdominal muscles along with flexibility of the lower back and hamstrings were considered to be

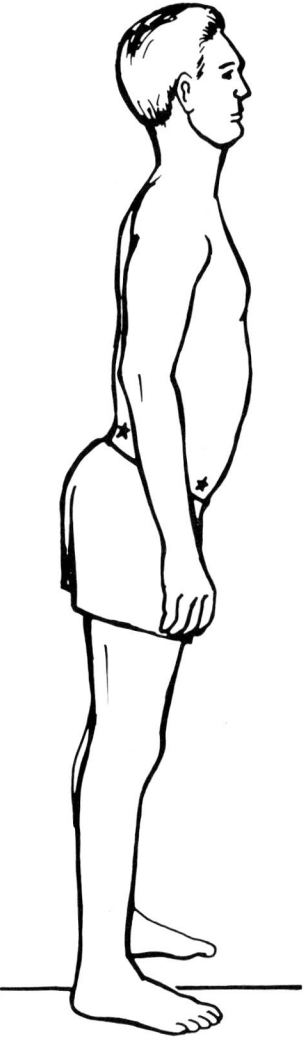

Figure 178. An exaggerated anterior pelvic tilt, with lower lumbar hyperextension. Reprinted from Gajdosik, R., Simpson, R., Smith, R., and Dontigny, R.L.: Pelvic tilt. *Physical Therapy,* 65:169–174, 1985. With permission of the American Physical Therapy Association.

important health-related fitness components. The rationale being that "substantial clinical evidence indicates that low back pain is associated with fitness deficiencies in the lower trunk region. Specifically, weakness of the abdominal muscles and lack of flexibility in the low back/hamstring musculature have been identified as precursors of low back pain" (Pate, 1983).

In the 1980 AAHPERD health-related physical fitness test, the strength and endurance of the abdominal muscles are determined by the number of bent knee sit-ups that can be performed in one minute. Flexibility of the low back and hamstrings is determined by the sit-and-reach test. While sitting with the knees extended, the student stretches his hands forward as far as possible toward or beyond the toes (Blair et al., 1983) (Figure 179).

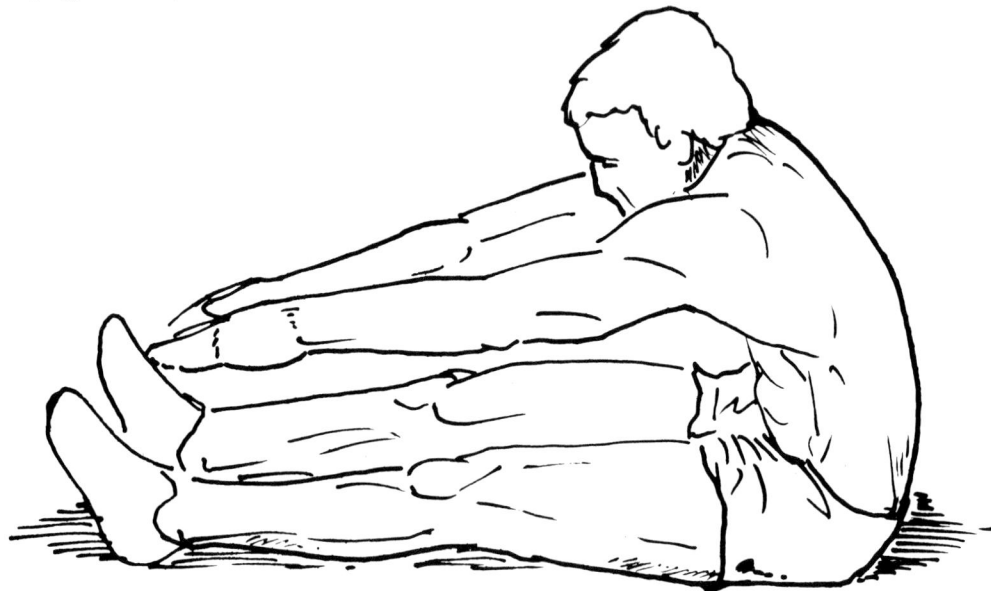

Figure 179. The sit-and-reach test. From Zacharkow, D.: *The Healthy Lower Back.* Springfield, Thomas, 1984. Reproduced with permission of Charles C Thomas, Publisher.

Besides their use in school fitness tests, sit-ups and toe-touching exercises are also considered to be good indicators of fitness for the general public. For example, in the January 1986 Reader's Digest Guide to Family Fitness (Kuntzleman, 1986), the number of bent knee sit-ups performed in one minute and the forward reach possible on the sit-and-reach test were both considered to be important measures of overall fitness for individuals from age six to over age sixty.

Unfortunately, both sit-ups and toe-touching exercises present many potential problems in regards to both spinal stress and the reinforcement of poor posture.

Sit-Up Exercises

The potential problems associated with sit-up exercises are as follows:

a. Sit-up exercises primarily strengthen the upper rectus abdominis muscle (Teshima, 1958; Lipetz and Gutin, 1970; Girardin, 1973; Rasch and Burke, 1978) (Figure 180). Based on subjective comments, the limiting factor in repetitive sit-ups is discomfort in the rectus abdominis muscle (Legg, 1981).

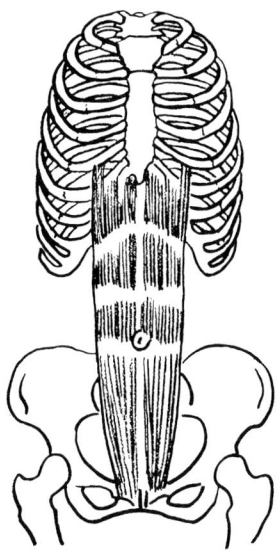

Figure 180. The rectus abdominis muscle. Sit-up exercises primarily strengthen the upper rectus.

According to Schultz (1983), "In a maximum voluntary attempted trunk flexion the rectus abdominis muscles provide about 40 per cent and the medial groups of the internal and external oblique muscles each provide about 25 percent of the strength." These "upper abdominals" should be grouped with the flexors of the body, whereas the important postural "lower abdominal" muscles are grouped with the extensors of the body (Rathbone, 1934, 1936).

b. Rectus muscle activity is not involved in maintaining or raising the intra-abdominal pressure (Floyd and Silver, 1950; Bartelink, 1957; Ono, 1958).

c. Sit-up exercises reinforce the movement pattern of spinal flexion with minimal hip flexion (Zacharkow, 1984b). This can contribute to ingraining poor technique when lifting, bending, or leaning forward in a chair.

The "postural depression" resulting from sit-up exercises, by approximating the front of the chest to the pelvis, is very similar to the "postural depression" in kyphotic sitting postures (Anderson, 1951; Posse, 1890) (Figures 181 and 182).

Figure 181. The "postural depression" in kyphotic sitting postures is very similar to the "postural depression" resulting from sit-up exercises. From Bradford, E.H., and Stone, J.S.: The seating of school children. *Transactions of the American Orthopaedic Association,* *12:*170–183, 1899.

d. Exercises that tend to increase the thoracic spinal flexion are detrimental to good posture (Powell, 1930). Among adolescent females, Toppenberg and Bullock (1986) found the greater degrees of thoracic kyphosis to be associated with the shorter lengths of the rectus abdominis muscle. Kendall (1965) noted the tendency for overdeveloped upper abdominal muscles in the common postural fault of round upper back.

e. Overemphasis on the rectus abdominis muscle can have an adverse effect on postural correction.

"Often the upper abdomen is constricted by overdevelopment of the upper rectus abdominus. This is especially true in women who are conscious of a prominent abdomen and who are expending misdirected effort to overcome the condition. If continuous effort is made to flatten

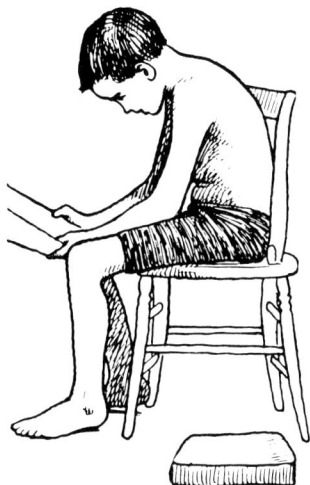

Figure 182. Another example of the "postural depression" in a kyphotic sitting posture. From Mosher, E.M.: *Health and Happiness.* New York, Funk and Wagnalls, 1913. Reproduced with permission of Harper and Row, Publishers, Inc.

the abdomen and the result is a constriction or deep crease just below the level of the ribs, the lower abdominal wall is made weaker and less effective in supporting organs, and the effort expended to make the abdomen less prominent does not produce the desired results" (Drew, 1945).

Frost (1938) found that with forced flattening of the abdomen, "most of the hollowing effect is in the upper abdomen, the sub-costal region, which involves definite interference with the free functioning of the diaphragm."

f. As the spine is forced into maximum flexion, full sit-ups will cause a marked rise in lumbar intradiscal pressure, comparable to bending forward twenty degrees with ten kilograms in each hand (Nachemson and Elfström, 1970). Halpern and Bleck (1979) commented that repeated sit-up activity "may as a side effect be causing or at least contributing to degenerative changes of the lumbar disk." White and Panjabi (1978) hypothesized that "the loads that are created on the discs by sit-ups can be expected to contribute to degeneration and failure of annular fibers."

Mutoh et al. (1983) reported twenty-nine cases of low back pain at a Tokyo hospital over a one year period related to sit-up exercises with the knees extended and the ankles supported. Of the twenty-nine cases, the diagnosis included twenty-two cases of lumbosacral strain (75.9 percent) and four cases of a lumbar disc lesion (13.8 percent).

g. Sit-up exercises promote trunk mobility, whereas training for trunk and pelvic stabilization are more important with abdominal exercises (Watanabe and Avakian, 1960; Armstrong, 1965; Zacharkow, 1984b).

Armstrong (1965) stressed that "a full range of movement of the spinal column itself is very rarely necessary to or even compatible with mechanical efficiency of the body as a whole. One of the main features of the development of a good active posture is the acquisition of the ability to brace the spine at or nearly at the neutral position so that it provides a stable centre from which the limbs can act with maximum efficiency."

h. The last point in regards to potential problems with sit-up exercises is rarely mentioned, but of major importance. McKenzie (1915) stressed that at the beginning of a supine trunk flexion movement, the lower abdominal obliques remain relaxed, and the lower inguinal regions tend to bulge. This is important to realize when advocating full or partial sit-ups for individuals who already have a weak, relaxed lower abdominal wall.

> "In the movements of straight flexion of the trunk the rectus muscle only is employed at the beginning and the relaxed oblique muscles are distended, forming two distinct pouches or weakened areas over the lower abdomen, and by the time they contract in self-protection the mischief may have been done.
>
> It is in such conditions and under such circumstances that hernia is likely to be acquired, because hernia, like other swellings, enlarges in the line of least resistance. Perhaps one of the most potent causes is a standing posture in which the abdomen is protruded and the chest sunken, forcing down the abdominal contents on the relaxed lower zone, and I have been struck with the number of cases in which hernia came on unconsciously, without apparent cause, other than perhaps a long walk or a fatiguing day's standing. Even repeated and violent effort seems less fruitful of cases than the dull and steady pressure on the relaxed abdominal walls" (McKenzie, 1915).

Bowen (1919) also stressed that with a weak abdominal wall, abdominal exercises should be avoided that do not bring the oblique muscles into action. "Direct flexion of the trunk by raising the head and shoulders from lying position is risky because it begins with isolated action of the rectus" (Bowen, 1919).

With a weak abdominal wall and sagging of the viscera, movements "that bring the lower parts of the abdominal wall into contraction first and most strongly are of course to be preferred, and for this reason leg-raising is better than trunk-raising" (Bowen, 1919).

Toe-Touching Exercises

The problems associated with both sitting and standing toe-touching exercises include the following:

a. The ability to touch one's toes will be influenced by the length of certain body segments (Broer and Galles, 1958; Wear, 1963; Adrichem and Korst, 1973). Individuals with a longer trunk-plus-arm measurement and relatively short legs will have an advantage in the performance of a toe-touch test. However, individuals with long legs and a relatively short trunk-plus-arm measurement will be at a disadvantage in a toe-touch test (Broer and Galles, 1958).

b. Toe-touching exercises can be extremely stressful during the adolescent growth spurt. According to Wiles (1937), "A great number of postural deformities commence in late childhood and adolescence, during periods of rapid growth."

The increase in height until approximately age thirteen or fourteen is due more to the lengthening of the legs, whereas after age fourteen or fifteen the trunk gains more rapidly in length (Tyler, 1907).

Kendall (1965) stressed that:

> " . . . there is a period between the years of ten and fourteen when a majority of children may not be able to touch the toes with knees straight. The inability to successfully perform this feat apparently results from a discrepancy between leg and trunk length during this growth period. To encourage or force children to accomplish this feat may be harmful in the sense that undue flexibility of the back may result" (Figure 183).

Kemper and Verschuur (1985) noted in their longitudinal study of teenagers in The Netherlands that the marked improvement in "sit-and-reach performance in girls at skeletal ages 12 and 13 and in boys at skeletal ages 14 and 15 also runs concurrently with the increase of trunk height in proportion to leg length." Branta et al. (1984) also felt that an increased score on a sit-and-reach test may be due to the growth spurt in trunk length, rather than an actual increase in hip flexibility.

c. Toe-touching exercises will tend to increase the thoracic and lumbar spinal flexion. The spinal stress from these movements will be the most severe in individuals with tight hamstrings (Lambrinudi, 1934; Milne and Mierau, 1979). Hamstring tightness is very prevalent throughout the school-age population (Milne and Mierau, 1979; Salminen, 1984).

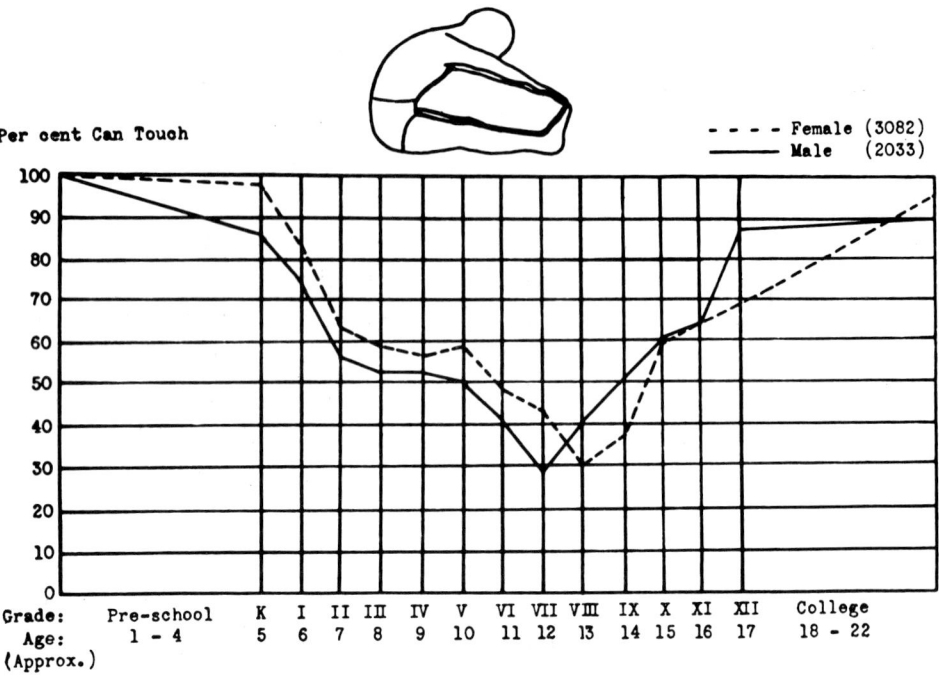

Figure 183. Ability to touch the toes, sitting with the knees straight, for different age groups. From Kendall, H.O., and Kendall, F.P.: Normal flexibility according to age groups. *The Journal of Bone and Joint Surgery, 30-A:*690–694, 1948. Reproduced with permission of The Journal of Bone and Joint Surgery.

According to Fisk (1987):

"When there has been hamstring tightness for a long time, say, since early schooldays, a characteristic pattern of bending becomes obvious: when bending to touch the toes with the knees straight, flexion starts in the upper spine, and there is an obvious limitation of flexion round the hips. To compensate for this loss of hip pivot there is an apparent increase in spinal flexion, obvious in the dorsal and upper lumbar spine.

Teaching such people how to stretch out their hamstrings is less than half the battle. They then have to be reprogrammed to bend with their hips and knees, not their back. This is extremely difficult. Their programmed pattern of bending is so ingrained on their computer circuits, and any attempt to alter this pattern takes a long time and a lot of effort. It would be simpler if such a pattern was not allowed to develop in the first place. An adequate training program in schools is needed."

Supine hamstring stretching techniques are preferable in order to reduce the thoracolumbar spinal stress (Zacharkow, 1984b; Milne and Mierau, 1979).

d. Hypermobility in spinal flexion is considered to be a major risk factor in low back pain (Biering-Sørensen, 1984a, 1984b). In regards to an increased flexibility in flexion, Kirby et al. (1981) found that female gymnasts with a history of low back pain had greater toe-touching ability than those gymnasts without low back symptoms.

e. Toe-touching exercises reinforce the movement pattern of spinal flexion with minimal hip flexion (Zacharkow, 1984b). A lifting posture similar to the standing toe-touch position can be very dangerous, as the fully flexed position involves passive vertebral support with the back muscles relaxed. Therefore, all the stress is on the posterior aspect of the discs and posterior ligaments of the back (Floyd and Silver, 1955; Wolf et al., 1979; Watanabe, 1981; Ekholm et al., 1982; Kippers and Parker, 1983; Tanii and Masuda, 1985).

MAINTAINING PROPER AXIAL RELATIONSHIP OF BODY SEGMENTS: PRACTICAL APPLICATIONS

Leaning Forward

Knudsen (1947) observed that from "daily life we are so used to carrying the body forward at work by a rounding of the back more than by a movement in the hips." He felt that such a working posture "deforms the body more than any other, and by its bad effect on the organs of the chest and the abdomen it undermines health" (Knudsen, 1947).

Checkley (1890) noted man's lack of awareness of joint movements, particularly in differentiating between hip flexion and spinal flexion. As a result, spinal flexion is overemphasized with minimal hip motion.

An erect trunk posture is characterized by the normal axial relationship of the pelvis and thorax, the ribs and chest raised to normal, and the head being held in its normal position. As a result, the abdominal and thoracic cavities are not constricted (Thomas and Goldthwait, 1922; Goldthwait, 1909). These characteristics of an erect trunk posture are important to maintain when leaning forward in both a standing and sitting position.

"In leaning forward the trunk must be kept straight; that is, the pelvis, thorax, and head remain in their relatively normal position. Bend forward and backward with motion only in the hip joint. Keep the chest high. Allow practically no motion in the sacro-lumbar or lumbo-dorsal junctures.

In bending thus, the pelvis moves with the spine as if the spine were not flexible" (Thomas and Goldthwait, 1922).

One needs to develop a "consciousness of control" over the back extensors of the thoracolumbar spine in order to prevent spinal flexion when leaning forward (Thomas and Goldthwait, 1922). Figures 184 to 186 illustrate the difference between good and bad forward leaning postures, in both sitting and standing positions.

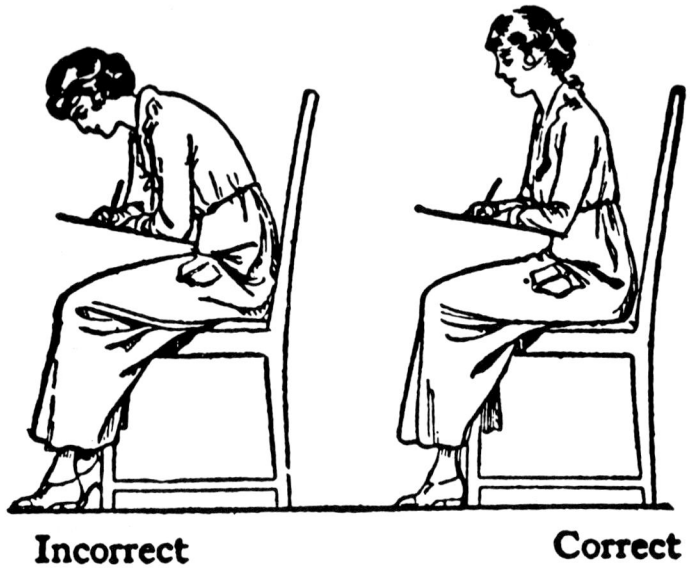

Incorrect   **Correct**

Figure 184. Leaning forward in a chair. From Wood, T.D., and Rowell, H.G.: *Health Supervision and Medical Inspection of Schools.* Philadelphia, Saunders, 1927. Reproduced with permission of the National Board of the Y.W.C.A. and W. B. Saunders Company.

Besides being limited by hamstring and hip extensor tightness, hip mobility in flexion will also be limited by restrictive clothing, such as tight trousers or skirts (Eklundh, 1965; Stubbs and Osborne, 1979; Stubbs et al., 1985; Fisk, 1986).

While maintaining a deep forward bending work posture, thoracolumbar spinal extension can be facilitated by keeping one forearm supported on the thigh (Suenaga, 1982; Maeda et al., 1980b).

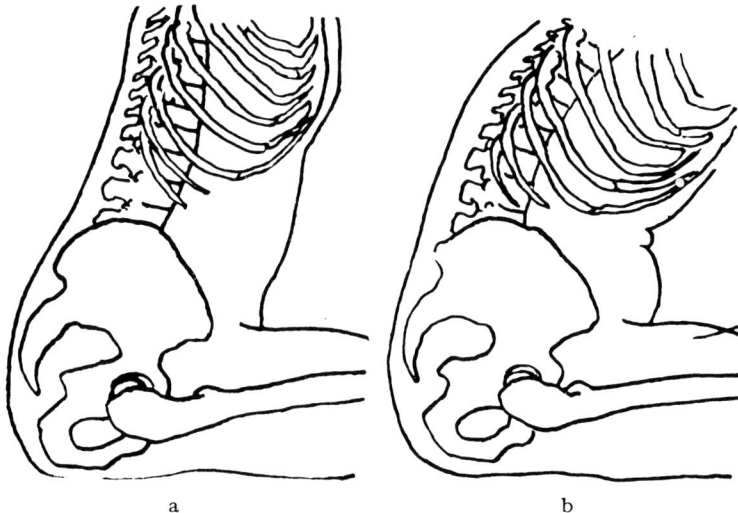

a   b

Figure 185.
 a. When leaning forward from the hips, the abdominal cavity is long and the chest and diaphragm are raised.
 b. When leaning forward by spinal flexion, the abdominal cavity is short and compressed. The ribs are lowered and the chest cavity is diminished. The upper trunk is partly resting on the viscera.

From Knudsen, K.A.: *A Textbook of Gymnastics, 2nd ed. Volume One. Form-Giving Exercises.* London, J. & A. Churchill Ltd., 1947. Reproduced with permission of Longman Group Ltd.

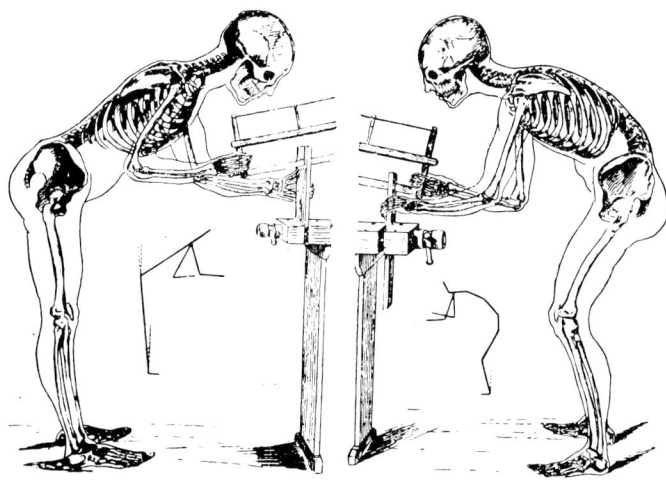

Figure 186. Leaning forward in a standing work posture.
 Left: Leaning forward by hip flexion.
 Right: Leaning forward by spinal flexion. The effect on the abdominal and thoracic cavities will be similar to Figure 185b. From Burgerstein, L.: *School Hygiene.* New York, Frederick A. Stokes Co., 1915. Reproduced with permission of Harper and Row, Publishers, Inc.

Lifting

The normal axial relationship of the pelvis, thorax, and head in an erect trunk posture is also critical in proper lifting technique. When lifting, the leaning forward of the trunk should occur at the hips, with the spine being stabilized by the back extensor musculature. Without the conscious muscular stabilization of the spine, the vertebral column will be more vulnerable to injury from the passive tensile stress on the posterior ligaments and discs (Asmussen et al., 1965; Keller, 1965; Poulsen and Jørgensen, 1971; Kippers and Parker, 1983; Zacharkow, 1984b).

Important points for maintaining an extended spinal posture with activation of the erector spinae are as follows:

a. The trunk posture should not be vertical, but leaning somewhat forwards. The exact trunk angle will depend on the size and height of the object being lifted. The trunk angle will also vary between individuals, depending on one's height and body proportions (Zacharkow, 1984b).

A leaning forward trunk posture will facilitate activation of the erector spinae, along with the reflex co-contraction of the lower abdominals (Rathbone, 1934, 1936; Carlsöö, 1961, 1980; Zacharkow, 1984b).

b. The body weight should be located just behind the balls of the feet at the start of the lift (Bancroft, 1913; Whitney, 1962; Bendix and Eid, 1983; Zacharkow, 1984b). The command "chest over toes" as advocated by Bancroft (1913) in obtaining proper standing posture will also be helpful in obtaining the proper starting position in lifting.

When getting in position to lift a heavy object from the floor, it will be helpful to first get as close to the object as possible. From this position, concentrate on holding the pelvis back and bringing the chest and upper trunk forwards. This movement will help facilitate extension of the lower thoracic spine and activation of the erector spinae and lower abdominals.

c. Keeping the chin tucked in at the start of the lift will lengthen the neck, thereby preventing a hyperextended head and upper cervical spine. This will facilitate extensor tone in the trunk and legs (Haynes, 1928; Jones, 1963; Anderson, 1960; Guthrie, 1963; Keller, 1965) (Figure 187).

d. With heavy lifts, the knees should not be flexed much beyond 90 degrees, as greater knee flexion angles will markedly reduce the quadriceps' efficiency (Keller, 1965; Schafer, 1983). Rising from a lifting posture with extreme hip and knee flexion will be similar to rising from a chair with a very low seat height. At the start of the lift, the hips will rise too

Figure 187. It is important to maintain the proper axial relationship of the head, thorax, and pelvis when bending and lifting. From Stransky, J., and Stone, R.B.: *The Alexander Technique.* New York, Beaufort Books, 1981. Reproduced with permission of Beaufort Books, Inc.

fast, and a kyphotic lifting posture will result (Davis et al., 1965; Grieve, 1977).

A lifting posture with a vertical trunk and full flexion at the hips and knees will also present other problems. It will be a very unstable lifting posture, as the individual will be forced to balance up on his toes (Davies, 1978; Grieve and Pheasant, 1982; Zacharkow, 1984b) (Figure 188). The lifter will be more concerned with keeping his balance than with performing the lift (Davies, 1978).

When a large object is lifted with a vertical trunk and a full squat posture, the arms must also be further extended towards the horizontal at the start of the lift. As such an arm position will produce a high torque at the shoulders, the individual will usually flex the trunk forward to lessen the load moment arm about the shoulders (Chaffin, 1975).

Lifting from a Seated Position

Although not often appreciated, lifting light objects from a seated position can be very stressful to the spine. With the trunk flexed forward and the arm extended forward, a seated lift can greatly increase the torque exerted on the lumbar spine (Tichauer, 1968, 1978).

Reaching Overhead

When reaching up for an object, the proper axial relationship of the pelvis, thorax, and head is just as critical as when lifting an object. A backward arching of the upper trunk can result in a forceful hyper-extension of the lower lumbar spine, along with relaxation of the erector spinae (Carlsöö, 1961; Stransky and Stone, 1981). With good technique, the hips and pelvis are held back and the upper trunk is brought slightly forwards, resulting in activation of both the erector spinae and lower abdominals (Figure 189).

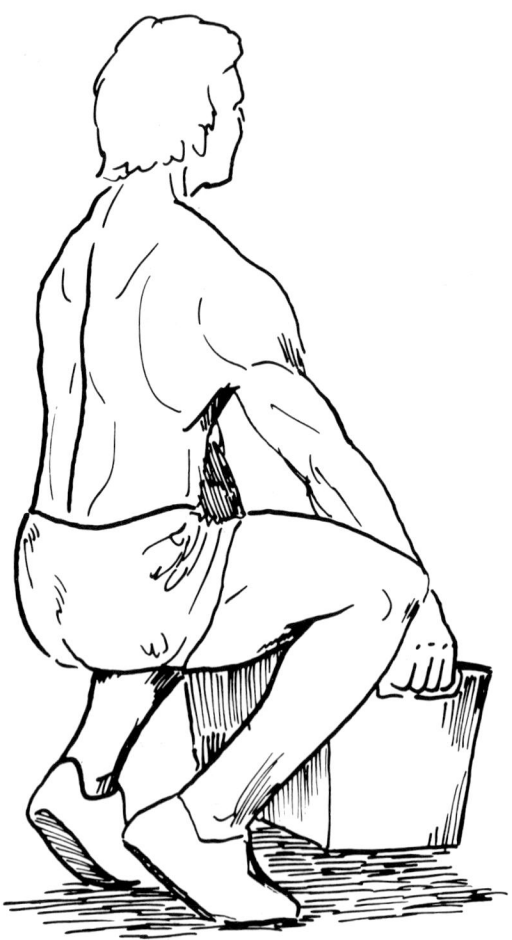

Figure 188. Balancing up on the toes with a vertical trunk—an unstable lifting posture. From Zacharkow, D.: *The Healthy Lower Back.* Springfield, Thomas, 1984. Reproduced with permission of Charles C Thomas, Publisher.

Sitting In/Rising From a Chair

As opposed to maintaining the proper axial relationship of the pelvis, thorax, and head, the vast majority of individuals will sit and rise with a hyperlordotic upper cervical and lower lumbar spine, excessive tension in the neck and lower back, a protruding lower abdomen, and a decreased intra-abdominal pressure (Alexander, 1918).

Standing to Sitting

One foot should be placed a very short step forward in advance of the other, with the back of the rear leg almost touching the front edge of the seat. Then, sway backward from the ankles until the body weight is mainly over the rear foot. While holding the hips and pelvis back and

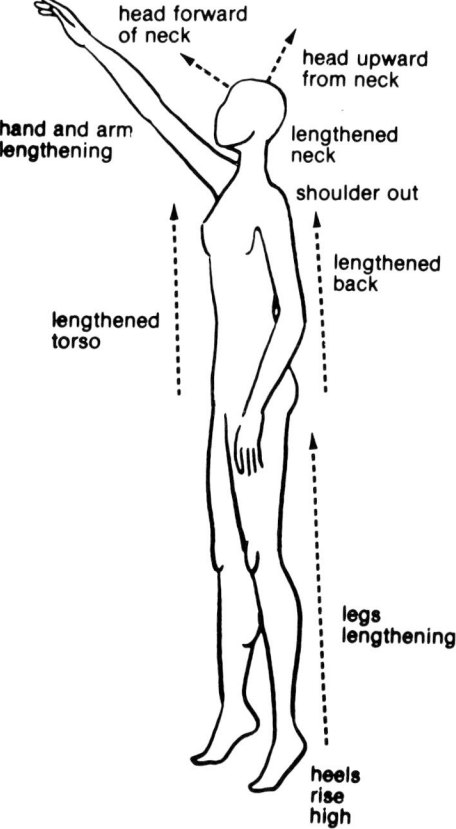

Figure 189. Proper technique for reaching upwards. From Stransky, J., and Stone, R.B.: *The Alexander Technique.* New York, Beaufort Books, 1981. Reproduced with permission of Beaufort Books, Inc.

keeping the chin drawn in, allow the hip and knee joints to slowly bend as hinges in order to achieve a sitting position.

Sitting to Standing

From a seated position at the front of the chair, draw the feet back so that one foot can be placed underneath the front edge of the seat. Similar to swaying backward from the ankles in a standing position, allow the pelvis to rock posteriorly over the ischial tuberosities (Haynes, 1928). Then, concentrate on holding the pelvis back and keeping the chin drawn in, and incline the trunk forwards from the hips. When the center of gravity falls over the feet, straighten the legs at the hips, knees, and ankles until the erect position is attained.

Supine Rest Position

A supine rest posture with a small roll or pad under the lower thorax can be helpful in promoting extension of the thoracic spine, expansion of the lower ribs, and a raised position of the diaphragm (Goldthwait and Brown, 1911; Thomas and Goldthwait, 1922; Sutton, 1925; Rathbone, 1934; Goldthwait et al., 1934; Hawley, 1937) (Figure 190).

While lying on the floor or a firm surface, the support should be placed so that its highest point is under the region of the tenth thoracic to eleventh thoracic vertebrae. The hands should be placed behind the neck. Based on individual comfort, the lower extremities can be kept extended, or the hips and knees can be flexed with the feet flat on the floor.

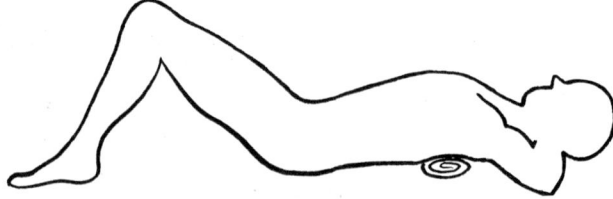

Figure 190. Supine rest posture. Adapted from Hawley, G.: *The Kinesiology of Corrective Exercise.* Philadelphia, Lea and Febiger, 1937. With permission of Lea and Febiger.

BIBLIOGRAPHY

Aarås, A.: Neck and shoulder pain. In Nelson, M. (Ed.): *Low Back Pain and Industrial and Social Disablement.* Middlesex, Back Pain Association, 1983, pp. 40–44.

Abildgaard, A.U., and Daugaard, K.: Videnskab og praksis. *Ugeskrift For Laeger, 141:*3147–3151, 1979.

Abramson, A.S.: Bone disturbances in injuries to the spinal cord and cauda equina (paraplegia). *The Journal of Bone and Joint Surgery, 30-A:*982–987, 1948.

Adams, M.A., and Hutton, W.C.: The effect of posture on the role of the apophysial joints in resisting intervertebral compressive forces. *The Journal of Bone and Joint Surgery, 62-B:*358–362, 1980.

Adams, M.A., and Hutton, W.C.: Prolapsed intervertebral disc. *Spine, 7:*184–191, 1982.

Adkins, H.V., Robins, V., Eckland, F., Perry, J., and Nickel, V.L.: Selective stretching for the paralytic patient. *The Physical Therapy Review, 40:*644–648, 1960.

Adrichem, J.A.M., and Korst, J.K.: Assessment of the flexibility of the lumbar spine. *Scandinavian Journal of Rheumatology, 2:*87–91, 1973.

Agarwal, N., Del Guercio, L.R.M., and Lee, B.Y.: The role of nutrition in the management of pressure sores. In Lee, B.Y. (Ed.): *Chronic Ulcers of the Skin.* New York, McGraw-Hill, 1985, pp. 133–145.

Agris, J., and Spira, M.: Pressure ulcers: prevention and treatment. *Clinical Symposia, 31:*1–32, 1979.

Ahlbäck, S-O., and Lindahl, O.: Sagittal mobility of the hip-joint. *Acta Orthopaedica Scandinavica, 34:*310–322, 1964.

Åkerblom, B.: *Standing and Sitting Posture.* Stockholm, Nordiska Bokhandeln, 1948.

Åkerblom, B.: Chairs and sitting. In Floyd, W.F., and Welford, A.T. (Eds.): *Symposium on Human Factors in Equipment Design.* London, Lewis, 1954, pp. 29–35.

Alden, D.G., Daniels, R.W., Kanarick, A.F.: Keyboard design and operation: a review of the major issues. *Human Factors, 14:*275–293, 1972.

Alexander, C.: The etiology of femoral epiphysial slipping. *The Journal of Bone and Joint Surgery, 48B:*299–311, 1966.

Alexander, C.J.: Scheuermann's disease. *Skeletal Radiology, 1:*209–221, 1977.

Alexander, F.M.: *Man's Supreme Inheritance.* New York, Dutton, 1918.

Allman, R.M., Laprade, C.A., Noel, L.B., Walker, J.M., Moorer, C.A., Dear, M.R., and Smith, C.R.: Pressure sores among hospitalized patients. *Annals of Internal Medicine, 105:*337–342, 1986.

Andersen, K.E., and Kvorning, S.A.: Medical aspects of the decubitus ulcer. *International Journal of Dermatology, 21:*265–270, 1982.

371

Anderson, J., and Breidahl, P.: Cartilage atrophy following spinal cord damage. *Australasian Radiology, 25:*98–103, 1981.

Anderson, T.: *Human Kinetics and Analysing Body Movements.* London, Heinemann, 1951.

Anderson, T. McC.: *Manual Lifting and Handling.* London, Industrial Society, 1960.

Andersson, B.J.G., Jonsson, B., and Örtengren, R.: Myoelectric activity in individual lumbar erector spinae muscles in sitting. A study with surface and wire electrodes. *Scandinavian Journal of Rehabilitation Medicine, Supplement 3:*91–108, 1974a.

Andersson, B.J.G., and Örtengren, R.: Lumbar disc pressure and myoelectric back muscle activity during sitting. III. Studies on a wheelchair. *Scandinavian Journal of Rehabilitation Medicine, 6:*122–127, 1974b.

Andersson, B.J.G., and Örtengren, R.: Myoelectric back muscle activity during sitting. *Scandinavian Journal of Rehabilitation Medicine, Supplement 3:*73–90, 1974c.

Andersson, B.J.G., Örtengren, R., Nachemson, A., and Elfström, G.: Lumbar disc pressure and myoelectric back muscle activity during sitting. Part one. Studies on an experimental chair. *Scandinavian Journal of Rehabilitation Medicine, 6:*104–114, 1974d.

Andersson, B.J.G., Örtengren, R., Nachemson, A., and Elfström, G.: Lumbar disc pressure and myoelectric back muscle activity during sitting. Part four. Studies on a car driver's seat. *Scandinavian Journal of Rehabilitation Medicine, 6:*128–133, 1974e.

Andersson, B.J.G., and Örtengren, R.: Lumbar disc pressure and myoelectric back muscle activity during sitting. II. Studies on an office chair. *Scandinavian Journal of Rehabilitation Medicine, 6:*115–121, 1974f.

Andersson, B.J.G., Örtengren, R., Nachemson, A.L., Elfström, G., and Broman, H.: The sitting posture: an electromyographic and discometric study. *Orthopedic Clinics of North America, 6:*105–120, 1975.

Andersson, G.B.J.: The load on the lumbar spine in sitting postures. In Oborne, D.J., and Levis, J.A. (Eds.): *Human Factors in Transport Research, vol. 2.* London, Academic Pr, 1980, pp. 231–239.

Andersson, G.B.J., Murphy, R.W., Örtengren, R., and Nachemson, A.L.: The influence of backrest inclination and lumbar support on lumbar lordosis. *Spine, 4:*52–58, 1979.

Antioch College Studies. *The Effects of Modern Shoes Upon Proper Body Mechanics.* Yellow Springs, Antioch College, 1931.

Appleton, A.B.: Posture. *The Practitioner, 156:*48–55, 1946.

Armstrong, J.R.: *Lumbar Disc Lesions,* 3rd ed. Edinburgh, Livingstone, 1965.

Arndt, R.: Body posture. In *Health and Ergonomic Considerations of Visual Display Units—Symposium Proceedings.* Denver, March 1–2, 1982, pp. 29–44.

Arndt, R.: Working posture and musculoskeletal problems of video display terminal operators-review and reappraisal. *American Industrial Hygiene Association Journal, 44:*437–446, 1983.

Aronovitz, R., Greenway, R., Lindan, O., Reswick, J., and Scanlan, J.: A pneumatic cell matrix to measure the distribution of contact pressure over the human body.

In *Proceedings of the Sixteenth Annual Conference on Engineering in Medicine and Biology*, 1963, pp. 62–63.

Asatekin, M.: Postural and physiological criteria for seating—a review. *M.E.T.U. Journal of the Faculty of Architecture*, 1:55–83, 1975.

Asmussen, E.: The weight-carrying function of the human spine. *Acta Orthopaedica Scandinavica, 29:*276–290, 1960.

Asmussen, E., and Klausen, K.: Form and function of the erect human spine. *Clinical Orthopaedics, 25:*55–63, 1962.

Asmussen, E., Poulsen, E., and Rasmussen, B.: Quantitative evaluation of the activity of the back muscles in lifting. *Communications from the Danish National Association for Infantile Paralysis, 21:*3–13, 1965.

Atherton, J.A., Chatfield, J., Clarke, A.K., and Harrison, R.A.: Static chairs for the arthritic. *Journal of Occupational Therapy (England), 43:*366–367, 1980.

Aveling, J.H.: *Posture in Gynecic and Obstetric Practice.* Philadelphia, Lindsay and Blakiston, 1879.

Avon, G., and Schmitt, L.: Electromyographie du trapèze dans diverses positions de travail à la machine à écrire. *Ergonomics, 18:*619–626, 1975.

Ayoub, M.M.: Sitting down on the job (properly). *Industrial Design, 19:*42–45, April 1972.

Ayoub, M.M.: Work place design and posture. *Human Factors, 15:*265–268, 1973.

Babbs, F.W.: A design layout method for relating seating to the occupant and vehicle. *Ergonomics, 22:*227–234, 1979.

Bailey, B.N.: *Bedsores.* London, Edward Arnold, 1967, p. 9.

Baird, R.A., DeBenedetti, M.J., and Eltorai, I.: Non-septic hip instability in the chronic spinal cord injury patient. *Paraplegia, 24:*293–300, 1986.

Bancroft, J.H.: *The Posture of School Children.* New York, Macmillan, 1913.

Barbenel, J.C.: Measurement of interface pressures. In Barbenel, J.C., Forbes, C.D., and Lowe, G.D.O. (Eds.): *Pressure Sores.* London, Macmillan, 1983, pp. 67–78.

Barbenel, J.C., Evans, J.H., and Jordan, M.M.: Tissue mechanics. *Engineering in Medicine*, 7:5–9, 1978.

Barbenel, J.C., Ferguson-Pell, M.W., and Evans, J.H.: Pressures produced on hospital mattresses. *Health Bulletin, 39:*62–68, 1981.

Barbenel, J.C., Jordan, M.M., Nicol, S.M., and Clark, M.O.: Incidence of pressure sores in the Greater Glasgow Health Board area. *The Lancet, 2:*548–550, September 10, 1977.

Bardsley, G.I., Bell, F., Black, R.C., and Barbenel, J.C.: Movements during sitting and their relationship to pressure sores. In Barbenel, J.C., Forbes, C.D., and Lowe, G.D.O. (Eds.): *Pressure Sores.* London, Macmillan, 1983, pp. 157–165.

Barker, V.: *Posture Makes Perfect.* Auckland, Fitworld, 1985.

Barkla, D.: The estimation of body measurements of British population in relation to seat design. *Ergonomics, 4:*123–132, 1961.

Barkla, D.M.: Chair angles, duration of sitting, and comfort ratings. *Ergonomics, 7:*297–304, 1964.

Barlow, W.: An investigation into kinaesthesia. *The Medical Press and Circular, 215:*60–63, 1946.

Barlow, W.: *The Alexander Technique.* New York, Warner, 1980.

Bartelink, D.L.: The role of abdominal pressure in relieving the pressure on the lumbar intervertebral discs. *The Journal of Bone and Joint Surgery, 39-B:* 718–725, 1957.

Barton, A., and Barton, M.: *The Management and Prevention of Pressure Sores.* London, Faber and Faber, 1981.

Basmajian, J.V.: *Surface Anatomy.* Baltimore, Williams and Wilkins, 1977.

Basmajian, J.V.: *Muscles Alive. Their Functions Revealed by Electromyography,* 4th ed. Baltimore, Williams and Wilkins, 1978.

Batchelor, K.W., and Farmelo, G.P.: A basic design of a chair for hemiplegics with special reference to the back shape. *Biomedical Engineering, 10:* 373–378, 1975.

Bedbrook, G.: *The Care and Management of Spinal Cord Injuries.* New York, Springer-Verlag, 1981, p. 99.

Bell Telephone Laboratories: *Video Display Terminals: Preliminary Guidelines for Selection, Installation, and Use.* Short Hills, Bell Telephone Laboratories, 1983.

Bendix, T.: Seated trunk posture at various seat inclinations, seat heights, and table heights. *Human Factors, 26:* 695–703, 1984.

Bendix, T., and Biering-Sørensen, F.: Posture of the trunk when sitting on forward inclining seats. *Scandinavian Journal of Rehabilitation Medicine, 15:* 197–203, 1983.

Bendix, T., and Eid, S.E.: The distance between the load and the body with three bi-manual lifting techniques. *Applied Ergonomics, 14:* 185–192, 1983.

Bendix, T., and Hagberg, M.: Trunk posture and load on the trapezius muscle whilst sitting at sloping desks. *Ergonomics, 27:* 873–882, 1984.

Bendix, T., Krohn, L., Jessen, F., and Aarås, A.: Trunk posture and trapezius muscle load while working in standing, supported-standing, and sitting positions. *Spine, 10:* 433–439, 1985b.

Bendix, T., Sørensen, S.S., and Klausen, K.: Lumbar curve, trunk muscles, and line of gravity with different heel heights. *Spine, 9:* 223–227, 1984.

Bendix, T., Winkel, J., and Jessen, F.: Comparison of office chairs with fixed forwards or backwards inclining, or tiltable seats. *European Journal of Applied Physiology, 54:* 378–385, 1985a.

Bennett, H.E.: Some requirements of good school seating. *Elementary School Journal, 23:* 203–214, 1922.

Bennett, H.E.: A study of school posture and seating. *Elementary School Journal, 26:* 50–57, 1925.

Bennett, H.E.: *School Posture and Seating.* Boston, Ginn and Company, 1928.

Bennett, L.: Transferring load to flesh, part eight. Stasis and stress. *Bulletin of Prosthetics Research, 10-23:* 202–210, Spring, 1975.

Bennett, L., Kavner, D., Lee, B.Y., Trainor, F.S., and Lewis, J.M.: Skin blood flow in seated geriatric patients. *Archives of Physical Medicine and Rehabilitation, 62:* 392–398, 1981.

Bennett, L., Kavner, D., Lee, B.Y., Trainor, F.S., and Lewis, J.M.: Skin stress and blood flow in sitting paraplegic patients. *Archives of Physical Medicine and Rehabilitation, 65:* 186–190, 1984.

Bennett, L., and Lee, B.Y.: Pressure and shear in pressure sores. In *Proceedings of the*

National Symposium on the Care, Treatment and Prevention of Decubitus Ulcers. Arlington, Virginia, November 1984, pp. 41–46.

Bennett, L., and Lee, B.Y.: Pressure versus shear in pressure sore causation. In Lee, B.Y. (Ed.): *Chronic Ulcers of the Skin.* New York, McGraw-Hill, 1985, pp. 39–56.

Bennett, R.L.: Classification and treatment of early lateral deviations of the spine following acute anterior poliomyelitis. *Archives of Physical Medicine and Rehabilitation, 36:*9–17, 1955.

Bennett, R.L.: Recognition and care of early scoliosis. *Archives of Physical Medicine and Rehabilitation, 42:*211–225, 1961.

Benninghoff, A.: *Lehrbuch der Anatomie des Menschen, vol. one.* München, 1939. Translated in Åkerblom, B.: *Standing and Sitting Posture.* Stockholm, Nordiska Bokhandeln, 1948, p. 23.

Benz, C., Grob, R., and Haubner, P.: *Designing VDU Workplaces.* Köln, Verlag TÜV Rheinland, 1983.

Berger, N., and Lusskin, R.: Orthotic components and systems. In American Academy of Orthopaedic Surgeons: *Atlas of Orthotics.* St. Louis, Mosby, 1975, pp. 344–363.

Bergqvist, U.: Video display terminals and health. *Scandinavian Journal of Work, Environment and Health, 10 (Supplement 2):*1–87, 1984.

Berjian, R.A., Douglass, H.O., Holyoke, E.D., Goodwin, P.M., and Priore, R.L.: Skin pressure measurements on various mattress surfaces in cancer patients. *American Journal of Physical Medicine, 62:*217–226, 1983.

Berry, R.B.: The late results of surgical treatment of pressure sores in paraplegics. *British Journal of Surgery, 67:*473–474, 1980.

Bhatnager, V., Drury, C.G., and Schiro, S.G.: The effect of time-at-task on posture and performance in inspection. In Attwood, D.A., and McCann, C. (Eds.): *Proceedings of the 1984 International Conference on Occupational Ergonomics, vol. 1.* Toronto, Ontario, Canada, May 7–9, 1984, pp. 289–293.

Bhatnager, V., Drury, C.G., and Schiro, S.G.: Posture, postural discomfort, and performance. *Human Factors, 27:*189–199, 1985.

Biering-Sørensen, F.: Physical measurements as risk indicators for low-back trouble over a one-year period. *Spine, 9:*106–119, 1984a.

Biering-Sørensen, F.: A one-year prospective study of low back trouble in a general population. *Danish Medical Bulletin, 31:*362–375, 1984b.

Bitterman, M.E.: Fatigue defined as reduced efficiency. *American Journal of Psychology, 57:*569–573, 1944.

Black, R., Filippone, A.F., and Hahn, G.: The thermographic study of skin after wheelchair sitting. In Stokes, I.A.F. (Ed.): *Mechanical Factors and the Skeleton.* London, Libbey, 1981, pp. 201–204.

Black, R.C., and Reed, L.D.: The use of thermography in the prevention of pressure sores. In Barbenel, J.C., Forbes, C.D., and Lowe, G.D.O. (Eds.): *Pressure Sores.* London, Macmillan, 1983, pp. 167–175.

Black, S.: *Man and Motor Cars.* New York, Norton, 1966.

Blair, S.N., Falls, H.B., and Pate, R.R.: A new physical fitness test. *The Physician and Sportsmedicine, 11:*87–95, April 1983.

Bogduk, N.: The anatomy and pathophysiology of whiplash. *Clinical Biomechanics, 1:*92–101, 1986.

Bolton, C.B.: Ventile, incompressible cushions. *Applied Ergonomics, 3:*101–105, 1972.

Bouisset, S., and Zattara, M.: A sequence of postural movements precedes voluntary movement. *Neuroscience Letters, 22:*263–270, 1981.

Bouissou, H., Pieraggi, M.T., and Julian, M.: Dermis ageing. *Pathology Research and Practice, 178:*515–517, 1984.

Bowen, J.R., MacEwen, G.D., and Mathews, P.A.: Treatment of extension contracture of the hip in cerebral palsy. *Developmental Medicine and Child Neurology, 23:*23–29, 1981.

Bowen, W.P.: *A Teacher's Course in Physical Training.* Ann Arbor, Wahr, 1917.

Bowen, W.P.: *Applied Anatomy and Kinesiology,* 2nd ed. Philadelphia, Lea and Febiger, 1919.

Bradford, E.H., and Stone, J.S.: The seating of school children. *Transactions of the American Orthopaedic Association, 12:*170–183, 1899.

Brand, P.W.: Panel discussion on pressure ulcerations. In Fredricks, S., and Brody, G.S. (Eds.): *Symposium on the Neurologic Aspects of Plastic Surgery.* St. Louis, Mosby, 1978, vol. 17, p. 248.

Brand, P.W.: Comments on the article "development of test methods for evaluation of wheelchair cushions." *Bulletin of Prosthetics Research, 17:*3–4, Spring 1980.

Branta, C., Haubenstricker, J., and Seefeldt, V.: Age changes in motor skills during childhood and adolescence. *Exercise and Sport Sciences Reviews, 12:*467–520, 1984.

Branton, P.: *The Comfort of Easy Chairs.* Stevenage, Hertfordshire, England, The Furniture Industry Research Association, 1966.

Branton, P.: Behaviour, body mechanics and discomfort. In Grandjean, E. (Ed.): *Proceedings of the Symposium on Sitting Posture.* London, Taylor and Francis, 1969, pp. 202–213.

Branton, P.: Seating in industry. *Applied Ergonomics, 1:*159–165, 1970.

Branton, P.: Backshapes of seated persons—how close can the interface be designed? *Applied Ergonomics, 15:*105–107, 1984.

Branton, P., and Grayson, G.: An evaluation of train seats by observation of sitting behaviour. *Ergonomics, 10:*35–51, 1967.

Brattgård, S-O: Design of wheelchairs and wheelchair service based on scientific research. *Réadaptation, 11:*162–172, 1969.

Brattgård, S.O., Carlsöö, S., and Severinsson, K.: Temperature and humidity in the sitting area. In Kenedi, R.M., Cowden, J.M., and Scales, J.T. (Eds.): *Bedsore Biomechanics.* Baltimore, Univ Park, 1976, pp. 185–188.

Brattgård, S-O., Lindström, I., Severinsson, K., and Wihk, L.: Wheelchair design and quality. *Scandinavian Journal of Rehabilitation Medicine, Supplement 9:*15–19, 1983.

Brattgård, S-O., and Severinsson, K.: Investigations of pressure, temperature, and humidity in the sitting area in a wheelchair. In Asmussen, E., and Jorgensen, K. (Eds.): *Biomechanics VI-B. International Series on Biomechanics.* Baltimore, Univ. Park, 1978, vol. 2-B, pp. 270–273.

Braus, H.: *Anatomie des Menschen, vol. one.* Berlin, 1921. Translated in Åkerblom, B.: *Standing and Sitting Posture.* Stockholm, Nordiska Bokhandeln, 1948, p. 25.

Braverman, I.M., and Fonferko, E.: Studies in cutaneous aging: 1. The elastic fiber network. *The Journal of Investigative Dermatology, 78:*434–443, 1982a.

Braverman, I.M., and Fonferko, E.: Studies in cutaneous aging: II. The microvasculature. *The Journal of Investigative Dermatology, 78:*444–448, 1982b.

Braverman, I.M., and Keh-Yen, A.: Ultrastructure of the human dermal microcirculation. III. The vessels in the mid- and lower dermis and subcutaneous fat. *The Journal of Investigative Dermatology, 77:*297–304, 1981.

Broer, M.R., and Galles, N.R.G.: Importance of relationship between various body measurements in performance of the toe-touch test. *The Research Quarterly, 29:*253–263, 1958.

Bromley, I.: *Tetraplegia and Paraplegia—A Guide for Physiotherapists.* Edinburgh, Churchill, 1976, p. 211.

Brown, J.C., Swank, S.M., Matta, J., and Barras, D.M.: Late spinal deformity in quadriplegic children and adolescents. *Journal of Pediatric Orthopedics, 4:*456–461, 1984.

Browne, T.J.: Habit and posture, part two. *American Physical Education Review, 21:*176–189, 1916.

Brubaker, C.E., McClay, I.S., and McLaurin, C.A.: Effect of seat position on wheelchair propulsion efficiency. In *Proceedings of the Second International Conference on Rehabilitation Engineering,* Ottawa, 1984, pp. 12–14.

Brunnstrom, S.: *Clinical Kinesiology,* 3rd ed. Philadelphia, Davis, 1972.

Brunswic, M.: Ergonomics of seat design. *Physiotherapy, 70:*40–43, 1984a.

Brunswic, M.: Seat design in unsupported sitting. In *Proceedings of the 1984 International Conference on Occupational Ergonomics,* vol. 1. Toronto, Ontario, Canada, May 7–9, 1984b, pp. 294–297.

Bryan, C.S., Dew, C.E., Reynolds, K.L.: Bacteremia associated with decubitus ulcers. *Archives of Internal Medicine, 143:*2093–2095, 1983.

Bulstrode, S., Harrison, R.A., and Clarke, A.K.: *Assessment of Back Rests for Use in Car Seats.* DHSS Aids Assessment Programme, Health Publications Unit, Lancashire, United Kingdom, 1983.

Bunch, W.H., and Keagy, R.D.: *Principles of Orthotic Treatment.* Saint Louis, Mosby, 1976.

Burandt, U.: Röntgenuntersuchung über die stellung von becken und wirbelsäule beim sitzen auf vorgeneigten flächen. In Grandjean, E. (Ed.): *Proceedings of the Symposium on Sitting Posture.* London, Taylor and Francis, 1969, pp. 242–250.

Burandt, U., and Grandjean, E.: Sitting habits of office employees. *Ergonomics, 6:*217–228, 1963.

Burdett, R.G., Habasevich, R., Pisciotta, J., and Simon, S.R.: Biomechanical comparison of rising from two types of chairs. *Physical Therapy, 65:*1177–1183, 1985.

Bureau of the Census, United States Department of Commerce. *Population Estimates and Projections.* Series P-25, No. 922, October, 1982.

Burgerstein, L.: *School Hygiene.* New York, Stokes, 1915.

Burke, J., Sharrard, W.J.W., and Sutcliffe, M.L.: Pelvic obliquity and sitting buttock pressure distribution in paralytic scoliosis. *Journal of Bone and Joint Surgery,* 62-B:119–120, 1980.

Burnham, W.H.: Outlines of school hygiene. *The Pedagogical Seminary, 2:*9–71, 1892.

Burr, R.G.: Blood zinc in the spinal patient. *Journal of Clinical Pathology, 26:*773–775, 1973.

Burton, A.K.: Electromyography and office-chair design; a pilot study. *Behaviour and Information Technology, 3:*353–357, 1984.

Bush, C.A.: Study of pressures on skin under ischial tuberosities and thighs during sitting. *Archives of Physical Medicine and Rehabilitation, 50:*207–213, 1969.

Buti, L.B., Cortili, G., DeNigris, F., and Moretti, E.: Ergonomic design of a workplace for VDU operators. In Grandjean, E., and Vigliani, E. (Eds.): *Ergonomic Aspects of Visual Display Terminals.* London, Taylor and Francis, 1983, pp. 283–288.

Butler, R.W.: The nature and significance of vertebral osteochondritis. *Proceedings of the Royal Society of Medicine, 48:*895–902, 1955.

Cailliet, R.: *Hand Pain and Impairment,* 2nd ed. Philadelphia, Davis, 1975.

Cakir, A.: Human factors and VDT design. In Kolers, P.A., Wrolstad, M.E., and Bouma, H. (Eds.): *Processing of Visible Language 2.* New York, Plenum Press, 1980, pp. 481–495.

Cakir, A., Hart, D.J., and Stewart, T.F.M.: *Visual Display Terminals.* Chichester, Wiley, 1980.

Campbell, D.G.: Posture: a gesture toward life. *The Physiotherapy Review, 15:*43–47, 1935.

Cantoni, S., Colombini, D., Occhipinti, E., Grieco, A., Frigo, C., and Pedotti, A.: Posture analysis and evaluation at the old and new work place of a telephone company. In Grandjean, E. (Ed.): *Ergonomics and Health in Modern Offices.* London, Taylor and Francis, 1984, pp. 456–464.

Carlson, J.M., Lonstein, J., Beck, K.O., and Wilkie, D.C.: Seating for children and young adults with cerebral palsy. *Clinical Prosthetics and Orthotics, 10:*137–158, 1986.

Carlsöö, S.: The static muscle load in different work positions: an electromyographic study. *Ergonomics, 4:*193–211, 1961.

Carlsöö, S.: Influence of frontal and dorsal loads on muscle activity and on the weight distribution in the feet. *Acta Orthopaedica Scandinavica, 34:*299–309, 1964.

Carlsöö, S.: *How Man Moves.* London, Heinemann, 1972.

Carlsöö, S.: A back and lift test. *Applied Ergonomics, 11:*66–72, 1980.

Carlsöö, S.: The effect of vibration on the skeleton, joints and muscles. *Applied Ergonomics, 13:*251–258, 1982.

Casley-Smith, J.R.: Varying total tissue pressures and the concentration of initial lymphatic lymph. *Microvascular Research, 25:*369–379, 1983.

Center for Disease Control: Working with video display terminals: a preliminary health-risk evaluation. *Morbidity and Mortality Weekly Report, 29:*307–308, June 27, 1980.

Cerrato, P. L.: How diet helps the skin fight pressure sores. *RN, 49:*67–68, 1986.

Chaffin, D.B.: Localized muscle fatigue-definition and measurement. *Journal of Occupational Medicine, 15:*346–354, 1973.

Chaffin, D.B.: Biomechanics of manual materials handling and low-back pain. In Zenz, C. (Ed.): *Occupational Medicine.* Chicago, Year Book, 1975, pp. 443–467.

Chandler, F.A.: Stresses in a curved column. *The Journal of Bone and Joint Surgery, 15:*214, 1933.

Chantraine, A.: Actual concept of osteoporosis in paraplegia. *Paraplegia, 16:*51–58, 1978.

Chari, V.R., and Kirby, R.L.: Influence of the lower limbs on sitting balance while reaching forward. *Archives of Physical Medicine and Rehabilitation, 66:*529, 1985.

Chari, V.R., and Kirby, R.L.: Lower-limb influence on sitting balance while reaching forward. *Archives of Physical Medicine and Rehabilitation, 67:*730–733, 1986.

Checkley, E.: *A Natural Method of Physical Training.* Brooklyn, Bryant, 1890.

Checkley, E.: *A Natural Method of Physical Training,* revised ed. New York, Baker and Taylor, 1909.

Chow, W.W.: *Mechanical Properties of Gels and Other Materials with Respect to their Use in Pads Transmitting Forces to the Human Body.* Technical report no. 13, Medical School, Department of Physical Medicine and Rehabilitation, University of Michigan, September 1974.

Chow, W.W., Juvinall, R.C., and Cockrell, J.L.: Effects and characteristics of cushion covering membranes. In Kenedi, R. M., Cowden, J.M., and Scales, J.T. (Eds.): *Bedsore Biomechanics.* Baltimore, Univ. Park, 1976, pp. 95–102.

Chow, W.W., and Odell, E.I.: Deformations and stresses in soft body tissues of a sitting person. *Journal of Biomechanical Engineering, 100:*79–87, 1978.

Clark, M.C., and Faletti, M.V.: The role of seating types in egress difficulty experienced by older adults: a biomechanical analysis. In Swezey, R.W. (Ed.): *Proceedings of the Human Factors Society 29th Annual Meeting.* Baltimore, Sept 29–Oct 3, 1985, pp. 343–346.

Clark, M.O., Barbenel, J.C., Jordan, M.M., and Nicol, S.M.: Pressure sores. *Nursing Times, 74:*363–366, March 2, 1978.

Clark, R.P.: Micro-environmental air exchange rates in patient support systems. *Engineering in Medicine, 3:*6–7, 1974.

Clark, W.E.: The anatomy of work. In Floyd, W.F., and Welford, A.T. (Eds.): *Symposium on Human Factors in Equipment Design.* London, H.K. Lewis, 1954, pp. 5–15.

Clauser, C.E., McConville, J.T., and Young, J.W.: *Weight, Volume, and Center of Mass of Segments of the Human Body.* Technical Report AMRL–TR–69–70. Aerospace Medical Research Laboratory, Wright-Patterson Air Force Base, Ohio. August 1969.

Clemmesen, S.: Some studies on muscle tone. *Proceedings of the Royal Society of Medicine, 44:*637–646, 1951.

Cochran, G.V.B., and Palmieri, V.: Development of test methods for evaluation of wheelchair cushions. *Bulletin of Prosthetics Research, 10–33:*9–30, Spring 1980.

Cochran, G.V.B., and Slater, G.: Experimental evaluation of wheelchair cushions: report of a pilot study. *Bulletin of Prosthetics Research, 10–20:*29–61, Fall 1973.

Cochrane, G.M., and Wilshere, E.R. (Eds.): *Equipment for the Disabled. Wheelchairs,* 5th ed. Oxford, Oxfordshire Health Authority, 1982.

Coe, J.: Proof and practice: the design of a V.D.U. work-station. In Fisher, A.J., and Croft, P. (Eds.): *Proceedings of the 17th Annual Conference of the Ergonomics Society of Australia and New Zealand.* Sydney, Nov 27–28, 1980, pp. 187–197.

Coe, J.B.: Ergonomics and visual display units: work station design. In McPhee, B., and Howie, A. (Eds.) *Ergonomics and Visual Display Units.* Sydney, The Ergonomics Society of Australia and New Zealand, 1979, pp. 59–72.

Coe, J.B.: The influence of community status on work station design. In Shinnick, T., and Hill, G. (Eds.): *Proceedings of the 20th Annual Conference of the Ergonomics Society of Australia and New Zealand.* Adelaide, South Australia, December 1–2, 1983, pp. 185–190.

Coe, J.B.: Posture as a factor in repetitive strain injuries—the technological epidemic. In Adams, A.S., and Stevenson, M.G. (Eds.): *Proceedings of the 21st Annual Conference of the Ergonomics Society of Australia and New Zealand.* Sydney, 1984, pp. 119–123.

Coghill, G.E.: The educational methods of F. Matthias Alexander. In Alexander, F.M.: *The Universal Constant in Living.* New York, Dutton, 1941, pp. xxi–xxviii.

Cohen, L.A.: Role of eye and neck proprioceptive mechanisms in body orientation and motor coordination. *Journal of Neurophysiology, 24:*1–11, 1961.

Cohn, H.: *The Hygiene of the Eye in Schools.* London, Simpkin, Marshall and Co., 1886.

Collins, G.A., Cohen, M.J., Naliboff, B.D., and Schandler, S.L.: Comparative analysis of paraspinal and frontalis EMG, heart rate and skin conductance in chronic low back pain patients and normals to various postures and stress. *Scandinavian Journal of Rehabilitation Medicine, 14:*39–46, 1982.

Congleton, J.J., Ayoub, M.M., and Smith, J.L.: The design and evaluation of the neutral posture chair for surgeons. *Human Factors, 27:*589–600, 1985.

Constantian, M.B., and Jackson, H.S.: The ischial ulcer. In Constantian, M.B. (Ed.): *Pressure Ulcers—Principles and Techniques of Management.* Boston, Little, 1980, pp. 215–246.

Constantian, M.B., and Jones, M.V.: General nursing care of the patient with pressure ulcers. In Constantian, M.B. (Ed.): *Pressure Ulcers—Principles and Techniques of Management.* Boston, Little, 1980, pp. 123–139.

Cooper, D.G.: *Pelvic Stabilization for the Elderly.* Paper presented at Third International Seating Symposium, Memphis, Tennessee, February 26–28, 1987.

Cooper, K.H., and Holmstrom, F.M.G.: Injuries during ejection seat training. *Aerospace Medicine, 34:*139–141, 1963.

Corlett, E.N.: Human factors in the design of manufacturing systems. *Human Factors, 15:*105–110, 1973.

Corlett, E.N.: Pain, posture and performance. In Corlett, E.N., and Richardson, J.: *Stress, Work Design, and Productivity.* Chichester, John Wiley and Sons, 1981, pp. 27–42.

Corlett, E.N., and Bishop R.P.: A technique for assessing postural discomfort. *Ergonomics, 19:*175–182, 1976.

Corlett, E.N., and Eklund, J.A.E.: How does a backrest work? *Applied Ergonomics, 15:*111–114, 1984.

Corlett, E.N., Eklund, J.A., Houghton, C.S., and Webb, R.: A design for a sit-stand stool. In Coombes, K. (Ed.): *Proceedings of the Ergonomics Society's Conference 1983.* New York, Taylor and Francis, 1983, p. 157.

Cornelius, W.L., and Hinson, M.M.: The relationship between isometric contractions of hip extensors and subsequent flexibility in males. *Journal of Sports Medicine, 20:*75–80, 1980.

Cornell, W.S.: *Health and Medical Inspection of School Children.* Philadelphia, Davis, 1912.

Cotton, F.J.: School furniture for Boston schools. *American Physical Education Review, 9:*267–284, 1904.

Cotton, F.J.: *Report on School Furniture.* Appendix 6, Boston City Document No. 35, 1905, pp. 81–86.

Cousins, S.J., Jones, K.N., and Ackerley, K.E.: Aids in prevention and treatment of pressure sores: contoured cushion fabrication using the shapeable matrix. In Barbenel, J.C., Forbes, C.D., and Lowe, G.D.O. (Eds.): *Pressure Sores.* London, Macmillan, 1983, pp. 151–156.

Cowan, N.R.: The frontal cardiac silhouette in older people. *British Heart Journal, 27:*231–235, 1965.

Crewe, R.: The role of the occupational therapist in pressure sore prevention. In Barbenel, J.C., Forbes, C.D., and Lowe, G.D.O. (Eds.): *Pressure Sores.* London, Macmillan, 1983, pp. 121–132.

Croney, J.: *Anthropometry for Designers,* revised edition. New York, Van Nostrand Reinhold, 1981.

Cummine, J.L., Lonstein, J.E., Moe, J.H., Winter, R.B., and Bradford, D.S.: Reconstructive surgery in the adult for failed scoliosis fusion. *The Journal of Bone and Joint Surgery, 61-A:*1151–1161, 1979.

Cunningham, D.J.: The lumbar curve in man and apes. *Nature, 33:*378–379, 1886.

Cureton, T.K., and Wickens, J.S.: The center of gravity of the human body in the antero-posterior plane and its relation to posture, physical fitness, and athletic ability. *Research Quarterly, Supplement 6:*93–105, 1935.

Cushman, W.H.: Data entry performance and operator preferences for various keyboard heights. In Grandjean, E. (Ed.): *Ergonomics and Health in Modern Offices.* London, Taylor and Francis, 1984, pp. 495–504.

Cyriax, J.: *The Slipped Disc,* 2nd edition. Epping, Gower, 1975.

Dahn, I., Lassen, N.A., and Westling, H.: Blood flow in human muscles during external pressure or venous stasis. *Clinical Science, 32:*467–473, 1967.

Dail, C.W., and Affeldt, J.E.: Effect of body position on respiratory muscle function. *Archives of Physical Medicine and Rehabilitation, 38:*427–434, 1957.

Dainoff, M.: VDU work task categories. In *Health and Ergonomic Considerations of Visual Display Units—Symposium Proceedings.* Denver, American Industrial Hygiene Association, March 1–2, 1982, pp. 103–118.

Dainoff, M.J.: Video display terminals. The relationship between ergonomic design, health complaints and operator performance. *Occupational Health Nursing, 31:*29–33, 1983.

Dainoff, M.J.: Ergonomics of office automation—a conceptual overview. In Matthews,

M.L., and Webb, R.D.G. (Eds.): *Proceedings of the 1984 International Conference on Occupational Ergonomics, vol. 2.* Toronto, Ontario, Canada, May 7–9, 1984a, pp. 72–80.

Dainoff, M.J.: A model for human efficiency: relating health, comfort and performance in the automated office workstation. In Salvendy, G. (Ed.): *Human-Computer Interaction.* Amsterdam, Elsevier Science, 1984b, pp. 355–360.

Dainoff, M.: Some issues surrounding the design of ergonomic office chairs. In Cohen, B.G.F. (Ed.): *Human Aspects in Office Automation.* Amsterdam, Elsevier, 1984c, pp. 143–151.

Dainoff, M.J.: Ergonomics—more than just furniture. *Modern Office Technology,* :71–80, June 1984d.

Dainoff, M.J., Fraser, L., and Taylor, B.J.: Visual, musculoskeletal, and performance differences between good and poor VDT workstations. In Edwards, R.E. (Ed.): *Proceedings of the Human Factors Society 26th Annual Meeting,* Seattle, Washington, October 25–29, 1982, p. 144.

Dally, J.F.H.: An inquiry into the physiological mechanism of respiration with especial reference to the movements of the vertebral column and diaphragm. *Journal of Anatomy and Physiology (London), 43:*93–114, 1908.

Dalton, J.J., Hackler, R.H., and Bunts, R.C.: Amyloidosis in the paraplegic; incidence and significance. *The Journal of Urology, 93:*553–555, 1965.

Daly, C.H., and Odland, G.F.: Age-related changes in the mechanical properties of human skin. *The Journal of Investigative Dermatology, 73:*84–87, 1979.

Damkot, D.K., Pope, M.H., Lord, J., and Frymoyer, J.W.: The relationship between work history, work environment and low-back pain in men. *Spine, 9:*395–399, 1984.

Damodaran, L., Simpson, A., and Wilson, P.: *Designing Systems for People.* Manchester, NCC Publications, 1980.

Damon, A., Stoudt, H.W., and McFarland, R.A.: *The Human Body in Equipment Design.* Cambridge, Harvard University Press, 1966.

Daniel, R.K., and Faibisoff, B.: Muscle coverage of pressure points—the role of myocutaneous flaps. *Annals of Plastic Surgery, 8:*446–452, 1982.

Daniel, R.K., Priest, D.L., and Wheatley, D.C.: Etiologic factors in pressure sores: an experimental model. *Archives of Physical Medicine and Rehabilitation, 62:*492–498, 1981.

Daniel, R.K., Wheatley, D., and Priest, D.: Pressure sores and paraplegia: an experimental model. *Annals of Plastic Surgery, 15:*41–49, 1985.

Daniels, G.S.: *The "Average Man"?* Technical note WCRD 53-7, Wright Air Development Center, Wright-Patterson Air Force Base, Dayton, Ohio, December 1952.

Darcus, H.D., and Weddell, A.G.M.: Some anatomical and physiological principles concerned in the design of seats for naval war-weapons. *British Medical Bulletin, 5:*31–37, 1947.

David, J.A., Chapman, E.J., Chapman, R.G., and Lockett, B.: A survey of prescribed nursing treatment for patients with established pressure sores. *Care, The British Journal of Rehabilitation and Tissue Viability, 1:*18–20, Spring 1985.

Davies, B.T.: Training in manual handling and lifting. In Drury, C.G. (Ed.): *Safety in Manual Materials Handling.* Cincinnati, NIOSH, 1978, pp. 175–178.

Davis, P.R.: The thoraco-lumbar mortice joint. *Journal of Anatomy, 89:*370–377, 1955.

Davis, P.R., Troup, J.D.G., and Burnard, J.H.: Movements of the thoracic and lumbar spine when lifting: a chrono-cyclophotographic study. *Journal of Anatomy, London, 99:*13–26, 1965.

DeLateur, B.J., Berni, R., Hongladarom, T., and Giaconi, R.: Wheelchair cushions designed to prevent pressure sores: an evaluation. *Archives of Physical Medicine and Rehabilitation, 57:*129–135, 1976.

Dempsey, C.A.: Body support/restraint. *Product Engineering, 33:*106–115, 1962.

✓Dempster, W.T.: *Space Requirements of the Seated Operator.* WADC Technical Report 55-159. Wright Air Development Center, Dayton, Ohio, July 1955a.

Dempster, W.T.: The anthropometry of body action. *Annals of the New York Academy of Sciences, 63:*559–585, 1955b.

Denne, W.A.: An objective assessment of the sheepskins used for decubitus sore prophylaxis. *Rheumatology and Rehabilitation, 18:*23–29, 1979a.

Denne, W.A.: Some properties of sheep-skin and polyurethane foam. In *Seating Systems for the Disabled.* London, The Biological Engineering Society, 1979b, pp. 9–12.

Department of Education and Science: *School Furniture: Standing and Sitting Postures.* Building Bulletin 52. London, HMSO, 1976.

Department of Health and Social Security: *Pressure Sore Prevention.* The *Clinical Assessment of Cushions, Mattresses and Beds in the Spinal Injuries Department, Robert Jones and Agnes Hunt Orthopaedic Hospital, Oswestry.* London, DHSS, 1980.

Derry, D.E.: The influence of sex on the position and composition of the human sacrum. *Journal of Anatomy and Physiology, 46:*184–192, 1912.

DeTroyer, A.: Mechanical role of the abdominal muscles in relation to posture. *Respiration Physiology, 53:*341–353, 1983.

Deusinger, R.H., and Rose, S.J.: Analysis of external oblique EMG activity during back extension. *Physical Therapy, 65:*673–674, 1985.

Dhami, L.D., Gopalakrishna, A., and Thatte, R.L.: An objective study of the dimensions of the ischial pressure point and its correlation to the occurrence of a pressure sore. *British Journal of Plastic Surgery, 38:*243–251, 1985.

Dick, J.C.: The tension and resistance to stretching of human skin and other membranes, with results from a series of normal and oedematous cases. *Journal of Physiology, 112:*102–113, 1951.

Diebschlag, W., and Müller-Limmroth, W.: Physiological requirements on car seats: some results of experimental studies. In Oborne, D.J., and Levis, J.A. (Eds.): *Human Factors in Transport Research, vol. 2.* London, Academic Press, 1980, pp. 223–230.

Diffrient, N.: Design with backbone. *Industrial Design, 17:*44–47, October 1970.

Diffrient, N.: The Diffrient difference. *Leading Edge, 5:*41–59, June 1984.

Diffrient, N., Tilley, A.R., and Bardagjy, J.C.: *Humanscale 1/2/3.* Cambridge, MIT Pr, 1974.

Diffrient, N., Tilley, A.R., and Harman, D.: *Humanscale 7/8/9.* Cambridge, MIT Pr, 1981.

Dillon, J.: The role of ergonomics in the development of performance tests for furniture. *Applied Ergonomics, 12:*169–175, 1981.

Dinsdale, S.M.: Decubitus ulcers: role of pressure and friction in causation. *Archives of Physical Medicine and Rehabilitation, 55:*147–152, 1974.

Do, M.C., Bouisset, S., and Moynot, C.: Are paraplegics handicapped in the execution of a manual task? *Ergonomics, 28:*1363–1375, 1985.

Dresslar, F.B.: *School Hygiene.* New York, Macmillan, 1917.

Drew, L.C.: *Individual Gymnastics,* 3rd ed. Philadelphia, Lea and Febiger, 1926.

Drew, L.C.: *Individual Gymnastics,* 5th ed. Philadelphia, Lea and Febiger, 1945.

Drillis, R., and Contini, R.: *Body Segment Parameters.* Technical Report No. 1166.03, New York University, School of Engineering and Science, Research Division, New York, New York, September 1966.

Drummond, D., Breed, A.L., and Narechania, R.: Relationship of spine deformity and pelvic obliquity on sitting pressure distributions and decubitus ulceration. *Journal of Pediatric Orthopedics, 5:*396–402, 1985.

Drummond, D., Narechania, R., Breed, A., and Lange, T.: The relationship of unbalanced sitting and decubitus ulceration to spine deformity in paraplegic patients. In *Proceedings of the 17th Annual Meeting of the Scoliosis Research Society,* Denver, Colorado, September 22–25, 1982b, p. 94.

Drummond, D., Narechania, R., Breed, A., and Lange, T.: The relationship of unbalanced sitting and decubitus ulceration to spine deformity in paraplegic patients. *Orthopaedic Transactions, 7:*512, 1983.

Drummond, D.S., Narechania, R.G., Rosenthal, A.N., Breed, A.L., Lange, T.A., and Drummond, D.K.: A study of pressure distributions measured during balanced and unbalanced sitting. *The Journal of Bone and Joint Surgery, 64-A:*1034–1039, 1982a.

Drury, C.G. and Coury, B.G.: A methodology for chair evaluation. *Applied Ergonomics, 13:*195–202, 1982.

Drury, C.G., and Francher, M.: Evaluation of a forward-sloping chair. *Applied Ergonomics, 16:*41–47, 1985.

DuToit, G., and Gillespie, R.G.: Scoliosis in paraplegia. *The Journal of Bone and Joint Surgery, 61-B:*258–259, 1979.

Dyson, R.: Bed sores—the injuries hospital staff inflict on patients. *Nursing Mirror, 146:*30–32, June 15, 1978.

Eastman, M.C., and Kamon, E.: Posture and subjective evaluation at flat and slanted desks. *Human Factors, 18:*15–26, 1976.

Ek, A–C., and Boman, G.: A descriptive study of pressure sores: the prevalence of pressure sores and the characteristics of patients. *Journal of Advanced Nursing, 7:*51–57, 1982.

Ek, A–C., Gustavsson, G., and Lewis, D.H.: The local skin blood flow in areas at risk for pressure sores treated with massage. *Scandinavian Journal of Rehabilitation Medicine, 17:*81–86, 1985.

Ekholm, J., Arborelius, U.P., and Nemeth, G.: The load on the lumbosacral joint and trunk muscle activity during lifting. *Ergonomics, 25:*145–161, 1982.

Eklund, G., and Hagbarth, K.-E.: Normal variability of tonic vibration reflexes in man. *Experimental Neurology, 16:*80–92, 1966.

Eklund, J., and Corlett, E.N.: Shrinkage as a measure of the effect of load on the spine. *Spine, 9:*189–194, 1984.

Eklund, J.A.E., and Corlett, E.N.: Shrinkage—a measure for the evaluation of ergonomic designs. In Brown, I.D., Goldsmith, R., Coombes, K., and Sinclair, M.A.: *Ergonomics International 85. Proceedings of the Ninth Congress of the International Ergonomics Association,* London, Taylor and Francis, 1985, pp. 415–417.

Eklundh, M.: *Spare Your Back.* London, Duckworth, 1965.

Elias, R., Cail, F., Tisserand, M., and Christmann, H.: Investigations in operators working with CRT display terminals: relationships between task content and psychophysiological alterations. In Grandjean, E., and Vigliani, E. (Eds.): *Ergonomic Aspects of Visual Display Terminals.* London, Taylor and Francis, 1983, pp. 211–217.

Eliot, S.A.: 1833. Quoted in Hartwell (1895), pp. 195–196.

Ellis, H.: *Clinical Anatomy,* 6th ed. Oxford, Blackwell Sci, 1977.

Ellis, M.I., Seedhom, B.B., Amis, A.A., Dowson, D., and Wright, V.: Forces in the knee joint whilst rising from normal and motorized chairs. *Engineering in Medicine, 8:*33–40, 1979.

El-Toraei, I., and Chung, B.: The management of pressure sores. *The Journal of Dermatologic Surgery and Oncology, 3:*507–511, 1977.

Emerson, H.: Intra-abdominal pressures. *The Archives of Internal Medicine, 7:*754–784, 1911.

Emmons, W.H., and Hirsch, R.S.: Thirty millimeter keyboards: how good are they? In Edwards, R.E. (Ed.) *Proceedings of the Human Factors Society 26th Annual Meeting.* Seattle, October 25–29, 1982, pp. 425–429.

Enebuske, C.J.: *Progressive Gymnastic Day's Orders.* Boston, Silver, Burdett, 1892.

Engel, P., and Hildebrandt, G.: Wheelchair design—technological and physiological aspects. *Proceedings of the Royal Society of Medicine, 67:*409–413, 1974.

Eriksson, E.: Etiology: microcirculatory effects of pressure. In Constantian, M.B. (Ed.): *Pressure Ulcers—Principles and Techniques of Management.* Boston, Little, 1980, pp. 7–14.

Fahrner: *Das Kind und der Schultisch.* Zurich, Schulthess, 1865. Translated in Cohn (1886), pp. 94–98.

Feiss, H.O.: "School" lateral curvature. *The Cleveland Medical Journal, 4:*349–354, 1905.

Fellmann, Th., Bräuninger, U., Gierer, R., and Grandjean, E.: An ergonomic evaluation of VDTs. *Behaviour and Information Technology, 1:*69–80, 1982.

Fenwick, D.: *Wheelchairs and Their Users.* London, HMSO, 1977.

Ferguson, D., and Duncan, J.: Keyboard design and operating posture. *Ergonomics, 17:*731–744, 1974.

Ferguson-Pell, M., Wilkie, I.C., and Barbenel, J.C.: Pressure sore prevention for the wheelchair user. In *Proceedings of International Conference on Rehabilitation Engineering,* Toronto, 1980, pp. 167–171.

Fernand, R., and Fox, D.E.: Evaluation of lumbar lordosis. *Spine, 10:*799–803, 1985.

Fernie, G.R., Holden, J.M., and Lunau, K.: *Chair Design for the Elderly.* Paper presented at Third International Seating Symposium, Memphis, Tennessee, February 26–28, 1987.

Fiedler, R., and Fiedler, K.: Arbeitsstuhl und gesundheit. *Zeitschrift fur die Gesamte Hygiene, 23:*889–891, 1977.

Finlay, O.E.: Rehabilitation chair. *Physiotherapy, 67:*207, 1981.

Finlay, O.E., Bayles, T.B., Rosen, C., and Milling, J.: Effects of chair design, age and cognitive status on mobility. *Age and Ageing, 12:*329–335, 1983.

Finneson, B.E.: *Low Back Pain.* Philadelphia, Lippincott, 1973, p. 115.

Fiorini, G.T., and McCammond, D.: Forces on lumbo-vertebral facets. *Annals of Biomedical Engineering, 4:*354–363, 1976.

Fisher, S.V., and Patterson, P: Long term pressure recordings under the ischial tuberosities of tetraplegics. *Paraplegia, 21:*99–106, 1983.

Fisher, S.V., Szymke, T.E., Apte, S.Y., and Kosiak, M.: Wheelchair cushion effect on skin temperature. *Archives of Physical Medicine and Rehabilitation, 59:*68–72, 1978.

Fisk, J.: The 1982 Mennell-Travell distinguished lecture: the low back problem. *Manual Medicine, 2:*31–37, 1986.

Fisk, J.W.: *Medical Treatment of Neck and Back Pain.* Springfield, Thomas, 1987.

Fisk, J.W., and Baigent, M.L.: Hamstring tightness and Scheuermann's disease. *American Journal of Physical Medicine, 60:*122–125, 1981.

Fisk, J.W., Baigent, M.L., and Hill, P.D.: Scheuermann's disease. *American Journal of Physical Medicine, 63:*18–30, 1984.

Fitz, G.W.: Bed posture as an etiological factor in spinal curvature. *Transactions of the American Orthopaedic Association, 11:*249–251, 1898.

Floru, R., and Cail, F.: Psychophysiological investigations on VDU repetitive task. In Brown, I.D., Goldsmith, R., Coombes, K., and Sinclair, M.A. (Eds.): *Ergonomics International 85. Proceedings of the Ninth Congress of the International Ergonomics Association.* London, Taylor and Francis, 1985.

Floyd, W.F.: Postural factors in the design of motor car seats. *Proceedings of the Royal Society of Medicine, 60:*953–955, 1967.

Floyd, W.F., and Roberts, D.F.: Anatomical and physiological principles in chair and table design. *Ergonomics, 2:*1–16, 1958.

Floyd, W.F., and Silver, P.H.S.: Electromyographic study of patterns of activity of the anterior abdominal wall muscles in man. *Journal of Anatomy, 84:*132–145, 1950.

Floyd, W.F., and Silver, P.H.S.: The function of the erectores spinae muscles in certain movements and postures in man. *Journal of Physiology, 129:*184–203, 1955.

Floyd, W.F., and Ward, J.S.: Posture of schoolchildren and office workers. In *Proceedings of Second International Congress on Ergonomics,* Dortmund, 1964, pp. 351–360.

Floyd, W.F., and Ward, J.: Posture in industry. *The International Journal of Production Research, 5:*213–224, 1967.

Floyd, W.F., and Ward, J.S.: Anthropometric and physiological considerations in school, office and factory seating. In Grandjean, E. (Ed.): *Proceedings of the Symposium on Sitting Posture.* London, Taylor and Francis, 1969, pp. 18–25.

√ Fon, G.T., Pitt, M.J., and Thies, A.C.: Thoracic kyphosis: range in normal subjects. *American Journal of Roentgenology, 134:*979–983, 1980.

Forsberg, C–M., Hellsing, E., Linder-Aronson, S., and Sheikholeslam, A.: EMG activity in neck and masticatory muscles in relation to extension and flexion of the head. *European Journal of Orthodontics, 7:*177–184, 1985.

√ Frazier, L.M.: Coccydynia: a tail of woe. *North Carolina Medical Journal, 46:*209–212, 1985.

Fries, E.C., and Hellebrandt, F.A.: The influence of pregnancy on the location of the center of gravity, postural stability, and body alignment. *American Journal of Obstetrics and Gynecology, 46:*374–380, 1943.

Frost, L.H.: Individual structural differences in the orthopedic examination. *Journal of Health and Physical Education, 9:*90–93, 122, 1938.

√ Frymoyer, J.W., and Pope, M.H.: The role of trauma in low back pain: a review. *The Journal of Trauma, 18:*628–634, 1978.

Frymoyer, J.W., Pope, M.H., Clements, J.H., Wilder, D.G., MacPherson, B., and Ashikaga, T.: Risk factors in low-back pain. The *Journal of Bone and Joint Surgery, 65-A:*213–218, 1983.

Gajdosik, R.L., LeVeau, B.F., and Bohannon, R.W.: Effects of ankle dorsiflexion on active and passive unilateral straight leg raising. *Physical Therapy, 65:*1478–1482, 1985.

Galpin, J.E., Chow, A.W., Bayer, A.S., and Guze, L.B.: Sepsis associated with decubitus ulcers. *The American Journal of Medicine, 61:*346–350, 1976.

Ganong, W.F.: *Review of Medical Physiology,* 9th ed. Los Altos, Lange, 1979, p. 450.

Garber, S.L., and Krouskop, T.A.: Body build and its relationship to pressure distribution in the seated wheelchair patient. *Archives of Physical Medicine and Rehabilitation, 63:*17–20, 1982.

Garibaldi, R.A., Brodine, S., and Matsumiya, S.: Infections among patients in nursing homes. *The New England Journal of Medicine, 305:*731–735, 1981.

Garland, D.E., Hanscom, D.A., Keenan, M.A., Smith, C., and Moore, T.: Resection of heterotopic ossification in the adult with head trauma. *The Journal of Bone and Joint Surgery, 67-A:*1261–1269, 1985.

Garner, J.R.: Proper seating. *Industrial Medicine, 5:*324–327, 1936.

Garrett, A.L., Perry, J., and Nickel, V.L.: Paralytic scoliosis. *Clinical Orthopaedics, 21:*117–124, 1961.

Geisler, W.O., Jousse, A.T., and Wynne-Jones, M.: Survival in traumatic transverse myelitis. *Paraplegia, 14:*262–275, 1977.

Gibson, D.A., Albisser, A.M., and Koreska, J.: Role of the wheelchair in the management of the muscular dystrophy patient. *Canadian Medical Association Journal, 113:*964–966, 1975.

Gibson, D.A., Koreska, J., Robertson, D., Kahn, A., and Albisser, A.M.: The management of spinal deformity in Duchenne's muscular dystrophy. *Orthopedic Clinics of North America, 9:*437–450, 1978.

Gibson, D.A., and Wilkins, K.E.: The management of spinal deformities in Duchenne muscular dystrophy. *Clinical Orthopaedics and Related Research, 108:*41–51, 1975.

Girardin, Y.: EMG action potentials of rectus abdominis muscle during two types of abdominal exercises. *Medicine and Sport, 8:*301–308, 1973.

Giuliano, V.E.: The mechanization of office work. *Scientific American, 247:*149–164, Sept. 1982.

Glancy, J.: A new orthotic concept in the non-operative treatment of idiopathic scoliosis. *Orthotics and Prosthetics, 32:*15–31, December 1978.

Glassford, E.J.: *The Relationship of Hemodynamics to Seating Comfort.* Society of Automotive Engineers, Publication no. 770248, 1977, pp. 1–6.

Glassow, R.B.: *Fundamentals in Physical Education.* Philadelphia, Lea and Febiger, 1932.

Goldman, J.M., Rose, L.S., Williams, S.J., Silver, J.R., and Denison, D.M.: Effect of abdominal binders on breathing in tetraplegic patients. *Thorax, 41:*940–945, 1986.

Goldthwait, J.E.: The relation of posture to human efficiency and the influence of poise upon the support and function of the viscera. *The Boston Medical and Surgical Journal, 161:*839–848, 1909.

Goldthwait, J.E.: An anatomic and mechanistic conception of disease. *The Boston Medical and Surgical Journal, 172:*881–898, 1915.

Goldthwait, J.E.: The opportunity for the orthopedist in preventive medicine through educational work on posture. *The American Journal of Orthopedic Surgery, 14:*443–449, 1916.

Goldthwait, J.E.: The importance of correct furniture to assist in the best body function, as recognized by the Massachusetts Institute of Technology and Smith College. *The Journal of Bone and Joint Surgery, 5:*179–184, 1923.

Goldthwait, J.E., and Brown, L.T.: The recognition of congenital visceral ptosis in the treatment of the badly poised and poorly nourished child. *American Journal of Orthopaedic Surgery, 9:*253–267, 1911.

Goldthwait, J.E., Brown, L.T., Swaim, L.T., and Kuhns, J.G.: *Body Mechanics in the Study and Treatment of Disease.* Philadelphia, Lippincott, 1934.

Goldthwait, J.E., Brown, L.T., Swaim, L.T., and Kuhns, J.G.: *Essentials of Body Mechanics in Health and Disease,* 5th ed. Philadelphia, Lippincott, 1952.

Gould, G.M.: The optic and ocular factors in the etiology of the scoliosis of school children. *American Medicine, 9:*562–570, 1905.

Graebe, R.H.: *A Proposed Evaluation Method for Bio-Suspension Devices Using a Decubitus Threshold Pressure Concept.* Paper presented at the 54th Annual Session of the American Congress of Rehabilitation Medicine, October 30–November 2, 1977.

Graebe, R.H.: *Principles of Seating Tolerance and Tissue Protection.* Presented at the Second Annual Interagency Conference on Rehabilitation Engineering, Atlanta, August 30, 1979.

Grandjean, E.: Fatigue: its physiological and psychological significance. *Ergonomics, 11:*427–436, 1968.

Grandjean, E.: Fatigue. *American Industrial Hygiene Association Journal, 31:*401–411, 1970.

Grandjean, E.: *Ergonomics of the Home.* London, Taylor and Francis, 1973.

Grandjean, E.: *Fitting the Task to the Man,* 3rd ed. London, Taylor and Francis, 1980a.

Grandjean, E.: Sitting posture of car drivers from the point of view of ergonomics. In Oborne, D.J., and Levis, J.A. (Eds.): *Human Factors in Transport Research, vol. 2.* London, Academic Pr, 1980b, pp. 205–213.

Grandjean, E.: Preface. In Grandjean, E. (Ed.): *Ergonomics and Health in Modern Offices.* London, Taylor and Francis, 1984a.

Grandjean, E.: Postural problems at office machine work stations. In Grandjean, E. (Ed.): *Ergonomics and Health in Modern Offices.* London, Taylor and Francis, 1984b, pp. 445–455.

Grandjean, E.: Postures and the design of VDT workstations. *Behaviour and Information Technology, 3:*301–311, 1984c.

Grandjean, E., Boni, A., and Kretzschmar, H.: The development of a rest chair profile for healthy and notalgic people. In Grandjean, E. (Ed.): *Proceedings of the Symposium on Sitting Posture.* London, Taylor and Francis, 1969, pp. 193–201.

Grandjean, E., and Hünting, W.: Ergonomics of posture—review of various problems of standing and sitting posture. *Applied Ergonomics, 8:*135–140, 1977.

Grandjean, E., Hünting, W., Maeda, K., and Läubli, Th.: Constrained postures at office workstations. In Kvalseth, T.O. (Ed.): *Ergonomics of Workstation Design.* London, Butterworths, 1983b, pp. 19–27.

Grandjean, E., Hünting, W., and Nishiyama, K.: Preferred VDT workstation settings, body posture and physical impairments. *Journal of Human Ergology, 11:*45–53, 1982a.

Grandjean, E., Hünting, W., and Nishiyama, K.: Preferred VDT workstation settings, body posture and physical impairments. *Applied Ergonomics, 15:*99–104, 1984.

Grandjean, E., Hünting, W., and Pidermann, M.: VDT workstation design: preferred settings and their effects. *Human Factors, 25:*161–175, 1983a.

Grandjean, E., Hünting, W., Wotzka, G., and Schärer, R.: An ergonomic investigation of multipurpose chairs. *Human Factors, 15:*247–255, 1973.

Grandjean, E., Jenni, M., and Rhiner, A.: Eine indirekte methode zur erfassung des komfortgefühls beim sitzen. *Internationale Zeitschrift für Angewandte Physiologie Einschliesslich Arbeitsphysiologie, 18:*101–106, 1960.

Grandjean, E., Nishiyama, K., Hünting, W., and Piderman, M.: A laboratory study on preferred and imposed settings of a VDT workstation. *Behaviour and Information Technology, 1:*289–304, 1982b.

Granström, B., Kvarnström, S., and Tiefenbacher, F.: Electromyography as an aid in the prevention of excessive shoulder strain. *Applied Ergonomics, 16:*49–54, 1985.

Grant, L.J.: Interface pressure measurement between a patient and a support surface. *Care, The British Journal of Rehabilitation and Tissue Viability, 1:*7–9, Spring 1985.

Gray, F.E., Hanson, J.A., and Jones, F.P.: Postural aspects of neck muscle tension. *Ergonomics, 9:*245–256, 1966.

Grew, N.D., and Deane, G.: The physical effect of lumbar spinal supports. *Prosthetics and Orthotics International, 6:*79–87, 1982.

Grieve, D., and Pheasant, S.: Biomechanics. In Singleton, W.T. (Ed.): *The Body at Work.* Cambridge, Cambridge Univ., 1982, pp. 71–200.

Grieve, D.W.: The dynamics of lifting. *Exercise and Sport Sciences Reviews, 5:*157–179, 1977.

Griffith, B.H., and Schultz, R.C.: The prevention and surgical treatment of recurrent decubitus ulcers in patients with paraplegia. *Plastic and Reconstructive Surgery, 27:*248–260, 1961.

Griffiths, H.J., D'Orsi, C.J., and Zimmerman, R.E.: Use of ^{125}I photon scanning in the evaluation of bone density in a group of patients with spinal cord injury. *Investigative Radiology, 7:*107–111, 1972.

Grillner, S., Nilsson, J., and Thorstensson, A.: Intra-abdominal pressure changes during natural movements in man. *Acta Physiologica Scandinavica, 103:*275–283, 1978.

Groth, K.E.: Klinische beobachtungen und experimentelle studien über die entstehung des dekubitus. *Acta Chirurgica Scandinavica, 87 (Supplement 76):*1–209, 1942.

Grundy, D.J., and Silver, J.R.: Amputation for peripheral vascular disease in the paraplegic and tetraplegic. *Paraplegia, 21:*305–311, 1983.

Grundy, D.J., and Silver, J.R.: Major amputation in paraplegic and tetraplegic patients. *International Rehabilitation Medicine, 6:*162–165, 1984.

Guillon, F., Pignolet, F., Proteau, J., Contet, C.L., Somenzi, G., and Tarrière, C.L.: Variation de la pression intra-discale selon l'importance de l'appui lombaire en position assise, intérêt pour l'ergonomie du siège. *Archives des Maladies Professionnelles de Medecine due Travail et de Securite Sociale, 46:*106–108, 1985.

Gurfinkel, V.S., Lipshits, M.I., Mori, S., and Popov, K.E.: Stabilization of body position as the main task of postural regulation. *Fiziologiya Cheloveka, 7:*400–410, 1981.

Guthrie, D.I.: A new approach to handling in industry. *South African Medical Journal, 37:*651–655, 1963.

Guthrie, R.H., and Goulian, D.: Decubitus ulcers: prevention and treatment. *Geriatrics, 28:* 67–71, 1973.

Guttmann, L.: *Spinal Cord Injuries — Comprehensive Management and Research,* 2nd ed. Oxford, Blackwell Sci, 1976.

Guyton, A.C., Abernathy, B., Langston, J.B., Kaufmann, B.N., and Fairchild, H.M.: Relative importance of venous and arterial resistances in controlling venous return and cardiac output. *American Journal of Physiology, 196:*1008–1014, 1959.

Hackler, R.H.: A 25 year prospective mortality study in the spinal cord injured patient: comparison with the long-term living paraplegic. *The Journal of Urology, 117:*486–488, 1977.

Hall, E.C.: Seating for transportation. In Shinnick, T., and Hill, G. (Eds.): *Proceedings of the 20th Annual Conference of the Ergonomics Society of Australia and New Zealand.* Adelaide, South Australia, Dec. 1–2, 1983, pp. 191–211.

Hall, M.A.W.: Back pain and car-seat comfort. *Applied Ergonomics, 3:*82–90, 1972.

Halpern, A.A., and Bleck, E.E.: Sit-up exercises: an electromyographic study. *Clinical Orthopaedics and Related Research, 145:*172–178, 1979.

Handley, J.: Posture education: an essential component of the health based physical education programme. *The British Journal of Physical Education, 17:*37–38, 1986.

Hanson, J.A., and Jones, F.P.: Heart rate and small postural changes in man. *Ergonomics, 13:*483–487, 1970.

Hardy, L.: Improving active range of hip flexion. *Research Quarterly for Exercise and Sport, 56:*111–114, 1985.

Harmon, D.B.: *The Co-ordinated Classroom.* Grand Rapids, American Seating Company, 1949.

Harms-Ringdahl, K., and Ekholm, J.: Intensity and character of pain and muscular activity levels elicited by maintained extreme flexion position of the lower-cervical-upper-thoracic spine. *Scandinavian Journal of Rehabilitation Medicine, 18:*117–126, 1986.

Harms-Ringdahl, K., Ekholm, J., Schüldt, K., Németh, G., and Arborelius, U.P.: Load moments and myoelectric activity when the cervical spine is held in full flexion and extension. *Ergonomics, 29:*1539–1552, 1986.

Hartigan, J.D.: The dangerous wheelchair. *Journal of the American Geriatrics Society, 30:*572–573, 1982.

Hartnett, B.: Is the modern motor car ergonomically efficient? In Fisher, A.J., and Croft, P. (Eds.): *Ergonomics in Practice. Proceedings of the 17th Annual Conference of the Ergonomics Society of Australia and New Zealand.* Sydney, Nov. 27–28, 1980, pp. 103–106.

Hartwell, E.M.: *The Problem of School Seating.* In Annual Report of the Superintendent of Schools, School Document No. 4, Boston, 1895, pp. 169–236.

Hassard, G.H.: Heterotopic bone formation about the hip and unilateral decubitus ulcers in spinal cord injury. *Archives of Physical Medicine and Rehabilitation, 56:*355–358, 1975.

Hatami, T.: Electromyographic studies of influence of pregnancy on activity of the abdominal wall muscles. *The Tohoku Journal of Experimental Medicine, 75:*71–80, 1961.

Hawley, G.: *The Kinesiology of Corrective Exercise.* Philadelphia, Lea and Febiger, 1937.

Haynes, R.S.: Postural reflexes. *American Journal of Diseases of Children, 36:*1093–1107, 1928.

Health and Safety Executive: *Visual Display Units.* London, Her Majesty's Stationery Office, 1983.

Hedberg, G., Björkstén, M., Ouchterlony-Jonsson, E., and Jonsson, B.: Rheumatic complaints among Swedish engine drivers in relation to the dimensions of the driver's cab in the Rc engine. *Applied Ergonomics, 12:*93–97, 1981.

Helander, M.G., Billingsley, P.A., and Schurick, J.M.: An evaluation of human factors research on visual display terminals in the workplace. In Muckler, F.A. (Ed.): *Human Factors Review: 1984.* Santa Monica, Human Factors Society, 1984, pp. 55–129.

Helbig, K.: Sitzdruckverteilung beim ungepolsterten sitz. *Anthropologischer Anzieger, 36:*194–202, 1978.

Hellebrandt, F.A.: Standing as a geotropic reflex. *American Journal of Physiology, 121:*471–474, 1938.

Hellebrandt, F.A., Brogdon, E., and Tepper, R.H.: Posture and its cost. *American Journal of Physiology, 129:*773–781, 1940.

Hellebrandt, F.A., and Franseen, E.B.: Physiological study of the vertical stance of man. *Physiological Reviews, 23:*220–255, 1943.

Hellebrandt, F.A., Tepper, R.H., Braun, G.L., and Elliott, M.C.: The location of the cardinal anatomical orientation planes passing through the center of weight in young adult women. *American Journal of Physiology, 121:*465–470, 1938.

Hennessey, M.: Work posture and kinesiology: an ergonomic checklist. In Adams, A.S., and Stevenson, M.G. (Eds.): *Proceedings of the 21st Annual Conference of the Ergonomics Society of Australia and New Zealand,* Sydney, 1984, pp. 147–158.

Hentz, V.R.: Management of pressure sores in a specialty center. *Plastic and Reconstructive Surgery, 64:*683–691, 1979.

Hertel, A.: Zur steilschriftfrage. *Zeitschrift für Schulgesundheitspflege, 4:*672–675, 1891.

Hertzberg, H.T.E.: Seat comfort. In Hertzberg, H.T.E. (Ed.): *Annotated Bibliography of Applied Physical Anthropology in Human Engineering.* WADC Technical Report 56-30, Wright Air Development Center, Wright-Patterson Air Force Base, Dayton, Ohio, Appendix 1, pp. 297–300, 1958.

Hertzberg, H.T.E.: "Average" man is a fiction: range of sizes is key to efficient work places. *Contract, 11:*86–89, 1970.

Hertzberg, H.T.E.: *The Human Buttocks in Sitting: Pressures, Patterns, and Palliatives.* Society of Automotive Engineers, publication no. 720005, 1972.

Hiba, J.C.: Some ergonomic aspects in the design of a headrest for a passenger seat for coaches. In Oborne, D.J., and Levis, J.A. (Eds.): *Human Factors in Transport Research, vol. 2.* London, Academic Pr, 1980, pp. 257–265.

Hira, D.S.: An ergonomic appraisal of educational desks. *Ergonomics, 23:*213–221, 1980.

Hockenberry, J.: Seating design for worker efficiency. *Furniture Design and Manufacturing, 49:*42–46, December 1977.

Hockenberry, J.: A systems approach to long term task seating design. In Easterby, R., Kroemer, K.H.E., and Chaffin, D.B. (Eds.): *Anthropometry and Biomechanics. Theory and Application.* New York, Plenum, 1982, pp. 225–234.

Hodgson, V.R., Lissner, H.R., and Patrick, L.M.: Response of the seated human cadaver to acceleration and jerk with and without seat cushions. *Human Factors, 5:*505–523, 1963.

Hogan, B.J.: Seating design can cause lower back problems. *Design News, 38:*85–90, Sept. 13, 1982.

Hollinshead, W.H.: *Functional Anatomy of the Limbs and Back,* 3rd ed. Philadelphia, Saunders, 1969.

Holloway, G.A., Tolentino, G., and DeLateur, B.J.: Cutaneous blood flow responses to wheelchair cushion pressure loading measured by laser Doppler flowmetry. In Lee, B.Y. (Ed.): *Chronic Ulcers of the Skin.* New York, McGraw-Hill, 1985, pp. 57–67.

Holm, S., and Nachemson, A.: Nutrition of the intervertebral disc: effects induced by vibrations. *Orthopaedic Transactions, 9:*525, 1985.

Holstein, P.: Level selection in leg amputation for arterial occlusive disease. *Acta Orthopaedica Scandinavica, 53:*821–831, 1982.

Holstein, P., Lund, P., Larsen, B., and Schomacker, T.: Skin perfusion pressure measured as the external pressure required to stop isotope washout. *Scandinavian Journal of Clinical and Laboratory Investigation, 37:*649–659, 1977.

Holstein, P., Nielsen, P.E., and Barras, J.P.: Blood flow cessation at external pressure in the skin of normal human limbs. *Microvascular Research, 17:*71–79, 1979.

Homans, J.: Thrombosis of the deep leg veins due to prolonged sitting. *The New England Journal of Medicine, 250:*148–149, 1954.

✓ Hooton, E.A.: *A Survey in Seating.* Westport, Greenwood Press, 1945.

✓ Hope, E.W., Browne, E.A., and Sherrington, C.S.: *A Manual of School Hygiene.* Cambridge, Cambridge University, 1913.

Hoppenfeld, S.: *Physical Examination of the Spine and Extremities.* New York, Appleton-Century-Crofts, 1976.

Hosea, T.M., Simon, S.R., Delatizky, J., Wong, M.A., and Hsieh, C.-C.: Myoelectric analysis of the paraspinal musculature in relation to automobile driving. *Spine, 11:*928–936, 1986.

Houle, R.J.: Evaluation of seat devices designed to prevent ischemic ulcers in paraplegic patients. *Archives of Physical Medicine and Rehabilitation, 50:*587–594, 1969.

Howell, T.H.: Some terminal aspects of disease in old age: a clinical study of 300 patients. *Journal of the American Geriatrics Society, 17:*1034–1038, 1969.

Howland, I.S.: *The Teaching of Body Mechanics in Elementary and Secondary Schools.* New York, Barnes, 1936.

Howorth, B.: Dynamic posture. *Journal of the American Medical Association, 131:*1398–1404, 1946.

✓ Howorth, B.: The painful coccyx. *The Journal of the Western Pacific Orthopaedic Association, 15:*39–56, 1978.

Humphry, G.M.: *A Treatise on the Human Skeleton.* Cambridge, Macmillan, 1858.

Hünting, W., and Grandjean, E.: Sitzverhalten und subjektives wohlbefinden auf schwenkbaren und fixierten formsitzen. *Zeitschrift für Arbeitswissenschaft, 30:*161–164, 1976.

Hünting, W., Grandjean, E., and Maeda, K.: Constrained postures in accounting machine operators. *Applied Ergonomics, 11:*145–149, 1980.

Hünting, W., Läubli, Th., and Grandjean, E.: Postural and visual loads at VDT workplaces. 1. constrained postures. *Ergonomics, 24:*917–931, 1981.

Husain, T.: An experimental study of some pressure effects on tissues, with reference to the bed-sore problem. *Journal of Pathology and Bacteriology, 66:*347–358, 1953.

Hutchison, K.J., Overton, T.R., Biltek, K.B., Nixon, R., and Williams, H.T.G.: Skin blood flow during histamine flare using the clearance of epicutaneous applied Xenon-133 in diabetic and non-diabetic subjects. *Angiology, 34:*223–230, 1983.

IBM: *Human Factors of Workstations With Visual Displays.* San Jose, IBM, 1984.

Ingelmark, B.: Über schmerzhafte insuffizienzzustände im halse. *Acta Medica Scandinavica, 111:*172–189, 1942.

Ingelmark, B.E., and Lindström, J.: Asymmetries of the lower extremities and pelvis and their relations to lumbar scoliosis. *Acta Morphologica Neerlando-Scandinavica, 5:*221–234, 1963.

Institute for Consumer Ergonomics.: *Selecting Easy Chairs for Elderly and Disabled People.* University of Technology, Loughborough, Leicestershire, United Kingdom, 1983a.

Institute for Consumer Ergonomics: *Seating for Elderly and Disabled People. Report No. 2. Anthropometric Survey.* University of Technology, Loughborough, Leicestershire, United Kingdom, 1983b.

Institute for Consumer Ergonomics: *Seating for Elderly and Disabled People. Report No. 3. Desired Chair Dimension Trials.* University of Technology, Loughborough, Leicestershire, United Kingdom, 1983c.

Institute for Consumer Ergonomics: *Seating for Elderly and Disabled People. Report No. 4. Chair Rig Trials.* University of Technology, Loughborough, Leicestershire, United Kingdom, 1983d.

Institute for Consumer Ergonomics: *Seating for Elderly and Disabled People. Report No. 7. Technical Testing of Manufactured Chairs—Structural and Materials Testing.* University of Technology, Loughborough, Leicestershire, United Kingdom, 1983e.

Institute for Consumer Ergonomics: *Seating for Elderly and Disabled People. Report No. 8. Technical Testing of Manufactured Chairs-Stability Testing.* University of Technology, Loughborough, Leicestershire, United Kingdom, 1983f.

Institute for Consumer Ergonomics: *Seating for Elderly and Disabled People. Report No. 9. Chair Specifications and Guidelines for Chair Selection.* University of Technology, Loughborough, Leicestershire, United Kingdom, 1983g.

Jacobs, H., Sussman, D., Abernethy, C., Plank, G., and Stoklosa, J.: Please remain seated: Seat designs to help retain passengers during emergency stops. In *Proceedings of the Human Factors Society 24th Annual Meeting,* 1980.

James, J.I.P.: Paralytic scoliosis. *The Journal of Bone and Joint Surgery, 38-B:*660–685, 1956.

James, J.I.P.: *Scoliosis,* 2nd ed. Edinburgh, Churchill, 1976.

James, W.: What is an emotion? In Lange, C.G., and James, W: *The Emotions, vol. one.* Baltimore, Williams and Wilkins, 1922, p. 15.

Jay, P.: *Choosing the Best Wheelchair Cushion.* London, The Royal Association for Disability and Rehabilitation, 1983.

Johnson, P.H.: Coccygodynia. *The Journal of the Arkansas Medical Society, 77:*421–424, 1981.

Jokl, P.: Muscle and low back pain. In White, A.A., and Gordon, S.L. (Eds.): *American Academy of Orthopaedic Surgeons Symposium on Idiopathic Low Back Pain.* St. Louis, Mosby, 1982, pp. 456–462.

Jokl, P.: Keynote address: muscle and low-back pain. *Journal of the American Osteopathic Association, 84:*114–116, 1984.

Jones, F.P.: The influence of postural set on pattern of movement in man. *International Journal of Neurology, 4:*60–71, 1963.

Jones, F.P., Gray, F.E., Hanson, J.A., and Shoop, J.D.: Neck muscle tension and the postural image. *Ergonomics, 4:*133–142, 1961.

Jones, H.R.: *Postural Responses of Third Grade Children in Reading and Writing.* Tufts University, Medford, Massachusetts, August 1965.

Jones, H.W.F.: Teaching and equipping the patient at risk. In *The Prevention of*

Pressure Sores. Proceedings of a Conference on the Prevention of Pressure Sores. London, Department of Health and Social Security, 1979, pp. 23–25.

Jones, J.A.: Effect of posture work on the health of children. *American Journal of Diseases of Children, 46:*148–154, 1933.

Jones, J.C.: Methods and results of seating research. In Grandjean, E. (Ed.): *Proceedings of the Symposium on Sitting Posture.* London, Taylor and Francis, 1969, pp. 57–67.

Jordan, M.M., and Clark, M.O.: *Report on the Incidence of Pressure Sores in the Patient Community of the Greater Glasgow Health Board Area on 21st January, 1976.* The Bioengineering Unit, University of Strathclyde and The Greater Glasgow Health Board, February 1977.

Jordan, M.M., Nicol, S.M., and Melrose, A.L.: *Report on the Incidence of Pressure Sores in the Patient Community of the Borders Health Board Area on 13th October 1976.* The Bioengineering Unit, University of Strathclyde and The Borders Health Board, April 1977.

Joseph, J., and Nightingale, A.: Electromyography of muscles of posture: leg muscles in males. *Journal of Physiology, 117:*484–491, 1952.

Jürgens, H.W.: Die verteilung des körperdrucks auf sitzfläche und rückenlehne als problem der industrieanthropologie. In Grandjean, E. (Ed.): *Proceedings of the Symposium on Sitting Posture.* London, Taylor and Francis, 1969, pp. 84–91.

Jürgens, H.W.: Body movements of the driver in relation to sitting conditions in the car: a methodological study. In Oborne, D.J., and Levis, J.A. (Eds.): *Human Factors in Transport Research, vol. 2.* London, Academic Press, 1980, pp. 249–256.

Jürgens, H.W., and Helbig, K.: *Distribution of Body Pressure in a Sitting Posture on the Basis of Work Seat Design.* Research contract BMVg InSan No. 3571-V-072. Bonn, West Germany, 1973 (In German).

Kadaba, M.P., Ferguson-Pell, M.W., Palmieri, V.R., and Cochran, G.V.B.: Ultrasound mapping of the buttock-cushion interface contour. *Archives of Physical Medicine and Rehabilitation, 65:*467–469, 1984.

Kamijo, K., Tsujimura, H., Obara, H., and Katsumata, M.: *Evaluation of Seating Comfort.* Society of Automotive Engineers, paper 820761, 1982.

Kantowitz, B.H., and Sorkin, R.D.: *Human Factors: Understanding People-System Relationships.* New York, Wiley, 1983.

Kaplan, A.: Selecting a chair for the office. *Modern Office Procedures, 26:*130, 1981.

Karvonen, M.J., Koskela, A., and Noro, L.: Preliminary report on the sitting postures of school children. *Ergonomics, 5:*471–477, 1962.

Kaufman, T., Eichenlaub, E.H., Levin, M., Hurwitz, D.J., and Klain, M.: Tobacco smoking: impairment of experimental flap survival. *Annals of Plastic Surgery, 13:*468–472, 1984.

Kauzlarich, J.J.: The effect of wheelchair cushions on ride. In *Proceedings of the Fourth Annual Conference on Rehabilitation Engineering,* Washington, D.C., 1981, pp. 42–43.

Keagy, R.D., Brumlik, J., and Bergan, J. J.: Direct electromyography of the psoas major muscle in man. *The Journal of Bone and Joint Surgery, 48-A:*1377–1382, 1966.

Keegan, J.J.: Alterations of the lumbar curve related to posture and seating. *The Journal of Bone and Joint Surgery, 35A:*589–603, 1953.

Keegan, J.J.: Evaluation and improvement of seats. *Industrial Medicine and Surgery,* *31:*137–148, 1962.

Keegan, J.J.: *The Medical Problem of Lumbar Spine Flattening in Automobile Seats.* Society of Automotive Engineers, Publication 838A, 1964.

Keith, A.: Man's posture: its evolution and disorders. Lecture two. The evolution of the orthograde spine. *The British Medical Journal, 1:*499–502, 1923a.

Keith, A.: Man's posture: its evolution and disorders. Lecture four. The adaptations of the abdomen and of its viscera to the orthograde posture. *The British Medical Journal, 1:*587–590, 1923b.

Keller, J.: Preliminary exercises for lifting and placing. *Helse og Arbete, 1:*29–34, 1965.

Kellogg, J.H.: Physical deterioration resulting from school life; cause; remedy. *Proceedings of the National Educational Association:* 899–911, 1896.

Kellogg, J.H.: Observations on the relations of posture to health and a new method of studying posture and development. *The Bulletin of the Battle Creek Sanitarium and Hospital Clinic, 22:*193–216, 1927.

Kelly, E.D.: *Teaching Posture and Body Mechanics.* New York, Barnes, 1949.

Kelsey, J.L.: An epidemiological study of acute herniated lumbar intervertebral discs. *Rheumatology and Rehabilitation, 14:*144–159, 1975a.

Kelsey, J.L.: An epidemiological study of the relationship between occupations and acute herniated lumbar intervertebral discs. *International Journal of Epidemiology, 4:*197–205, 1975b.

Kelsey, J.L., Githens, P.B., O'Conner, T., Weil, U., Calogero, J.A., Holford, T.R., White, A.A., Walter, S.D., Ostfeld, A.M., and Southwick, W.O.: Acute prolapsed lumbar intervertebral disc. *Spine, 9:*608–613, 1984.

Kelsey, J.L., and Hardy, R.J.: Driving of motor vehicles as a risk factor for acute herniated lumbar intervertebral disc. *American Journal of Epidemiology, 102:*63–73, 1975.

Kemper, H.C.G., and Verschuur, R.: Motor performance fitness tests. *Medicine and Sport Science, 20:*96–106, 1985.

Kendall, F.P.: A criticism of current tests and exercises for physical fitness. *Journal of the American Physical Therapy Association, 45:*187–197, 1965.

Kendall, F.P., and McCreary, E.K.: *Muscles. Testing and Function,* 3rd ed. Baltimore, Williams and Wilkins, 1983.

Kendall, H. O., and Kendall, F.P.: Normal flexibility according to age groups. *The Journal of Bone and Joint Surgery, 30-A:*690–694, 1948.

Kendall, H.O., Kendall, F.P., and Boynton, D.A.: *Posture and Pain.* Huntington, Krieger, 1970.

Kendall, H.O., Kendall, F.P., and Wadsworth, G.E.: *Muscles — Testing and Function,* 2nd ed. Baltimore, Williams and Wilkins, 1971.

Kendall, P. H., and Underwood, C.S.: Seats and sitting. *Occupational Therapy (London), 31:*20–29, 1968.

Kerr, J.: *The Fundamentals of School Health.* New York, Macmillan, 1928.

King, G.R.: *A Study of Relationships Between the Changes in the Position of Body*

Segments When "Natural" Standing Position Is Altered to "Best" Standing Position. Thesis, Wellesley College, Wellesley, Massachusetts, June 1932.

Kippers, V., and Parker, A.W.: Hand positions at possible critical points in the stoop-lift movement. *Ergonomics, 26:*895–903, 1983.

Kira, A.: *The Bathroom.* New York, Viking Press, 1976.

Kirby, N.A., Barnerias, M.J., and Siebens, A.A.: An evaluation of assisted cough in quadriplegic patients. *Archives of Physical Medicine and Rehabilitation, 47:*705–710, 1966.

Kirby, R.L., Simms, F.C., Symington, V.J., and Garner, J.B.: Flexibility and musculo-skeletal symptomatology in female gymnasts and age-matched controls. *The American Journal of Sports Medicine, 9:*160–164, 1981.

Kirk, J.E., and Chieffi, M.: Variation with age in elasticity of skin and subcutaneous tissue in human individuals. *Journal of Gerontology, 17:*373–380, 1962.

Klausen, K.: The form and function of the loaded human spine. *Acta Physiologica Scandinavica, 65:*176–190, 1965.

Klausen, K.: The shape of the spine in young males with and without back complaints. *Clinical Biomechanics, 1:*81–84, 1986.

Klausen, K., and Rasmussen, B.: On the location of the line of gravity in relation to L5 in standing. *Acta Physiologica Scandinavica, 72:*45–52, 1968.

Kleeman, W.: The FAA chair study. *Design, 21 (No.278):*72,74,76,81, 1980.

Kleeman, W., and Prunier, T.: Evaluation of chairs used by air traffic controllers of the U.S. Federal Aviation Administration — implications for design. In Easterby, R., Kroemer, K.H.E., and Chaffin, D.B. (Eds.): *Anthropometry and Biomechanics. Theory and Application.* New York, Plenum, 1982, pp. 235–239.

Knapp, R., and Bradley, P.L.: Medical applications of an electrical pressure gauge for external use: preliminary report. *Bio-medical Engineering, 5:*116–119, 124, 1970.

Knave, B.G.: The visual display unit. In Ledin, H. (Ed.): *Ergonomic Principles in Office Automation,* 2nd ed. Bromma, Ericsson Information Systems AB, 1983, pp. 11–41.

Knudsen, K.A.: *A Text-Book of Gymnastics.* Philadelphia, Lippincott, 1920.

Knudsen, K.A.: *A Textbook of Gymnastics, vol. one.* London, Churchill, 1947.

Kohara, J.: *The Application of Human Engineering to Design.* Chicago, Institute of Design, Illinois Institute of Technology, November 1965.

Kohara, J., and Hoshi, A.: Fitting the seat to the passenger. *Industrial Design, 13(10):*54–59, 1966.

Kohara, J., and Sugi, T.: *Development of Biomechanical Manikins for Measuring Seat Comfort.* Society of Automotive Engineers, publication no. 72006, 1972.

Konz, S.: *Work Design: Industrial Ergonomics,* 2nd ed. Columbus, Grid, 1983.

Koreska, J., Gibson, D.A., and Albisser, A.M.: Structural support system for unstable spines. In Komi, P.V. (Ed.): *Biomechanics V-A. International Series on Biomechanics.* Baltimore, Univ. Park, 1976, Vol. 1-A, pp. 474–483.

Kosiak, M.: Etiology and pathology of ischemic ulcers. *Archives of Physical Medicine and Rehabilitation, 40:*62–69, 1959.

Kosiak, M., Kubicek, W.G., Olson, M., Danz, J.N., and Kottke, F.J.: Evaluation of pressure as a factor in the production of ischial ulcers. *Archives of Physical Medicine and Rehabilitation, 39:*623–629, 1958.

Kostuik, J.P., and Hall, B.B.: Spinal fusions to the sacrum in adults with scoliosis. *Spine, 8:*489–500, 1983.

Kotelmann, L.: *School Hygiene.* Syracuse, Bardeen, 1899.

Kottke, F.J.: Evaluation and treatment of low back pain due to mechanical causes. *Archives of Physical Medicine and Rehabilitation, 42:*426–440, 1961.

Kovar, M.G.: Health of the elderly and use of health services. *Public Health Reports, 92:*9–19, 1977.

Krämer, J.: Pressure dependent fluid shifts in the intervertebral disc. *Orthopedic Clinics of North America, 8:*211–216, 1977.

Krämer, J.: *Intervertebral Disk Diseases.* Chicago, Year Book, 1981.

Krämer, J., Kolditz, D., and Gowin, R.: Water and electrolyte content of human intervertebral discs under variable load. *Spine, 10:*69–71, 1985.

Krebs, M., Ragnarrson, K.T., and Tuckman, J.: Orthostatic vasomotor response in spinal man. *Paraplegia, 21:*72–80, 1983.

Krjukova, D.N.: Electromyographic analysis of fatigue of postural muscles as a function of seated working posture. *Gigiena Truda i Professional'nye Zabolevanija, 4:*12–16, April 1977.

Kroemer, K.H.E.: Seating in plant and office. *American Industrial Hygiene Association Journal, 32:*633–652, 1971.

Kroemer, K.H.E.: Human engineering the keyboard. *Human Factors, 14:*51–63, 1972.

Kroemer, K.H.E.: How to fit the equipment to the operator. In *Health and Ergonomic Considerations of Visual Display Units—Symposium Proceedings.* Denver, American Industrial Hygiene Association, March 1–2, 1982, pp. 45–71.

Kroemer, K.H.E., and Hill, S.G.: Preferred line of sight angle. *Ergonomics, 29:*1129–1134, 1986.

Kroemer, K.H.E., and Price, D.L.: Ergonomics in the office: comfortable work stations allow maximum productivity. *Industrial Engineering, 14:*24–32, July 1982.

Kroemer, K.H.E., and Robinette, J.C.: *Ergonomics in the Design of Office Furniture. A Review of European Literature.* Technical Report AMRL–TR-68-80. Aerospace Medical Research Laboratories, Wright-Patterson Air Force Base, Ohio, July 1968.

Kroemer, K.H.E., and Robinette, J.C.: Ergonomics in the design of office furniture. *Industrial Medicine, 38:*115–125, 1969.

Kroese, A.J.: The contribution of muscle and skin circulation to reactive hyperaemia in the human lower limb. *Vasa, 6:*9–14, 1977.

Krouskop, T.A., Noble, P.C., Garber, S.L., and Spencer, W.A.: The effectiveness of preventive management in reducing the occurrence of pressure sores. *Journal of Rehabilitation Research and Development, 20:*74–83, 1983.

Krouskop, T.A., Reddy, N.P., Spencer, W.A., and Secor, J.W.: Mechanisms of decubitus ulcer formation—an hypothesis. *Medical Hypotheses, 4:*37–39, 1978.

Kukkonen, R., Huuhtanen, P., and Hakala, P.: Prevalence of data operators' musculoskeletal symptoms during the workday and work week. *Behavior and Information Technology, 3:*347–351, 1984.

Kuntzleman, C.T.: The Reader's Digest guide to family fitness. *Reader's Digest, 128:*F1–F12, January 1986.

Kuorinka, I.: Bodily discomfort. In Salvendy, G. (Ed.): *Handbook of Industrial Engineering.* New York, John Wiley and Sons, 1982, pp. 6.5.1–6.5.8.

Kvarnström, S.: Occurrence of musculoskeletal disorders in a manufacturing industry, with special attention to occupational shoulder disorders. *Scandinavian Journal of Rehabilitation Medicine, Supplement 8:*1–114, 1983.

Laging, B.: Furniture design for the elderly. *Rehabilitation Literature, 27:*130–140, 1966.

Lambrinudi, C.: Adolescent and senile kyphosis. *The British Medical Journal, 2:*800–804, 1934.

Lamid, S., and El Ghatit, A.Z.: Smoking, spasticity and pressure sores in spinal cord injured patients. *American Journal of Physical Medicine, 62:*300–306, 1983.

Lancourt, J.E., Dickson, J.H., and Carter, R.E.: Paralytic spinal deformity following traumatic spinal cord injury in children and adolescents. *The Journal of Bone and Joint Surgery, 63-A:*47–53, 1981.

Lander, C., Korbon, G.A., DeGood, D.E., and Rowlingson, J.C.: The Balans chair and its semi-kneeling position: an ergonomic comparison with the conventional sitting position. *Spine, 12:*269–272, 1987.

Landin, S., Hagenfeldt, L., Saltin, B., and Wahren, J.: Muscle metabolism during exercise in hemiparetic patients. *Clinical Science and Molecular Medicine, 53:*257–269, 1977.

Landis, E.M.: Micro-injection studies of capillary blood pressure in human skin. *Heart, 15:*209–228, 1930.

Larsen, B., Holstein, P., and Lassen, N.A.: On the pathogenesis of bedsores. *Scandinavian Journal of Plastic and Reconstructive Surgery, 13:*347–350, 1979.

Latham, F.: A study in body ballistics: seat ejection. *Proceedings of the Royal Society of London, Series B, 147:*121–139, 1957.

Launis, M.: Design of a VDT work station for customer service. In Grandjean, E. (Ed.): *Ergonomics and Health in Modern Offices.* London, Taylor and Francis, 1984, pp. 465–470.

Lauridsen, K.V., and Lund, T.: Wheelchairs. *Communications from the Testing and Observation Institute of the Danish National Association for Infantile Paralysis, 17:*3–19, 1964.

Laville, A.: Postural reactions related to activities on VDU. In Grandjean, E., and Vigliani, E. (Eds.): *Ergonomic Aspects of Visual Display Terminals.* London, Taylor and Francis, 1983, pp. 167–174.

Laville, A.: Postural stress in high-speed precision work. *Ergonomics, 28:*229–236, 1985.

Lay, W.E., and Fisher, L.C.: Riding comfort and cushions. *SAE Journal (Transactions), 47:*482–496, 1940.

Le, K.M., Madsen, B.L., Barth, P.W., Ksander, G.A., Angell, J.B., and Vistnes, L.M.: An in-depth look at pressure sores using monolithic silicon pressure sensors. *Plastic and Reconstructive Surgery, 74:*745–754, 1984.

Le Carpentier, E.F.: Easy chair dimensions for comfort-a subjective approach. In

Grandjean, E. (Ed.): *Proceedings of the Symposium on Sitting Posture.* London, Taylor and Francis, 1969, pp. 214–223.

Lee, G.S.: *Invisible Exercise.* New York, Dutton, 1922.

Lee, M., and Wagner, M.M.: *Fundamentals of Body Mechanics and Conditioning.* Philadelphia, Saunders, 1949.

Lee, R.I., and Brown, L.T.: A new chart for the standardization of body mechanics. *The Journal of Bone and Joint Surgery, 5:*753–756, 1923.

Le Floch, P.: Les conditions anatomiques des positions assises. *Bulletin de L'Association des Anatomistes, 65:*447–457, 1981.

Le Floch, P., and Guillaumat, M.: The orthopaedic consequences of the sitting position. *Orthopaedic Transactions, 6:*57, 1982.

Legg, S.J.: The effect of abdominal muscle fatigue and training on the intra-abdominal pressure developed during lifting. *Ergonomics, 24:*191–195, 1981.

Leibowitz, J.: For the victims of our culture: the Alexander technique. *Dance Scope, 4:*32–37, 1967.

Less, M., and Eickelberg, W.W.B.: Force changes in neck vertebrae and muscles. In Komi, P.V. (Ed.): *Biomechanics V-A. International Series on Biomechanics, vol. 1A.* Baltimore, Univ Park, 1976, pp. 530–536.

Life, M.A., and Pheasant, S.T.: An integrated approach to the study of posture in keyboard operation. *Applied Ergonomics, 15:*83–90, 1984.

Lincoln, D.F.: *The Sanitary Conditions and Necessities of School-Houses and School Life.* Concord, Republican Press, 1886.

Lincoln, D.F.: *School and Industrial Hygiene.* Philadelphia, Blakiston, 1896.

Lindan, O., Greenway, R.M., and Piazza, J.M.: Pressure distribution on the surface of the human body: 1. Evaluation in lying and sitting positions using a "bed of springs and nails." *Archives of Physical Medicine and Rehabilitation, 46:*378–385, 1965.

Lindh, M.: Biomechanics of the lumbar spine. In Frankel, V.H., and Nordin, M. (Eds.): *Basic Biomechanics of the Skeletal System.* Philadelphia, Lea and Febiger, 1980, pp. 255–290.

Lipetz, S., and Gutin, B.: An electromyographic study of four abdominal exercises. *Medicine and Science in Sports, 2:*35–38, 1970.

Loane, T.D., and Kirby, R.L.: Static rear stability of conventional and lightweight variable-axle-position wheelchairs. *Archives of Physical Medicine and Rehabilitation, 66:*174–176, 1985.

Lopez, E.M.: The role of musculocutaneous flaps in the closure of pressure sores: present status. *Journal of the American Paraplegia Society, 6:*87–89, 1983.

Lorenz, A.: *Die Heutige Schulbankfrage.* Wien, 1888. Condensed translation in Hartwell (1895).

Loud, P., and Gladwin, J.: Domestic accidents involving easy chairs. In Institute for Consumer Ergonomics: *Seating for Elderly and Disabled People. Report No. 8, Appendix.* University of Technology, Loughborough, Leicestershire, United Kingdom, 1983f.

Lovett, R.W.: The mechanics of lateral curvature of the spine. *Boston Medical and Surgical Journal, 142:*622–627, 1900.

Lovett, R.W.: Round shoulders and faulty attitude: a method of observation and record, with conclusions as to treatment. *Boston Medical and Surgical Journal, 147:*510–520, 1902.

Lovett, R.W.: *Lateral Curvature of the Spine and Round Shoulders,* 3rd ed. Philadelphia, P. Blakiston's Son and Co., 1916.

Lowman, C.L.: The relation of the abdominal muscles to paralytic scoliosis. *The Journal of Bone and Joint Surgery, 14:*763–771, 1932.

Lowthian, P.: Pressure sore prevalence. *Nursing Times, 75:*358–360, 1979.

Lowthian, P.: A review of pressure sore pathogenesis. *Nursing Times, 78:*117–121, 1982.

Lowthian, P.T.: Pressure sores 1. Practical prophylaxis. *Modern Geriatrics, 5:*25–30, November 1975.

Lowthian, P.T.: Underpads in the prevention of decubiti. In Kenedi, R.M., Cowden, J.M., and Scales, J.T. (Eds.): *Bedsore Biomechanics.* Baltimore, Univ Park, 1976, pp. 141–145.

Lucas, D.B.: Mechanics of the spine. *Bulletin of the Hospital for Joint Disease, 31:*115–131, 1970.

Lucas, D.B., and Bresler, B.: *Stability of the Ligamentous Spine.* Report Number 40, Biomechanics Laboratory, University of California, San Francisco and Berkeley, January 1961.

Luciani, L.: *Human Physiology, vol. 3.* London, Macmillan, 1915.

Lueder, R.K.: Seat comfort: a review of the construct in the office environment. *Human Factors, 25:*701–711, 1983.

Lueder, R.: The art and science of ergonomics. *Industry In-Depth, 9:*1–4, 1984.

Lundervold, A.: Electromyographic investigations of position and manner of working in typewriting. *Acta Physiologica Scandinavica, 24 (Supplementum 84):1*–171, 1951a.

Lundervold, A.: Electromyographic investigations during sedentary work, especially typewriting. *The British Journal of Physical Medicine, 14:*32–36, 1951b.

Lundervold, A.: Electromyographic investigations during typewriting. *Ergonomics, 1:*226–233, 1958.

Luque, E.R.: Segmental spinal instrumentation of the lumbar spine. *Clinical Orthopaedics and Related Research, 203:*126–134, 1986.

Luque, E.R., Cassis, N., and Ramírez-Wiella, G.: Segmental spinal instrumentation in the treatment of fractures of the thoracolumbar spine. *Spine, 7:*312–317, 1982.

Macaulay, M.A.: Seating and visibility. In Giles, J.G. (Ed.): *Vehicle Operation and Testing.* London, Iliffe, 1969, pp. 28–46.

Macgibbon, B., and Farfan, H.F.: A radiologic survey of various configurations of the lumbar spine. *Spine, 4:*258–266, 1979.

Macnab, I.: *Backache.* Baltimore, Williams and Wilkins, 1977.

Maeda, K.: Occupational cervicobrachial disorder and its causative factors. *Journal of Human Ergology, 6:*193–202, 1977.

Maeda, K., Hünting, W., and Grandjean, E.: Localized fatigue in accounting machine operators. *Journal of Occupational Medicine, 22:*810–816, 1980a.

Maeda, K., Hünting, W., and Grandjean, E.: Factor analysis of localized fatigue complaints of accounting-machine operators. *Journal of Human Ergology, 11:*37–43, 1982.

Maeda, K., Okazaki, F., Suenaga, T., Sakurai, T., and Takamatsu, M.: Low back pain related to bowing posture of greenhouse farmers. *Journal of Human Ergology, 9:*117–123, 1980b.

Mahlamäki, S., Rauhala, E., Remes, A., and Hänninen, O.: Effect of arm supports on EMG activity of trapezius muscles in typists. *Acta Physiologica Scandinavica, 126:*18A, 1986.

Majeske, C., and Buchanan, C.: *A Quantitative Description of Two Sitting Postures: with and without a Lumbar Roll.* Paper presented at the Virginia Physical Therapy Association state conference, Williamsburg, Virginia, 1983.

Majeske, C., and Buchanan, C.: Quantitative description of two sitting postures. *Physical Therapy, 64:*1531–1533, 1984.

Makris, P.E., Louizou, C., Markakis, C., Tsakiris, D.A., and Mandalaki, T.: Long lasting sitting position and haemostasis. *Thrombosis and Haemostasis, 55:*119–121, 1986.

Mandal, A.C.: Work-chair with tilting seat. *Ergonomics, 19:*157–164, 1976.

Mandal, A.C.: The seated man (homo sedens). *Applied Ergonomics, 12:*19–26, 1981.

Mandal, A.C.: The correct height of school furniture. *Human Factors, 24:*257–269, 1982.

Mandal, A.C.: The correct height of school furniture. *Physiotherapy, 70:*48–53, 1984.

Manley, M.T., Wakefield, E., and Key, A.G.: The prevention and treatment of pressure sores in the sitting paraplegic. *South African Medical Journal, 52:*771–774, 1977.

Mantle, M.J., Greenwood, R.M., and Currey, H.L.F.: Backache in pregnancy. *Rheumatology and Rehabilitation, 16:*95–101, 1977.

Marchant, R.P.: Comparison of polyurethane and latex foams for furniture. *Journal of Cellular Plastics, 8:*85–89, 1972.

Marek, T., Noworol, C., Gedliczka, A., and Matuszek, L.: A study of a modified VDT-stand arrangement. *Behaviour and Information Technology, 3:*405–409, 1984.

Margolis, S.A., Jones, R.M., and Brown, B.E.: The subasis bar: an effective approach to pelvic stabilization in seated positioning. In *Proceedings of the Eighth Annual Conference of the Rehabilitation Engineering Society of North America,* Memphis, Tennessee, 1985, pp. 45–47.

Mark, L.S., Vogele, D.C., Dainoff, M.J., Cone, S., and Lassen, K.: Measuring movement at ergonomic workstations. In Eberts, R.E., and Eberts, C.G. (Eds.): *Trends in Ergonomics/Human Factors II.* Amsterdam, Elsevier, 1985, pp. 431–438.

Marriott, I.A., and Stuchly, M.A.: Health aspects of work with visual display terminals. *Journal of Occupational Medicine, 28:*833–848, 1986.

Marsk, A.: Studies on weight-distribution upon the lower extremities in individuals working in a standing position. *Acta Orthopaedica Scandinavica, Supplement 31:*1–64, 1958.

Martin, F.H.: Gymnastics and other mechanical means in the treatment of visceral prolapse and its complications. *Surgery, Gynecology and Obstetrics, 15:*150–165, 1912.

Matsen, F.A.: *Compartmental Syndromes.* New York, Grune and Stratton, 1980.

Matsen, F.A., Wyss, C.R., King, R.V., Barnes, D., and Simmons, C.W.: Factors affecting the tolerance of muscle circulation and function for increased tissue pressure. *Clinical Orthopaedics and Related Research, 155:*224–230, 1981.

Matsen, F.A., Wyss, C.R., Krugmire, R.B., Simmons, C.W., and King, R.V.: The effects of limb elevation and dependency on local arteriovenous gradients in normal human limbs with particular reference to limbs with increased tissue pressure. *Clinical Orthopaedics and Related Research, 150:*187–195, 1980.

Mayer, L.: Further studies of fixed paralytic pelvic obliquity. *The Journal of Bone and Joint Surgery, 18:*87–100, 1936.

McClelland, I., and Ward, J.S.: Ergonomics in relation to sanitary ware design. *Ergonomics, 19:*465–478, 1976.

McConnel, J.K.: *The Adjustment of Muscular Habits.* London, Lewis, 1933.

McCormick, H.G.: *The Metabolic Cost of Maintaining a Standing Position.* Morningside Heights, King's Crown, 1942.

McCraw, J.B., Dibbell, D.G., and Carraway, J.H.: Clinical definition of independent myocutaneous vascular territories. *Plastic and Reconstructive Surgery, 60:*341–352, 1977.

McFarland, R.A., and Stoudt, H.W.: *Human Body Size and Passenger Vehicle Design.* Society of Automotive Engineers, publication no. SP-142A, 1961.

McGregor, J.C.: Surgical treatment of ischial pressure sores. *Journal of the Royal College of Surgeons of Edinburgh, 29:*242–245, 1984.

McKenzie, M.W., and Rogers, J.E.: Use of trunk supports for severely paralyzed people. *The American Journal of Occupational Therapy, 27:*147–148, 1973.

McKenzie, R.A.: *The Lumbar Spine. Mechanical Diagnosis and Therapy.* Waikanae, Spinal Publications, 1981.

McKenzie, R.T.: Influence of school life on curvature of the spine. *Proceedings of the National Educational Association:*939–948, 1898.

McKenzie, R.T.: *Exercise in Education and Medicine,* 2nd ed. Philadelphia, Saunders, 1915.

McMillan, M.: *Massage and Therapeutic Exercise,* 3rd ed. Philadelphia, Saunders, 1932.

Meisenbach, R.: In abstract of discussion. *Journal of the American Medical Association, 68:*329–330, 1917.

Merbitz, C.T., King, R.B., Bleiberg, J., and Grip, J.C.: Wheelchair push-ups: measuring pressure relief frequency. *Archives of Physical Medicine and Rehabilitation, 66:*433–439, 1985.

Mercier, C.: *The Nervous System and the Mind.* London, Macmillan, 1888.

Meyer, G.H.: *Die Statik und Mechanik des Menschlichen Knochengerüstes.* Leipzig, Engelmann, 1873.

Meyer, H.: Das aufrechte stehen. *Archiv für Anatomie und Physiologie:* 2–45, 1853.

Michocki, R.J., and Lamy, P.P.: The problem of pressure sores in a nursing home population: statistical data. *Journal of the American Geriatrics Society, 24:*323–328, 1976.

Mierau, D.R., Cassidy, J.D., Hamin, T., and Milne, R.A.: Sacroiliac joint dysfunction and low back pain in school aged children. *Journal of Manipulative and Physiological Therapeutics, 7:*81–84, 1984.

Miller, G.E., and Seale, J.: Lymphatic clearance during compressive loading. *Lymphology, 14:*161–166, 1981.

Miller, G.E., and Seale, J.L.: The mechanics of terminal lymph flow. *Journal of Biomechanical Engineering, 107:*376–380, 1985.

Miller, I., and Suther, T.W.: Preferred height and angle settings of CRT and keyboard for a display station input task. In *Proceedings of the Human Factors Society 25th Annual Meeting,* 1981, pp. 492–496.

Miller, M.E., and Sachs, M.L.: *About Bedsores.* Philadelphia, Lippincott, 1974.

Miller, W., and Suther, T.W.: Display station anthropometrics: preferred height and angle settings of CRT and keyboard. *Human Factors, 25:*401–408, 1983.

Mills, L.: The effects of faulty cranio-spinal form and alignment upon the eyes. *American Journal of Ophthalmology, 2:*493–499, 1919.

Milne, J.S., and Lauder, I.J.: Age effects in kyphosis and lordosis in adults. *Annals of Human Biology, 1:*327–337, 1974.

Milne, J.S., and Williamson, J.: A longitudinal study of kyphosis in older people. *Age and Ageing, 12:*225–233, 1983.

Milne, R.A., and Mierau, D.R.: Hamstring distensibility in the general population: relationship to pelvic and low back stress. *Journal of Manipulative and Physiological Therapeutics, 2:*146–150, 1979.

Minami, R.T., Mills, R., and Pardoe, R.: Gluteus maximus myocutaneous flaps for repair of pressure sores. *Plastic and Reconstructive Surgery, 60:*242–249, 1977.

Minns, R.J., and Sutton, R.A.: Pressures under the ischium detected by a pedobarograph. *Engineering in Medicine, 11:*111–115, 1982.

Minns, R.J., Sutton, R.A., Duffus, A., and Mattinson, R.: Underseat pressure distribution in the sitting spinal injury patient. *Paraplegia, 22:*297–304, 1984.

Moe, J.H., Winter, R.B., Bradford, D.S., and Lonstein, J.E.: *Scoliosis and Other Spinal Deformities.* Philadelphia, Saunders, 1978.

Montagna, W., and Carlisle, K.: Structural changes in aging human skin. *The Journal of Investigative Dermatology, 73:*47–53, 1979.

Moolten, S.E.: Bedsores in the chronically ill patient. *Archives of Physical Medicine and Rehabilitation, 53:*430–438, 1972.

Moolten, S.E.: Bedsores. *Hospital Medicine, 13:*83–103, May 1977.

Mooney, V., Einbund, M.J., Rogers, J.E., and Stauffer, E.S.: Comparison of pressure distribution qualities in seat cushions. *Bulletin of Prosthetics Research, 10-15:*129–143, Spring 1971.

Moore, M.A., and Hutton, R.S.: Electromygraphic investigation of muscle stretching techniques. *Medicine and Science in Sports and Exercise, 12:*322–329, 1980.

Morimoto, S.: Effect of sitting posture on human body. *The Bulletin of Tokyo Medical and Dental University, 20:*19–34, 1973.

Morris, J.M.: Biomechanics of the spine. *Archives of Surgery, 107:*418–423, 1973.

Morris, J.M., Lucas, D.B., and Bresler, B.: Role of the trunk in stability of the spine. *The Journal of Bone and Joint Surgery, 43:*327–351, 1961.

Mosely, L.H., Finseth, F., and Goody, M.: Nicotine and its effect on wound healing. *Plastic and Reconstructive Surgery, 61:*570–575, 1978.

Mosher, E.M.: Habitual postures of school children. *Educational Review, 4:*339–349, 1892.

Mosher, E.M.: Hygienic desks for school children. *Educational Review, 18:*9–14, 1899.

Mosher, E.M.: *Health and Happiness.* New York, Funk and Wagnalls, 1913.

Mosher, E.M.: Faulty habits of posture, a cause of enteroptosis. *International Journal of Surgery, 27:*174–179, 1914.

Mosher, E.M.: Habits of posture as related to health and efficiency. *International Journal of Surgery, 32:*40–45, 1919.

Motloch, W.M.: *Analysis of Medical Costs Associated with Healing of Pressure Sores in Adolescent Paraplegic.* Rehabilitation Engineering Center, Children's Hospital at Stanford, Palo Alto, 1978.

Moustafa, H.F., and Hopewell, J.W.: Blood flow clearance changes in pig skin after single doses of X rays. *British Journal of Radiology, 52:*138–144, 1979.

Mulholland, J.H., Tui, C., Wright, A.M., Vinci, V., and Shafiroff, B.: Protein metabolism and bed sores. *Annals of Surgery, 118:*1015–1023, 1943.

Mulliner, M.R.: *Mechano-therapy.* Philadelphia, Lea and Febiger, 1929.

Munton, J.S., Ellis, M.I., Chamberlain, M.A., and Wright, V.: An investigation into the problems of easy chairs used by the arthritic and the elderly. *Rheumatology and Rehabilitation, 20:*164–173, 1981.

Murrell, K.F.H.: *Ergonomics. Man in His Working Environment.* London, Chapman and Hall, 1965.

Mutoh, Y., Mori, T., Nakamura, Y., and Miyashita, M.: The relation between sit-up exercises and the occurrence of low back pain. In Matsui, H., and Kobayashi, K. (Eds.): *Biomechanics VIII-A.* Champaign, Human Kinetics, 1983, pp. 180–185.

NASA (National Aeronautics and Space Administration): *Anthropometric Source Book. Volume I: Anthropometry for Designers.* NASA Reference Publication 1024, 1978.

Nachemson, A.: Electromyographic studies on the vertebral portion of the psoas muscle. *Acta Orthopaedica Scandinavica, 37:*177–190, 1966.

Nachemson, A., and Elfström, G.: Intravital dynamic pressure measurements in lumbar discs. A study of common movements, maneuvers and exercises. *Scandinavian Journal of Rehabilitation Medicine, Supplement 1:*1–40, 1970.

Nakaseko, M., Grandjean, E., Hünting, W., and Gierer, R.: Studies on ergonomically designed alphanumeric keyboards. *Human Factors, 27:*175–187, 1985.

Narsete, T.A., Orgel, M.G., and Smith, D.: Pressure sores. *American Family Physician, 28:*135–139, 1983.

National Institute for Occupational Safety and Health: *Potential Health Hazards of Video Display Terminals.* Cincinnati, NIOSH Report No. 81-129, June 1981.

National Research Council: *Video Displays, Work, and Vision.* Washington, National Academy Press, 1983.

Natow, A.B.: Nutrition in prevention and treatment of decubitus ulcers. *Topics in Clinical Nursing, 5:*39–44, July 1983.

Nelham, R.L.: Seating for the chairbound disabled person—a survey of seating equipment in the United Kingdom. *Journal of Biomedical Engineering, 3:*267–274, 1981.

Nelham, R.L.: Principles and practice in the manufacture of seating for the handicapped. *Physiotherapy, 70:*54–58, 1984.

Nickel, V.L.: The treatment of patients with severe paralysis. *Postgraduate Medicine, 21:*581–590, 1957.

Nicol, K., and Koerner, U.: Pressure distribution on a chair for disabled subjects. In Winter, D.A., Norman, R., Wells, R., Hayes, K., and Patla, A. (Eds.): *Biomechanics IX-A.* Champaign, Human Kinetics, 1985, pp. 274–280.

Nicolaisen, T., and Jørgensen, K.: Trunk strength, back muscle endurance and low-back trouble. *Scandinavian Journal of Rehabilitation Medicine, 17:*121–127, 1985.

Nielsen, H.V.: Effects of externally applied compression on blood flow in subcutaneous and muscle tissue in the human supine leg. *Clinical Physiology, 2:*447–457, 1982.

Nielsen, H.V.: External pressure-blood flow relations during limb compression in man. *Acta Physiologica Scandinavica, 119:*253–260, 1983a.

Nielsen, H.V.: Effects of externally applied compression on blood flow in the human dependent leg. *Clinical Physiology, 3:*131–140, 1983b.

Noble, P.C.: *Some Contributions of Rehabilitation Engineering to the Pressure Sore Problem.* In Proceedings of a Rehabilitation Workshop, Royal Australasian College of Surgeons, Perth, Western Australia, 1977, pp. 169–180.

Noble, P.C.: *An Examination of Sitting Pressures in the Physically Disabled.* Bioengineering Division, Royal Perth Hospital, Perth, Western Australia, November 1978a.

Noble, P.C.: *A Comparative Assessment of Three Commercial Wheelchair Cushions.* Bioengineering Division, Royal Perth Hospital, Perth, Western Australia, December 1978b.

Noble, P.C.: *Rehabilitation Engineering in the Management of Tissue Trauma.* Paper presented at the Second Far East and South Pacific Spinal Injuries Conference, November 1979.

Noble, P.C.: *The Prevention of Pressures Sores in Persons with Spinal Cord Injuries. Monograph Number Eleven.* New York, World Rehabilitation Fund, 1981.

Nola, G.T., and Vistnes, L.M.: Differential response of skin and muscle in the experimental production of pressure sores. *Plastic and Reconstructive Surgery, 66:*728–733, 1980.

Norton, D., McLaren, R., and Exton-Smith, A.N.: *An Investigation of Geriatric Nursing Problems in Hospital.* Edinburgh, Churchill, 1975.

Nubar, Y., and Contini, R.: A minimal principle in biomechanics. *Bulletin of Mathematical Biophysics, 23:*377–391, 1961.

Oborne, D.J.: *Ergonomics at Work.* Chichester, Wiley, 1982.

Occhipinti, E., Colombini, D., Frigo, C., Pedotti, A., and Grieco, A.: Sitting posture: analysis of lumbar stresses with upper limbs supported. *Ergonomics, 28:*1333–1346, 1985.

O'Connell, A.L.: Electromyographic study of certain leg muscles during movements of the free foot and during standing. *American Journal of Physical Medicine, 37:*289–301, 1958.

Ohry, A., Heim, M., Steinbach, T.V., and Rozin, R.: The needs and unique problems facing spinal cord injured persons after limb amputation. *Paraplegia, 21:*260–263, 1983.

Ohuchi, K., and Hayashi, Y.: A study of the cushioning of seat assembly. *Japanese Journal of Ergonomics, 5:*251–256, 1969.

Okada, M.: An electromyographic estimation of the relative muscular load in different human postures. *Journal of Human Ergology, 1:*75–93, 1972.

Okada, M., and Fujiwara, K.: Muscle activity around the ankle joint as correlated with the center of foot pressure in an upright stance. In Matsui, H., and Kobayashi, K. (Eds.): *International Series on Biomechanics, Vol. 4A.* Champaign, Human Kinetics, 1983, pp. 209–216.

Okushima, H.: Study on hydrodynamic pressure of lumbar intervertebral disc. *Archiv Fur Japanische Chirurgie, 39:*45–57, 1970.

Olsen, F.G.: Comfort Packages for Diverse Passenger Car Objectives. Society of Automobile Engineers, publication no. 650463, 1965.

Olszewski, W.L., and Engeset, A.: Intrinsic contractility of prenodal lymph vessels and lymph flow in human leg. *American Journal of Physiology, 239:*H775–H783, 1980.

O'Neill, N.E.: A study of the anthropometric differences of the lordotic curvature in adult men and women. *Ergonomics, 25:*494–495, 1982.

Ong, C.N.: VDT work place design and physical fatigue: a case study in Singapore. In Grandjean, E. (Ed.): *Ergonomics and Health in Modern Offices.* London, Taylor and Francis, 1984, pp. 484–494.

Ong, C.N., Hoong, B.T., and Phoon, W.O.: Visual and muscular fatigue in operators using visual display terminals. *Journal of Human Ergology, 10:*161–171, 1981.

Onishi, N., Sakai, K., and Kogi, K.: Arm and shoulder muscle load in various keyboard operating jobs of women. *Journal of Human Ergology, 11:*89–97, 1982.

Ono, K.: Electromyographic studies of the abdominal wall muscles in visceroptosis. *The Tohoku Journal of Experimental Medicine, 68:*347–354, 1958.

O'Reilly, A.: Scoliosis and school seating. *American Physical Education Review, 19:*571–575, 1914.

Osebold, W.R., Mayfield, J.K., Winter, R.B., and Moe, J.H.: Surgical treatment of paralytic scoliosis associated with myelomeningocele. *The Journal of Bone and Joint Surgery, 64-A:*841–856, 1982.

Oshima, M.: Optimum conditions of seat design. In *Proceedings of the Fourth International Ergonomics Association Congress,* Strasbourg, July 6–10, 1970, pp. 1–9.

Östberg, O.: CRTs pose health problems for operators. *International Journal of Occupational Health and Safety, 44:*24–26, 46, 50, 52, Nov.–Dec. 1975.

Östberg, O.: The health debate. *Reprographics Quarterly, 12:*80–83, 1979.

Östberg, O.: Work environment issues of Swedish office workers: a union perspective. In Cohen, B.G.F. (Ed.): *Human Aspects in Office Automation.* Amsterdam, Elsevier, 1984, pp. 127–141.

Oxford, H.W.: Factors in the design of seats used in public transport. In *Proceedings of the Tenth Annual Conference of the Ergonomics Society of Australia and New Zealand,* Sydney, November 1973, pp. 13.1–13.16.

Palmieri, V.R., Haelen, G.T., and Cochran, G.V.B.: A comparison of sitting pressures on wheelchair cushions as measured by air cell transducers and miniature electronic transducers. *Bulletin of Prosthetics Research, 10–33:*5–8, Spring 1980.

Panero, J., and Zelnik, M.: *Human Dimension and Interior Space.* New York, Whitney Library of Design, 1979.

Panjabi, M.M., Andersson, G.B.J., Jorneus, L., Hult, E., and Mattsson, L.: In vivo measurements of spinal column vibrations. *The Journal of Bone and Joint Surgery, 68-A:*695–702, 1986.

Parsons, K.C., and Griffin, M.J.: Predicting the vibration discomfort of seated passengers. In Oborne, D.J., and Levis, J.A. (Eds.): *Human Factors in Transport Research, vol. 2.* London, Academic Pr, 1980, pp. 114–122.

Pate, R.R.: A new definition of youth fitness. *The Physician and Sportsmedicine, 11:*77–83, April 1983.

Patterson, R.: Is pressure the most important parameter? In *Proceedings of the National Symposium on the Care, Treatment and Prevention of Decubitus Ulcers.* Arlington, Virginia, November 1984, pp. 69–72.

Patterson, R.P., and Fisher, S.V.: The accuracy of electrical transducers for the measurement of pressure applied to the skin. *IEEE Transactions on Biomedical Engineering, BME-26:*450–456, 1979.

Patterson, R.P., and Fisher, S.V.: Pressure and temperature patterns under the ischial tuberosities. *Bulletin of Prosthetics Research, 17:*5–11, Fall 1980.

Pearson, D.A., and Wetle, T.T.: Long-term care. In Jonas, S. (Ed.). *Health Care Delivery in the United States,* 2nd ed. New York, Springer, 1981, pp. 218–234.

Perkash, I., O'Neill, H., Politi-Meeks, D., and Beets, C.L.: Development and evaluation of a universal contoured cushion. *Paraplegia, 22:*358–365, 1984.

Petersen, C.M., Amundsen, L.R., and Schendel, M.J.: Comparison of the effectiveness of two pelvic stabilization systems on pelvic movement during maximal isometric trunk extension and flexion muscle contractions. *Physical Therapy, 67:*534–539, 1987.

Petersen, N.C., and Bittmann, S.: The epidemiology of pressure sores. *Scandinavian Journal of Plastic and Reconstructive Surgery, 5:*62–66, 1971.

Pheasant, S.T.: *Anthropometrics. An Introduction for Schools and Colleges.* London, British Standards Institution, 1984.

Pheasant, S.: *Bodyspace.* London, Taylor and Francis, 1986.

Phelps, W.M., and Kiphuth, R.J.H.: *The Diagnosis and Treatment of Postural Defects.* Springfield, Thomas, 1932.

Pile, J.F.: *Modern Furniture.* New York, John Wiley and Sons, 1979.

Pinchcofsky-Devin, G.D., and Kaminski, M.V.: Correlation of pressure sores and nutritional status. *Journal of the American Geriatrics Society, 34:*435–440, 1986.

Platts, E.A.: Wheelchair design—survey of users' views. *Proceedings of the Royal Society of Medicine, 67:*414–416, 1974.

Pollack, A.A., and Wood, E.H.: Venous pressure in the saphenous vein at the ankle in man during exercise and changes in posture. *Journal of Applied Physiology, 1:*649–662, 1949.

Pope, M.H., Bevins, T., Wilder, D.G., and Frymoyer, J.W.: The relationship between anthropometric, postural, muscular, and mobility characteristics of males ages 18–55. *Spine, 10:*644–648, 1985.

Pope, M.H., Wilder, D.G., and Frymoyer, J.W.: Vibration as an aetiologic factor in low back pain. In *Engineering Aspects of the Spine.* London, Mechanical Engineering Publications Ltd., 1980, pp. 11–17.

Pope, P.M.: A study of instability in relation to posture in the wheelchair. *Physiotherapy, 71:*124–129, 1985a.

Pope, P.M.: Proposals for the improvement of the unstable postural condition and some cautionary notes. *Physiotherapy, 71:*129–131, 1985b.

Porreca, R.C., and Chagares, R.M.: Op-Site®: a treatment for pressure sores in the orthopaedic patient population. *Orthopaedic Nursing, 2:*30–36, September/October 1983.

Porter, C.: *School Hygiene and the Laws of Health.* London, Longmans, Green, and Co., 1906.

Porter, J.M., and Davis, G.N.: An assessment of alternative seating. In Coombes, K. (Ed.): *Proceedings of the Ergonomics Society's Conference 1983.* New York, Taylor and Francis, 1983, pp. 202–203.

Posse, N.: *The Swedish System of Educational Gymnastics,* 2nd ed. Boston, Lee and Shepard, 1890.

Pottier, M., Dubreuil, A., and Monod, H.: Les variations de volume du pied au cours de la station assise prolongée. *Le Travail Humain, 30:*111–122, 1967.

Pottier, M., Dubreuil, A., and Monod, H.: The effects of sitting posture on the volume of the foot. *Ergonomics, 12:*753–758, 1969.

Poulsen, E., and Jørgensen, K.: Back muscle strength, lifting, and stooped working postures. *Applied Ergonomics, 2:*133–137, 1971.

Powell, M.E.: *Trunk Strength and Flexibility as Factors in Posture.* Thesis, Wellesley College, Wellesley, Massachusetts, June 1930.

Pustinger, C., Dainoff, M.J., and Smith, M.: VDT workstation adjustability: effects on worker posture, productivity, and health complaints. In Eberts, R.E., and Eberts, C.G. (Eds.): *Trends in Ergonomics/Human FactorsII.* Amsterdam, Elsevier, 1985, pp. 445–451.

Pyle, W.L.: Hygiene of the eye. In Pyle, W. L. (Ed.): *A Manual of Personal Hygiene.* Philadelphia, Saunders, 1913, pp. 169–273.

Radke, A.O.: *The Importance of Seating in Driver Comfort and Performance.* Society of Automotive Engineers, publication no. 838B, 1964.

Radl, G.W.: Experimental investigations for optimal presentation-mode and colours of symbols on the CRT-screen. In Grandjean, E., and Vigliani, E. (Eds.): *Ergonomic Aspects of Visual Display Terminals.* London, Taylor and Francis, 1983, pp. 127–135.

Radl, G.W.: Optimal presentation mode and colours of symbols on VDUs. In Pearce, B.G. (Ed.): *Health Hazards of VDTs?* Chichester, Wiley, 1984, pp. 157–168.

Rang, M., Douglas, G., Bennet, G.C., and Koreska, J.: Seating for children with cerebral palsy. *Journal of Pediatric Orthopedics, 1:*279–287, 1981.

Rasch, P.J., and Burke, R.K.: *Kinesiology and Applied Anatomy. The Science of Human Movement,* 6th ed. Philadelphia, Lea and Febiger, 1978, pp. 242–243.

Rathbone, J.L.: *Corrective Physical Education.* Saunders, Philadelphia, 1934.

Rathbone, J.L.: *Residual Neuromuscular Hypertension: Implications for Education.* New York, Columbia University, 1936.

Rathbone, J.L., and Hunt, V.V.: *Corrective Physical Education,* 7th ed. Saunders, Philadelphia, 1965.

Rebiffé, R.: An ergonomic study of the arrangement of the driving position in motor cars. *Proceedings of the Institution of Mechanical Engineers, 181 (part 3D):*43–50, 1966.

Rebiffé, R.: General reflections on the postural comfort of the driver and passengers; consequences on seat design. In Oborne, D.J., and Levis, J.A. (Eds.): *Human Factors in Transport Research, vol. 2.* London, Academic Pr, 1980, pp. 240–248.

Reddy, M.P.: Management of common "pinched nerves" in the elderly. *Geriatrics, 41:*61–74, 1986.

Reed, J.W.: Pressure ulcers in the elderly: prevention and treatment utilizing the team approach. *Maryland State Medical Journal, 30:*45–50, November 1981.

Reichel, S.M.: Shearing force as a factor in decubitus ulcers in paraplegics. *Journal of the American Medical Association, 166:*762–763, 1958.

Reswick, J.B., Lindan, O., and Lippay, A.: *A Device to Measure Pressure Distribution Between the Human Body and Various Supporting Surfaces.* Report No. EDC 4-64-7. Case Institute of Technology and Highland View Hospital, Cleveland, Ohio, June 1964.

Reus, W.F., Robson, M.C., Zachary, L., and Heggers, J.P.: Acute effects of tobacco smoking on blood flow in the cutaneous micro-circulation. *British Journal of Plastic Surgery, 37:*213–215, 1984.

Reynolds, E., and Hooton, E.A.: Relation of the pelvis to erect posture. *American Journal of Physical Anthropology, 21:*253–278, 1936.

Reynolds, E., and Lovett, R.W.: An experimental study of certain phases of chronic backache. *Journal of the American Medical Association, 54:*1033–1043, 1910.

Ridder, C.A.: *Basic Design Measurements for Sitting.* Agricultural Experiment Station, University of Arkansas, Fayetteville, Bulletin 616, October 1959.

Ringsdorf, W.M., and Cheraskin, E.: Vitamin C and human wound healing. *Oral Surgery, 53:*231–236, 1982.

Riskind, J.H., and Gotay, C.C.: Physical posture: could it have regulatory or feedback effects on motivation and emotion? *Motivation and Emotion, 6:*273–298, 1982.

Rizzi, M.: Entwicklung eines verschiebbaren rückenprofils für auto-und ruhesitze. In Grandjean, E. (Ed.): *Proceedings of the Symposium on Sitting Posture.* London, Taylor and Francis, 1969, pp. 112–119.

Roaf, R.: The causation and prevention of bed sores. In Kenedi, R.M., Cowden, J.M., and Scales, J.T. (Eds.): *Bedsore Biomechanics.* Baltimore, Univ Park, 1976, pp. 5–9.

Roaf, R.: *Posture.* London, Academic Press, 1977.

Robbins, S.E., and Hanna, A.M.: Running-related injury prevention through bare-foot adaptations. *Medicine and Science in Sports and Exercise, 19:*148–156, 1987.

Roberts, D.F.: Passenger comfort-seating. *Bus and Coach, 35:*214–219, 1963.

Robinson, C.A.: Cervical spondylosis and muscle contraction headache. In Dalessio,

D.J. (Ed.): *Wolff's Headache and Other Head Pain,* 4th ed. New York, Oxford Univ. Press, 1980, pp. 362–380.

Robson, M.C., and Krizek, T.J.: The role of infection in chronic pressure ulcerations. In Fredricks, S., and Brody, G.S. (Eds.): *Symposium on the Neurologic Aspects of Plastic Surgery.* St. Louis, Mosby, 1978, pp. 242–246.

Rockey, J., and Nelham, R.L.: Seating for the chairbound child. In McCarthy, G.T. (Ed.): *The Physically Handicapped Child.* London, Faber and Faber, 1984, pp. 255–273.

Rohmert, W., and Luczak, H.: Ergonomics in the design and evaluation of a system for "postal video letter coding." *Applied Ergonomics, 9:*85–95, 1978.

Rose, S.J., Norton, B.J., and Kelly, D.O.: An electromyographic analysis of the McKenzie press-up and extension-in-standing exercises. *Physical Therapy, 61:*684, 1981.

Rosegger, R., and Rosegger, S.: Health effects of tractor driving. *Journal of Agricultural Engineering Research, 5:*241–275, 1960.

Rosemeyer, B.: Eine methode zur beckenfixierung im arbeitssitz. *Zeitschrift für Orthopädie und ihre Grenzgebiete, 110:*514–517, 1972.

Rosemeyer, B.: Der einflub von schäden des haltungs-und bewegungssystems auf die sitzhaltung. *Arbeitsmedizin Sozialmedizin Präventivmedizin, 8:*273–279, 1973.

Rosemeyer, B.: Die aufrechten Körperhaltungen des menschen eine vergleichende untersuchung. *Zeitschrift für Orthopädie und ihre Grenzgebiete, 112:*151–159, 1974.

Roth, B.: *The Treatment of Lateral Curvature of the Spine,* 2nd ed. London, Lewis, 1899.

Rowell, L.B., Detry, J–M.R., Blackmon, J.R., and Wyss, C.: Importance of the splanchnic vascular bed in human blood pressure regulation. *Journal of Applied Physiology, 32:*213–220, 1972.

Rudd, T.N.: The pathogenesis of decubitus ulcers. *Journal of the American Geriatrics Society, 10:*48–53, 1962.

Rühmann, H–P.: Basic data for the design of consoles. In Schmidtke, H. (Ed.): *Ergonomic Data for Equipment Design.* New York, Plenum Press, 1984, pp. 115–144.

Rupp, B.A.: Visual display standards: a review of issues. *Proceedings of the Society for Information Display, 22:*63–72, 1981.

Ruschhaupt, W.F.: Differential diagnosis of edema of the lower extremities. *Cardiovascular Clinics, 13:*307–320, 1983.

Salminen, J.J.: The adolescent back. *Acta Paediatrica Scandinavica, Supplement 315:*1–122, 1984.

Sargent, D.A.: *Physical Education.* Boston, Ginn, 1906.

Sather, M.R., Weber, C.E., and George, J.: Pressure sores and the spinal cord injury patient. *Drug Intelligence and Clinical Pharmacy, 11:*154–169, 1977.

Sauter, S.L., Chapman, L.J., and Knutson, S.J.: *Improving VDT Work. Causes and Control of Health Concerns in VDT Use.* Madison, University of Wisconsin, 1984.

Sauter, S.L., Gottlieb, M.S., Rohrer, K.M., and Dodson, V.N.: *The Well-Being of Video Display Terminal Users: An Exploratory Study.* Cincinnati, National Institute for Occupational Safety and Health, 1983.

Scales, J.T.: Pressures on the patient. *Journal of Drug Research:*5–12, July 1980 (special issue).

Schaedel, S.F.: *Human Factors in Design of Passenger Seats for Commercial Aircraft—A Review.* University of Virginia, Charlottesville, March 1977. NASA Grant No. NGR 47-005-181.

Schafer, R.C.: *Clinical Biomechanics. Musculoskeletal Actions and Reactions.* Baltimore, Williams and Wilkins, 1983.

Schell, V.C., and Wolcott, L.E.: The etiology, prevention and management of decubitus ulcers. *Missouri Medicine, 63:*109–112, 119, 1966.

Schenk, F.: Zur schulbankfrage. *Zeitschrift für Schulgesundheitspflege, 7:*529–545, 1894.

Schlegel, K.F.: Sitzschaeden und deren vermeidung durch eine neuartige sitzkonstruktion. *Medizinische Klinik, 51:*1940–1942, 1956.

Schmidtke, H.: Ergonomic data for the design of body support. In Schmidtke, H. (Ed.): *Ergonomic Data for Equipment Design.* New York, Plenum Press, 1984, pp. 159–178.

Schmoyer, M.: Low back pain following spinal cord injury. *Physical Therapy, 64:*731, 1984.

Schoberth, H.: *Sitzhaltung, Sitzschaden, Sitzmöbel.* Berlin, Springer-Verlag, 1962.

Schoberth, H.: Die wirbelsäule von schulkindern-orthopädische forderungen an schulsitze. In Grandjean, E. (Ed.): *Proceedings of the Symposium on Sitting Posture.* London, Taylor and Francis, 1969, pp. 98–111.

Schüldt, K., Ekholm, J., Harms-Ringdahl, K., Németh, G., and Arborelius, U.P.: Effects of changes in sitting work posture on static neck and shoulder muscle activity. *Ergonomics, 29:*1525–1537, 1986.

Schultz, A.B.: Biomechanics of the spine. In Nelson, M. (Ed.): *Low Back Pain and Industrial and Social Disablement.* Middlesex, Back Pain Association, 1983, pp. 20–25.

Schurmeier, H.L.: A consideration of posture and its relation to bodily mechanics. *American Medicine, 33:*143–150, 1927.

Schwatt, H.: School furniture with special reference to lateral curvature of the spine. *International Clinics 2, (20th Series):*195–210, 1910.

Scudder, C.L.: *Seating of Pupils in the Public Schools.* Boston School Document No. 9, 1892.

Scull, E.R., and Noble, P.C.: *The Involvement of the Bioengineering Division in the Management of Pressure Sore Problems Within the Acute and Rehabilitation Areas of the Hospital.* Bioengineering Division, Department of Medical Physics, Royal Perth Hospital, Perth, Western Australia, June 1977.

Secrest, D., and Dainoff, M.: Performance changes resulting from ergonomic factors in VDT data entry workstation design. In Mital, A. (Ed.): *Advances in Ergonomics/Human Factors I. Proceedings of the First Mid-Central Ergonomics/Human Factors Conference.* Cincinnati, April 12–14, 1984.

Seedhom, B.B., and Terayama, K.: Knee forces during the activity of getting out of a chair with and without the aid of arms. *Biomedical Engineering, 11:*278–282, 1976.

Severy, D.M., Brink, H.M., and Baird, J.D.: Rigid seats with 28 in. seatback effectively reduce injuries in 30+ mph rear-end impacts. *SAE Journal, 77:*20–25, April 1969.

Shackel, B.: Workstation design. In Ledin, H. (Ed.): *Ergonomic Principles in Office Automation,* 2nd ed. Bromma, Ericsson Information Systems AB, 1983, pp. 59–77.

Shackel, B., Chidsey, K.D., and Shipley, P.: The assessment of chair comfort. In Grandjean, E. (Ed.): *Proceedings of the Symposium on Sitting Posture.* London, Taylor and Francis, 1969, pp. 155–192.

Shah, J.: The choice of cushion. *Science and Practice of Physical Care, 1:*61, July 1981.

Shannon, M.L.: Five famous fallacies about pressure sores. *Nursing 84, 14:*34–41, October 1984.

Shaw, B.H., and Snowdon, C.: *The Mechanics of Mattresses.* Biomechanical Research and Development Unit, Department of Health and Social Security, Roehampton, London, 1979.

Shaw, E.R.: *School Hygiene.* New York, Macmillan, 1902.

Shea, J.D.: Pressure sores. Classification and management. *Clinical Orthopaedics and Related Research, 112:*89–100, 1975.

Shepard, M.A., Parker, D., and DeClercque, N.: The under-reporting of pressure sores in patients transferred between hospital and nursing home. *Journal of the American Geriatrics Society, 35:*159–160, 1987.

Shephard, R.J.: *Men at Work.* Springfield, Thomas, 1974.

Shields, R.K.: Effect of a lumbar support on seated buttock pressure. In Donath, M., Friedman, H., and Carlson, M. (Eds.): *Proceedings of the Ninth Annual Conference on Rehabilitation Technology.* Minneapolis, June 23–26, 1986, pp. 408–410.

Shipley, P.: Chair comfort for the elderly and infirm. *Nursing, 20:*858–860, 1980.

Shute, S.J., and Starr, S.J.: Effects of adjustable furniture on VDT users. *Human Factors, 26:*157–170, 1984.

Shvartz, E., Reibold, R.C., White, R.T., and Gaume, J.G.: Hemodynamic responses in orthostasis following five hours of sitting. *Aviation, Space, and Environmental Medicine, 53:*226–231, 1982.

Sieker, H.O., Burnum, J.F., Hickam, J.B., and Penrod, K.E.: Treatment of postural hypotension with a counter-pressure garment. *Journal of the American Medical Association, 161:*132–135, 1956.

Simons, D.G., and Travell, J.G.: Myofascial origins of low back pain. 2. Torso muscles. *Postgraduate Medicine, 73:*81–92, 1983.

Skarstrom, W.: Kinesiology of the trunk, shoulder and hip, applied to gymnastics. *American Physical Education Review, 13:*199–208, 1908.

Skarstrom, W.: *Gymnastic Kinesiology.* Springfield, Bassette, 1909.

Slechta, R.F., Forrest, J., Carter, W.K., and Wade, E.A.: *Comfort Evaluation of the C-118 Pilot Seat (Aerotherm).* WADC Technical Report 58-312, Wright Air Development Center, Wright-Patterson Air Force Base, Dayton, Ohio, 1959a.

Slechta, R.F., Forrest, J., Carter, W.K., and Wade, E.A.: *Comfort Evaluation of the C-124 Crew Seat (Weber).* WADC Technical Report 58-316. Wright Air Development Center, Wright-Patterson Air Force Base, Dayton, Ohio, 1959b.

Slechta, R.F., Wade, E.A., Carter, W.K., and Forrest, J.: *Comparative Evaluation of Aircraft Seating Accommodation.* WADC Technical Report 57-136, Wright Air Development Center, Wright-Patterson Air Force Base, Dayton, Ohio, 1957.

Smeathers, J.E., and Biggs, W.D.: Mechanics of the spinal column. In *Engineering Aspects of the Spine.* London, Mechanical Engineering Publications Ltd., 1980, pp. 103–109.

Smidt, G., Herring, T., Amundsen, L., Rogers, M., Russell, A., and Lehmann, T.: Assessment of abdominal and back extensor function. *Spine, 8:*211–219, 1983.

Smith, J.W.: The act of standing. *Acta Orthopaedica Scandinavica, 23:*159–168, 1953.

Smith, M.J.: Human factors issues in VDT use: environmental and workstation design considerations. *IEEE Computer Graphics and Applications, 4:*56–63, November 1984a.

Smith, M.J.: Health issues in VDT work. In Bennett, J., Case, D., Sandelin, J., and Smith, M. (Eds.): *Visual Display Terminals. Usability Issues and Health Concerns.* Englewood Cliffs, Prentice-Hall, 1984b, pp. 193–224.

Smith, M.J.: VDT strain: psychological strain or physical basis? In Swezey, R.W. (Ed.): *Proceedings of the Human Factors Society 29th Annual Meeting.* Baltimore, Sept. 29–Oct. 3, 1985, pp. 689–693.

Snell, R.S.: *Atlas of Clinical Anatomy.* Boston, Little, 1978.

Snijders, C.J.: On the form of the human spine and some aspects of its mechanical behaviour. *Acta Orthopaedica Belgica, 35:*584–594, 1969.

Snijders, C.J., Seroo, J.M., Snijder, J.G., and Hoedt, H.T.: Change in form of the spine as a consequence of pregnancy. In *Digest of the Eleventh International Conference on Medical and Biological Engineering,* Ottawa, 1976, pp. 670–671.

Snyder, H.L.: Keyboard design. In Ledin, H. (Ed.): *Ergonomic Principles in Office Automation.,* 2nd ed. Bromma, Ericsson Information Systems AB, 1983, pp. 43–58.

Snyder, L.H., and Ostrander, E.R.: *Spatial and Physical Considerations in the Nursing Home Environment: An Interim Report of Findings.* Department of Design and Environmental Analysis, College of Human Ecology, Cornell University, Ithaca, New York, August, 1972.

Soames, R.W., and Atha, J.: The role of the antigravity musculature during quiet standing in man. *European Journal of Applied Physiology, 47:*159–167, 1981.

Souther, S.G., Carr, S.D., and Vistnes, L.M.: Wheelchair cushions to reduce pressure under bony prominences. *Archives of Physical Medicine and Rehabilitation, 55:*460–464, 1974.

Spira, M., and Hardy, S.B.: Our experiences with high thigh amputations in paraplegics. *Plastic and Reconstructive Surgery, 31:*344–352, 1963.

Springer, T.J.: *Visual Display Terminal Workstations: A Comparative Evaluation of Alternatives.* State Farm Mutual Automobile Insurance Co., Bloomington, Illinois, 1982a.

Springer, T.J.: VDT workstations: a comparative evaluation of alternatives. *Applied Ergonomics, 13:*211–212, 1982b.

Springer, T.J.: Selecting the right furniture. *The Ergonomics Newsletter, 2:*17–19, 1983.

Staas, W.E., and La Mantia, J.G.: Rehabilitation approach. In Parish, L.C., Witkowski, J.A., and Crissey, J.T. (Eds.): *The Decubitus Ulcer.* New York, Masson, 1983, pp. 83–91.

Staffel, F.: Zur hygiene des sitzens. *Zentralblatt für Allgemeine Gesundheitspflege, 3:*403–421, 1884.

Staffel, F.: *Die Menschlichen Haltungstypen und ihre Beziehungen zu den Rückgratsverkrümmungen.* Wiesbaden, Bergmann, 1889. Translated in Bonne, A.J.: On the shape of the human vertebral column. *Acta Orthopaedica Belgica, 35:*567–582, 1969.

Stafford, G.T.: *Preventive and Corrective Physical Education.* New York, Barnes, 1928.

Stark, R.B.: *Plastic Surgery.* New York, Harper and Row, 1962.

Starr, S.J., Thompson, C.R., and Shute, S.J.: Effects of video display terminals on telephone operators. *Human Factors, 24:*699–711, 1982.

Stauffer, E.S.: Trauma. In Hardy, J.H. (Ed.): *Spinal Deformity in Neurological and Muscular Disorders.* St. Louis, Mosby, 1974, pp. 219–236.

Stecher, W.A.: An inquiry into the problem of desks for school children. *American Physical Education Review, 16:*453–458, 1911.

Steelcase, Inc.: *The Steelcase National Study of Office Environments, No. II. Comfort and Productivity in the Office of the 80's.* Conducted by Louis Harris and Associates, Inc., 1980.

Steen, B.: The function of certain neck muscles in different positions of the head with and without loading of the cervical spine. *Acta Morphologica Neerlando-Scandinavica, 6:*301–310, 1966.

Stewart, S.F.: Physiology of the unshod and shod foot with an evolutionary history of footgear. *American Journal of Surgery, 68:*127–138, 1945.

Stewart, S.F.C., Palmieri, V., and Cochran, G.V.B.: Wheelchair cushion effect on skin temperature, heat flux, and relative humidity. *Archives of Physical Medicine and Rehabilitation, 61:*229–233, 1980.

Stewart, T.: Ergonomics and visual display units—is there a problem? In McPhee, B., and Howie, A. (Eds.): *Ergonomics and Visual Display Units. Proceedings of a Conference.* Melbourne, September 5, 1979 and Sydney, September 7, 1979, pp. 3–23.

Stewart, T.: Problems caused by the continuous use of visual display units. *Lighting Research and Technology, 12:*26–36, 1980.

Stewart, T.F.M.: Practical experiences in solving VDU ergonomics problems. In Grandjean, E., and Vigliani, E. (Eds.): *Ergonomic Aspects of Visual Display Terminals.* London, Taylor and Francis, 1983, pp. 233–240.

Stockton, C.G.: Hygiene of the digestive apparatus. In Pyle, W.L. (Ed.): *A Manual of Personal Hygiene,* 5th ed. Philadelphia, Saunders, 1913, pp. 44–48.

Stoddard, A., and Osborn, J.F.: Scheuermann's disease or spinal osteochondrosis. *The Journal of Bone and Joint Surgery, 61-B:*56–58, 1979.

Stokes, I.A.F., and Abery, J.M.: Influence of the hamstring muscles on lumbar spine curvature in sitting. *Spine, 5:*525–528, 1980.

Stone, J.S.: School furniture in relation to lateral curvature. *American Physical Education Review, 5:*142–148, 1900.

Stone, P.T.: Issues in vision and lighting for users of VDUs. In Pearce, B.G. (Ed.): *Health Hazards of VDTs?* Chichester, Wiley, 1984, pp. 77–88.

Stoshak, M.L., and Mortimer, E.A.: Jean seam coccygodynia. *Pediatrics, 76:*138, 1985.

Stranden, E., Aarås, A., Anderson, D.M., Myhre, H.O., and Martinsen, K.: The effects of working posture on muscular-skeletal load and circulatory condition. In Coombes, K. (Ed.): *Proceedings of the Ergonomics Society's Conference 1983.* New York, Taylor and Francis, 1983, pp. 163–167.

Stransky, J., and Stone, R.B.: *The Alexander Technique*. New York, Beaufort, 1981.

Strassburg Commission: 1882. Quoted in Cohn (1886), pp. 104–105.

Strasser, H.: *Lehrbuch der Muskel und Gelenkmechanik*. Berlin, Springer, 1913.

Strohl, K.P., Mead, J., Banzett, R.B., Loring, S.H., and Kosch, P.C.: Regional differences in abdominal muscle activity during various maneuvers in humans. *Journal of Applied Physiology, 51:*1471–1476, 1981.

Stubbs, D.A., Buckle, P.W., Hudson, M.P., Barton, P.E., and Rivers, P.M.: Nurses uniform: an investigation of mobility. *International Journal of Nursing Studies, 22:*217–229, 1985.

Stubbs, D.A., and Osborne, C.M.: How to save your back. *Nursing, 3:*116–124, 1979.

Suenaga, T.: Studies on body postures of farmers during harvesting strawberries from greenhouses and their workloads. *Journal of the Kurume Medical Association, 45:*760–777, 1982.

Sulzberger, M.B., Cortese, T.A., Fishman, L., and Wiley, H.S.: Studies on blisters produced by friction. *The Journal of Investigative Dermatology, 47:*456–465, 1966.

Suther, T.W., and McTyre, J.H.: Effect on operator performance at thin profile keyboard slopes of 5, 10, 15, and 25 degrees. In Edwards, R.E. (Ed.): *Proceedings of the Human Factors Society 26th Annual Meeting*. Seattle, October 25–29, 1982, pp. 430–434.

Sutton, L.P.: Importance of considering the body mechanics of children. *American Journal of Diseases of Children, 30:*550–557, 1925.

Swartout, R., and Compere, E.L.: Ischiogluteal bursitis. *Journal of the American Medical Association, 227:*551–552, 1974.

Swearingen, J.J., Wheelwright, C.D., and Garner, J.D.: *An Analysis of Sitting Areas and Pressure of Man*. Civil Aeromedical Research Institute, Federal Aviation Agency, Aeronautical Center, Oklahoma City, January, 1962.

Sweet, C.D.: Improvement of body mechanics in adolescent children. *American Journal of Diseases of Children,* 74:503–506, 1947.

Szabó, G., Pósch, E., and Magyar, Z.: Interstitial fluid, lymph and oedema formation. *Physiologica Academiae Scientiarum Hungaricae, 56:*367–378, 1980.

Szalay, E.A., Roach, J.W., Houkom, J.A., Wenger, D.R., and Herring, J.A.: Extension-abduction contracture of the spastic hip. *Journal of Pediatric Orthopedics, 6:*1–6, 1986.

Taft, L.T.: The care and management of the child with muscular dystrophy. *Developmental Medicine and Child Neurology, 15:*510–518, 1973.

Tanii, K., and Masuda, T.: A kinesiologic study of erectores spinae activity during trunk flexion and extension. *Ergonomics, 28:*883–893, 1985.

Tawast-Rancken, S.: *Lasten Ryhtivirheiden Määrä ja Luonne. Mannerheimin Lastensuojeluliiton Vuosikirja*. Helsinki, 1960, pp 6–17.

Taylor, H.L.: Results of research on conditions affecting posture. *Journal of the American Medical Association, 68:*327–330, 1917.

Taylor, J.M.: Physical culture in children—the objects to be attained. *Pediatrics, 12:*237–244, 1901.

Tepperman, P.S., DeZwirek, C.S., Chiarcossi, A.L., and Jimenez, J.: Pressure sores—prevention and step-up management. *Postgraduate Medicine, 62:*83–89, 1977.

Teshima, H.: Electromyographic study of patterns of activity of the abdominal wall muscles in man. *The Tohoku Journal of Experimental Medicine, 67:*281–292, 1958.

Thier, R. H.: Measurement of seat comfort. *Automobile Engineer, 53:*64–66, 1963.

Thiyagarajan, C., and Silver, J.R.: Aetiology of pressure sores in patients with spinal cord injury. *British Medical Journal, 289:*1487–1490, 1984.

Thomas, D.P., and Whitney, R.J.: Postural movements during normal standing in man. *Journal of Anatomy, 93:*524–539, 1959.

Thomas, L.C., and Goldthwait, J.E.: *Body Mechanics and Health.* Boston, Houghton Mifflin, 1922.

Thompson, D.A.: Where I stand on seating. *Human Factors Society Bulletin, 28:*1–2, Sept 1985.

Tibone, J., Sakimura, I., Nickel, V.L., and Hsu, J.D.: Heterotopic ossification around the hip in spinal cord-injured patients. *The Journal of Bone and Joint Surgery, 60-A:*769–775, 1978.

Tichauer, E.R.: The biomechanics of the arm-back aggregate under industrial working conditions. In Byars, E.F., Contini, R., and Roberts, V.L. (Eds.): *Biomechanics Monograph.* New York, The American Society of Mechanical Engineers, 1967, pp. 153–170.

Tichauer, E.R.: Industrial engineering in the rehabilitation of the handicapped. *The Journal of Industrial Engineering, 19:*96–104, 1968.

Tichauer, E.R.: *The Biomechanical Basis of Ergonomics.* New York, Wiley, 1978.

Tissot, J-C.: *Gymnastique Médicinale et Chirurgicale.* Paris, Bastien, 1780. Translated by Licht, E., and Licht, S. New Haven, Licht, 1964.

Toppenberg, R.M., and Bullock, M.I.: The interrelation of spinal curves, pelvic tilt and muscle lengths in the adolescent female. *The Australian Journal of Physiotherapy, 32:*6–12, 1986.

Torrance, C.: Pressure sores, part two. Predisposing factors: The at-risk patient. *Nursing Times, 77:*5–8, February 19, 1981.

Torrance, C.: *Pressure Sores: Aetiology, Treatment and Prevention.* London, Croom Helm, 1983.

Travell, J.: Chairs are a personal thing. *House Beautiful, 97:*190–193, 262–266, 269, 1955.

Travell, J.: Mechanical headache. *Headache, 7:*23–29, 1967.

Travers, P.H., and Stanton, B-A.: Office workers and video display terminals: physical, psychological and ergonomic factors. *Occupational Health Nursing, 32:*586–591, 1984.

Trefler, E., Hanks, S., Huggins, P., Chiarizzo, S., and Hobson, D.: A modular seating system for cerebral-palsied children. *Developmental Medicine and Child Neurology, 20:*199–204, 1978.

Trefler, E., and Taylor, S.: Decision making guidelines for seating and positioning children with cerebral palsy. In Trefler, E. (Ed.): *Seating for Children with Cerebral Palsy. A Resource Manual.* Memphis, University of Tennessee, 1984, pp. 55–76.

Trendafilov, A.G., Konstantinov, V.N., Popdimitrov, D.K., and Gradinarov, N.K.: Quantitative comfort rating of the sitting posture. In *Proceedings of a Symposium on Ergonomics in Machine Design,* Prague, October 2–7, 1967.

Troup, J.D.G.: Back pain and car-seat comfort (Appendix). *Applied Ergonomics, 3:*90–91, 1972.

Troup, J.D.G.: Driver's back pain and its prevention. *Applied Ergonomics, 9:*207–214, 1978.

Troup, J.D.G., and Edwards, F.C.: *Manual Handling and Lifting.* London, Her Majesty's Stationery Office, 1985.

Trumble, H.C.: The skin tolerances for pressure and pressure sores. *The Medical Journal of Australia, 2:*724–726, 1930.

Tyler, J.M.: *Growth and Education.* Boston, Houghton, Mifflin, 1907.

Varterasian, J.H., and Thompson, R.R.: *The Dynamic Characteristics of Automobile Seats with Human Occupants.* Society of Automotive Engineers, publication no. 770249, 1977.

Vidal, J.: Sagittal deviations of the spine. An attempted classification in relation to the balance of the pelvis. *Orthopaedic Transactions, 8:*34–35, 1984.

Vierordt, K.: *Grundriss der Physiologie des Menschen.* Tübingen, Laupp, 1862. Translated in Luciani, L.: *Human Physiology, vol. 3.* London, Macmillan, 1915, pp. 113–114.

deVries, H.A.: EMG fatigue curves in postural muscles. A possible etiology for idiopathic low back pain. *American Journal of Physical Medicine, 47:*175–181, 1968.

Vulcan, A.P., King, A.I., and Nakamura, G.S.: Effects of bending on the vertebral column during +Gz acceleration. *Aerospace Medicine, 41:*294–300, 1970.

Wachsler, R.A., and Learner, D.B.: An analysis of some factors influencing seat comfort. *Ergonomics, 3:*315–320, 1960.

Wadsworth, T.G., and Williams, J.R.: Cubital tunnel external compression syndrome. *British Medical Journal, 1:*662–666, 1973.

Wagenhäuser, F.J.: Epidemiology of postural disorders in young people. In Fehr, K., Huskisson, E.C., and Wilhelmi, E. (Eds.): *Rheumatological Research and the Fight Against Rheumatic Diseases in Switzerland.* Basel, Eular, 1978.

Wallin, D., Ekblom, B., Grahn, R., and Nordenborg, T.: Improvement of muscle flexibility. A comparison between two techniques. *The American Journal of Sports Medicine, 13:*263–268, 1985.

Walther, O.-E., Simon, E., and Jessen, C.: Thermoregulatory adjustments of skin blood flow in chronically spinalized dogs. *Pflügers Archives, 322:*323–335, 1971.

Ward, J.: Ward reservations. *Design (London), 333:*33–34, 1976.

Ward, R.J., Danziger, F., Bonica, J.J., Allen, G.D., and Tolas, A.G.: Cardiovascular effects of change of posture. *Aerospace Medicine, 37:*257–259, 1966.

Waris, P.: Occupational cervicobrachial syndromes. *Scandinavian Journal of Work, Environment and Health, 6 (Supplement 3):*3–14, 1980.

Warren, C.G., Ko, M., Smith, C., and Imre, J.V.: Reducing back displacement in the powered reclining wheelchair. *Archives of Physical Medicine and Rehabilitation, 63:*447–449, 1982.

Warren, J.C.: *Physical Education and the Preservation of Health.* Boston, Ticknor, 1846.

Warwick, R., and Williams, P.L. (Eds.): *Gray's Anatomy,* 35th British ed. Philadelphia, Saunders, 1973.

Watanabe, J., and Avakian, L.: *The Secrets of Judo.* Rutland, Tuttle, 1960.

Watanabe, K.: Biomechanical implications of EMG activity of erector spinae and gluteus maximus muscles in postural changes of the trunk. In Morecki, A., Fidelus, K., Kedzior, K., and Wit, A. (Eds.): *Biomechanics VII-B. International Series on Biomechanics.* Baltimore, Univ Park, 1981, Vol. 3B, pp. 23–30.

Watkin, B.: Are you sitting comfortably? *Health and Safety at Work, 5:*29–30, June 1983.

Watson, N.: Skin care and long-term rehabilitation. In Barbenel, J.C., Forbes, C.D., and Lowe, G.D.O. (Eds.): *Pressure Sores.* London, Macmillan, 1983, pp. 95–102.

Wear, C.L.: Relationship of flexibility measurements to length of body segments. *The Research Quarterly, 34:*234–238, 1963.

Weber, A., Sancin, E., and Grandjean, E.: The effects of various keyboard heights on EMG and physical discomfort. In Grandjean, E. (Ed.): *Ergonomics and Health in Modern Offices.* London, Taylor and Francis, 1984, pp. 477–483.

Weber, J., van der Star, A., and Snijders, C.J.: Development and evaluation of a new instrument for the measurement of work postures; in particular the inclination of the head and the spinal column. In Corlett, N., Wilson, J., and Manenica, I. (Eds.): *The Ergonomics of Working Postures.* London, Taylor and Francis, 1986, pp. 82–91.

Wells, R.P., Norman, R.W., Bishop, P., and Ranney, D.A.: Assessment of the static fit of automobile lap-belt systems on front-seat passengers. *Ergonomics, 29:*955–976, 1986.

Wells, T.J.: *Problems in Geriatric Nursing Care.* Edinburgh, Churchill, 1980.

Westgaard, R.H., and Aarås, A.: Postural muscle strain as a causal factor in the development of musculo-skeletal illnesses. *Applied Ergonomics, 15:*162–174, 1984.

Westgaard, R.H., and Aarås, A.: The effect of improved workplace design on the development of work-related musculo-skeletal illnesses. *Applied Ergonomics, 16:*91–97, 1985.

Wheeler, J., Woodward, C., Ucovich, R.L., Perry, J., and Walker, J.M.: Rising from a chair. Influence of age and chair design. *Physical Therapy, 65:*22–26, 1985.

White, A.A., and Panjabi, M.M.: *Clinical Biomechanics of the Spine.* Philadelphia, Lippincott, 1978.

Whitman, A.: Postural deformities in children. *New York State Journal of Medicine, 24:*871–874, 1924.

Whitney, R.J.: The stability provided by the feet during manoeuvres whilst standing. *Journal of Anatomy, 96:*103–111, 1962.

Wilder, D.G., Seroussi, R.E., Dimnet, J., and Pope, M.H.: The balance point of the lumbar motion segment. *Orthopaedic Transactions, 10:*413–414, 1986.

Wilder, D.G., Woodworth, B.B., Frymoyer, J.W., and Pope, M.H.: Vibration and the human spine. *Spine, 7:*243–254, 1982.

Wildnauer, R.H., Bothwell, J.W., and Douglass, A.B.: Stratum corneum biomechanical properties. 1. Influence of relative humidity on normal and extracted human stratum corneum. *The Journal of Investigative Dermatology, 56:*72–78, 1971.

Wiles, P.: Postural deformities of the anteroposterior curves of the spine. *The Lancet, 1:*911–919, April 17, 1937.

Wilkins, K.E., and Gibson, D.A.: The patterns of spinal deformity in Duchenne muscular dystrophy. *The Journal of Bone and Joint Surgery, 58-A:*24–32, 1976.

Williams, A.: A study of factors contributing to skin breakdown. *Nursing Research, 21:*238–243, 1972.

Winkel, J.: Swelling of the lower leg in sedentary work—a pilot study. *Journal of Human Ergology, 10:*139–149, 1981.

Winkel, J., and Jørgensen, K.: Evaluation of foot swelling and lower-limb temperatures in relation to leg activity during long-term seated office work. *Ergonomics, 29:*313–328, 1986.

Winter, R.B., and Pinto, W.C.: Pelvic obliquity. Its causes and its treatment. *Spine, 11:*225–234, 1986.

Wolf, S.L., Basmajian, J.V., Russe, C.T.C., and Kutner, M.: Normative data on low back mobility and activity levels. *American Journal of Physical Medicine, 58:*217–229, 1979.

Woodhull, A.M., Maltrud, K., and Mello, B.L.: Alignment of the human body in standing. *European Journal of Applied Physiology, 54:*109–115, 1985.

Woodhull-McNeal, A.P.: Activity in torso muscles during relaxed standing. *European Journal of Applied Physiology, 55:*419–424, 1986.

Woodson, W.E.: *Human Factors Design Handbook.* New York, McGraw-Hill, 1981.

Wotzka, G., Grandjean, E., Burandt, U., Kretzschmar, H., and Leonhard, T.: Investigations for the development of an auditorium seat. In Grandjean, E. (Ed.): *Proceedings of the Symposium on Sitting Posture.* London, Taylor and Francis, 1969, pp. 68–83.

van der Woude, L.H.V., de Groot, G., Hollander, A.P., van Ingen Schenau, G.J., and Rozendal, R.H.: Wheelchair ergonomics and physiological testing of prototypes. *Ergonomics, 29:*1561–1573, 1986.

Wright, G.R.: Overview of VDT ergonomics. *Occupational Health in Ontario, 3:*182–191, 1982.

Wyke, B.: Articular neurology—a review. *Physiotherapy, 58:*94–99, 1972.

Wyllie, F.J., McLean, N.R., and McGregor, J.C.: The problem of pressure sores in a regional plastic surgery unit. *Journal of the Royal College of Surgeons of Edinburgh, 29:*38–43, 1984.

Yamaguchi, Y., Umezawa, F., and Ishinada, Y.: Sitting posture: an electromyographic study on healthy and notalgic people. *Journal of Japanese Orthopaedic Association, 46:*277–282, 1972.

Yang, B.-J., Chen, C.-F., Lin, Y.-H., and Lien, I.-N.: Pressure measurement on the ischial tuberosity of the human body in sitting position and evaluation of the pressure relieving effect of various cushions. *Journal of the Formosan Medical Association, 83:*692–698, 1984.

Young, J.S., and Burns, P.E.: Pressure sores and the spinal cord injured: *Model Systems' SCI Digest, 3:*9–18, 25, Fall 1981a

Young, J.S., and Burns, P.E.: Pressure sores and the spinal cord injured: part two. *Model Systems' SCI Digest, 3:*11–26, 48, Winter 1981b.

Young, K., McDonagh, M.J.N., and Davies, C.T.M.: The effects of two forms of isometric training on the mechanical properties of the triceps surae in man. *Pflügers Archiv, 405:*384–388, 1985.

Zacharkow, D.: *Wheelchair Posture and Pressure Sores.* Springfield, Thomas, 1984a.

Zacharkow, D.: *The Healthy Lower Back.* Springfield, Thomas, 1984b.

Zacharkow, D.: The relationship between seating and positioning and the prevention of pressure sores. In *Proceedings of the National Symposium on the Care, Treatment and Prevention of Decubitus Ulcers.* Arlington, Virginia, November 1984c, p. 27.

Zacharkow, D.: Effect of posture and distribution of pressure in the prevention of pressure sores. In Lee, B.Y. (Ed.): *Chronic Ulcers of the Skin.* New York, McGraw-Hill, 1985, pp. 197–202.

Zipp, P., Haider, E., Halpern, N., and Rohmert, W.: Keyboard design through physiological strain measurements: *Applied Ergonomics, 14:*117–122, 1983.

Zwahlen, H.T., Hartmann, A.L., and Rangarajulu, S.L.: Effects of rest breaks in continuous VDT work on visual and musculoskeletal comfort/discomfort and on performance. In Salvendy, G. (Ed.): *Human-Computer Interaction.* Amsterdam, Elsevier, 1984, pp. 315–319.

INDEX

Lanchester Library

2375